Magneto-Rheological Fluids

Magneto-Rheological Fluids

Editors

Antonio Concilio
Salvatore Ameduri
Ignazio Dimino
Vikram G Kamble
Rosario Pecora

MDPI • Basel • Beijing • Wuhan • Barcelona • Belgrade • Manchester • Tokyo • Cluj • Tianjin

Editors

Antonio Concilio
Adaptive Structures
CIRA, The Italian Aerospace Research Centre
Capua
Italy

Salvatore Ameduri
Adaptive Structures
CIRA, The Italian Aerospace Research Centre
Capua
Italy

Ignazio Dimino
Adaptive Structures
CIRA, The Italian Aerospace Research Centre
Capua
Italy

Vikram G Kamble
Polymer Competence Center Leoben GmbH
Leoben
Austria

Rosario Pecora
Department of Industrial Engineering
Univ. of Naples Federico II
Naples
Italy

Editorial Office
MDPI
St. Alban-Anlage 66
4052 Basel, Switzerland

This is a reprint of articles from the Special Issue published online in the open access journal *Applied Sciences* (ISSN 2076-3417) (available at: www.mdpi.com/journal/applsci/special_issues/Magneto-Rheological_Fluids).

For citation purposes, cite each article independently as indicated on the article page online and as indicated below:

LastName, A.A.; LastName, B.B.; LastName, C.C. Article Title. *Journal Name* **Year**, *Volume Number*, Page Range.

ISBN 978-3-0365-8491-1 (Hbk)
ISBN 978-3-0365-8490-4 (PDF)

Contents

About the Editors

Antonio Concilio

Antonio Concilio earned his degree in aeronautics engineering with honours at the University of Napoli "Federico II" (Italy) in 1989. He was awarded a PhD in aerospace engineering at the same university in 1995. In 2007, he completed the ECATA Master course in Aerospace Business Administration at ISAE-Supaero, Toulouse (France). Since 1989, he has been a researcher at the Italian Aerospace Research Centre (CIRA), Italy, where he is currently the Head of the Adaptive Structures Division (2023).

His main research interests focus on vehicles' noise and vibration control for increased comfort and protection of devices, smart structural system design, integrated sensor networks, and morphing of aircraft wing and control surfaces. He led CIRA activities within several European Union (EU) and nationally funded research contracts on smart structures and noise and vibration control. He was the European scientific coordinator of the Identification of an Aircraft Passenger Comfort Index Project (1998–2001) and the national project manager of the Design and Realisation Methods of Intelligent Systems for the Monitoring and Control of Aeronautic and Aerospace Structures project (2002–2006). He led an international team to develop, manufacture, and test an Adaptive Wing Trailing Edge within the EU Smart Aircraft Intelligent Structures project (2011–2015). Recently, he was the Italian Principal Investigator of the RESUME project, an Israeli–Italian cooperation on the development of Real-Time SHM Systems (2020–2023). In 2018, he earned the Italian National Qualification as a full professor (ASN). According to Scopus, his H-index rank is 24, with 183 indexed publications, 1835 citations in 973 publications, and 174 co-authors.

Salvatore Ameduri

Salvatore Ameduri, after having defended his PhD thesis (funded by EREA and within the cooperation between the University of Naples "Federico II" and CIRA) on the design of smart structure systems, became a senior researcher at CIRA in 2002 in the Smart Structures and Materials Department. He is an author and a co-author of two International Patents on a drop-nose LE and a SMA-based variable chamber TE. He is an author of about 120 Journal and Conference papers (smart structures and materials, morphing, noise and vibration, and acoustics), a reviewer of multiple publications, a Guest Editor of Special Issues of various journals, an award winner of numerous project competitions, an invited speaker, and a scientific reviewer for financing proposals. He has taken coordination and scientific roles for many scientific projects. He was the project manager for CIRA for the "Fuselage and Empennage with Cabin and Cargo Architecture for H2 Integration, FASTER-H2", Clean Aviation Project (2023–2025), and the "Shape Adaptive Blades for Rotorcraft Efficiency, SABRE", H2020 Project (2017–2020); a designer of a weight and balance demonstrator for the project "Design, Manufacturing, and Qualification of a New Generation of Landing Gear for a Compound Fast Rotorcraft", ANGELA (2017–2021, Clean Sky 2); a project manager for "Graphene-Polymeric Spray Sensor for Shape Recognition of Super-Deformable Structures, GRAPSS" (2016–2018, CIRA); a technical coordinator for "AIRGREEN 2" (2015–2023, Clean Sky 2 Programme), which focused on the development and validation of technologies for new-generation wings in wind tunnel and flight tests; a designer of a piezo-based de-icing system for the project "Green Airframe Icing Novel System", GAINS (Clean Sky 2 2015–2020); and a designer of active noise control systems for the LOWNOISE project (National Operational Programme, PON, for Research and Competitiveness, 2007–2013).

Ignazio Dimino

Dr. Ignazio Dimino graduated with honours in Aeronautical Engineering from the University of Palermo in 2004. He was a visiting researcher at the Centre of Acoustics and Vibration of Penn State University, State College, USA, in 2008. He earned his Ph.D. in Aeronautics from the Imperial College of London in 2012. He earned the qualification of Associate Professor from the National Scientific Habilitation in the field of aerospace structures and design in 2019. His technical background includes adaptive structures, wing morphing, smart actuation, and active noise and vibration control, ranging from conceptual design to final fabrication and testing. In 2020, he served as a mentor in a seven-week internship programme held at the Sandia National Labs NOMAD (Nonlinear Mechanics and Dynamics) Research Institute, University of New Mexico, Albuquerque, NM, USA. Currently, he is the Head of a Research Unit at the Adaptive Structures Division of the Italian Aerospace Research Centre (CIRA) and a Project Manager in the field of Smart Structures.

Dr. Dimino is the author of more than one hundred peer-reviewed papers on adaptive structures, structural dynamics, and vibro-acoustic control, and his H-index is 19. He is an Editor-in-chief of the *Journal of Adaptive Structures*, and he is a co-author of two books and three book chapters. In 2016 and 2019, he served as a programme committee member and chairman of the Greener Aviation conference in Brussels, which was organised by the 3AF Association (Aéronautique et Astronautique de France). In 2014, Dr. Dimino received the "Best Paper Award" at the 3AF/CEAS conference in Brussels. In 2021, he received the prize for the EREA Best Paper Award 2021, organised by EREA, the Association of European Research Establishments in Aeronautics. He is currently a member of the IFAR international working group on composites, focusing on new challenges for composite materials for the 2025–2045 timeframe.

Vikram G Kamble

Vikram G. Kamble is an accomplished researcher with a proven track record in both the academia and industry. With a rich academic journey spanning across multiple continents, Kamble holds an MS in Materials Engineering and Nanotechnology from the Politecnico di Milano, Italy, and a BE in Mechanical Engineering from Visvesvaraya Technological University, India. His diverse professional experience includes his role as a scientist at esteemed institutions such as the Leibniz Institute of Polymer Research in Germany and the Defence Research and Development Organisation in India. Since May 2023, he has been a dedicated researcher at the Polymer Competence Center Leoben GmbH in Austria.

Vikram's expertise extends beyond his scientific achievements. He has completed a range of professional certifications from renowned institutions, showcasing his proficiency in areas like project management, negotiation, and human resources. Notably, he has excelled in positions involving leadership, teamwork, and integrity. With a wealth of knowledge about scientific instruments and software skills under his belt, Vikram has made significant contributions to the field of polymers (sensors and actuators) and rheology of fluids. He has authored numerous research articles, book chapters, and edited books, solidifying his reputation as an expert in the discipline. Through his work, leadership, and continuous pursuit of knowledge, Vikram G. Kamble remains a dedicated contributor to the advancement in materials research and development.

Rosario Pecora

Rosario Pecora obtained his Master's degree in Aeronautical Engineering and Ph.D. in Transport Engineering from the University of Naples "Federico II" in 2002 and 2005, respectively. He has been serving as an Assistant Professor of Aircraft Structure Stability and a Lecturer of Aircraft Structures at the same university since 2011. His research primarily focuses on aero-servo-elasticity of unconventional structural systems, structure dynamics, and smart structures, with significant contributions to major European and pan-European projects.

Alongside his extensive research and academic work, Dr. Pecora has been actively involved with numerous aircraft manufacturing companies and research centres as a technical advisor. His expertise lies in areas such as loads, aeroelasticity, aircraft structure design, and certification (EASA CS-23 and -25 standards). Additionally, he held the role of Compliance Verification Engineer for structures, loads, and aeroelasticity at three different EASA-approved design organisations.

Dr. Pecora is the author of 98 indexed papers, an Editor of a book on morphing wing technologies for large civil aeroplanes, an Associate Editor of six indexed journals, a technical board member of four international congresses in aerospace engineering, and an inventor of European and US patents on SMA-based architectures for morphing wing trailing edges.

Preface to "Magneto-Rheological Fluids"

This paper collection is the result of an editorial initiative aimed at providing the most recent updates in the field of magneto-rheological fluids. Such materials may have a wide use, especially in optimising the dissipation in mechanical systems, but suffer from some drawbacks like their weight and, generally, the need to be continuously re-mixed. These aspects slowed down their actual implementation, somehow restricting their use to fields where those problems are easily solved. For instance, in the case of the ground vehicle suspensions, the impact is very limited, the fluid being aimed at substituting substances of the same consistency and being constantly shaken. Either of the mentioned aspects has limited consequences for seismic protection, and the additional weight percentage is definitely negligible with respect to the host system. It is therefore interesting to have a contemporary picture of how scientific and technological research is proceeding on the practical implementation of MRF-based systems, trying to overcome the classical shortcomings. Seismic protection, brake, and suspension optimisation are addressed in this collection, which is enriched with non-frequent and highly specialist topics and includes a conclusive review dealing with magneto-rheological fluid-based valves.

Antonio Concilio, Salvatore Ameduri, Ignazio Dimino, Vikram G Kamble, and Rosario Pecora
Editors

applied sciences

MDPI

Editorial

Magneto-Rheological Fluids

Antonio Concilio [1,*], Salvatore Ameduri [1], Ignazio Dimino [1] and Rosario Pecora [2]

[1] Adaptive Structures Department, The Italian Aerospace Research Centre, 81043 Capua, Italy; i.dimino@cira.it (I.D.)
[2] Industrial Engineering Department, University of Naples "Federico II", 80125 Napoli, Italy
* Correspondence: a.concilio@cira.it

Magneto-rheological fluids, or MRF, have been known for a long time in the technological and scientific community. They have great potential and, as well, have important showstoppers. This combination is always conflicting and may explain why applications of such excellent materials are not yet widespread, despite their market value having already topped the impressive figure of almost 1.5 billion EUR worldwide, and being projected to about 6 billion EUR in 2030.

An MRF is a liquid that can change its rheological state when subjected to a magnetic field. Usually, viscosity is the main objective of such changes; its characteristic may span from values typical of water up to becoming almost solid. Since viscosity is targeted, the main uses concentrate on phenomena wherein this property plays an important role. Therefore, dynamic aspects are mainly considered, with the primary aim to variate the overall viscous damping function to increase the dissipation at the highest level. For instance, controlling viscosity may be essential for vehicle suspensions or other devices dealing with the impact of a system on the ground. Having almost no damping would cause the target structure to vibrate indefinitely, so a certain amount of dissipation capability is necessary. However, it cannot exceed a certain threshold, otherwise, the time response is dramatically affected, in turn determining the system to approach equilibrium in a very long time, virtually infinite. In other words, having infinite damping is almost equivalent, macroscopically, to have infinite local stiffness, as an additional constraint is added. Additionally, dissipation capability depends on velocity, so the damper should be tuned with its current value to attain the best performance.

The potential of such an incredible instrument is hampered by its construction. Usually, a MRF is produced as an oil solution containing small metal particles. The magnetic-forced orientation of such particles hinders the fluid movement. Since metal is heavy, the particles tends to drop down, so that a continuous mixing shall be actuated even though the use of some additives can limit the issue. The drawbacks of this necessity are evident since it would require some external energy devoted to this operation which, in the case of a transport drum, is sometimes generated by a helix constantly mindling the fluid. Luckily, some applications are self-consistent, as the structures that enjoy the presence of the MRF are themselves subjected to continuous oscillations. In the case of cars or bikes, this operation is strictly connected to continuous hits with road irregularities; for this reason, it is not surprising that many patents and several common implementations of magneto-rheological fluids are for road vehicles.

Other major applications of MRF are related to seismic response attenuation, where the need to maximize the energy dissipation is crucial for guaranteeing the device's effectiveness and safeguarding the reference structure. In this case, the system may stand for a long time, silent, without anything to do. However, it should be ready to use at unpredictable moments. Therefore, its state should be preserved at each instant, and an automatic monitoring system is essential. This fact enlarges the costs of implementing such devices, even with the perspective of an exceptional response in danger.

This Special Issue presents two papers that deal with the application of MRF on buildings. The first one [1] is an extensive review of many publications focusing on devices

Citation: Concilio, A.; Ameduri, S.; Dimino, I.; Pecora, R. Magneto-Rheological Fluids. *Appl. Sci.* **2023**, *13*, 5044. https://doi.org/10.3390/app13085044

Received: 31 March 2023
Accepted: 11 April 2023
Published: 18 April 2023

fitted to certain structures. It deals with friction dampers (FD), tuned mass dampers (TMD), and viscous dampers (VD), further classifying them into passive, active, and semi-active systems. The second one is more directed to a specific application and deals with a vibration-suppression system based on MRF-based integrated dissipative bracings [2]. A simulation model is proposed and validated through experimental measures on a dynamically scaled model representative of a five-storey slender building, thus making the article extremely interesting for further investigations.

The other important cluster of papers deals instead with the use of the MRF as a device for enhancing the performance of brake or suspension systems. The first of the series faces the design optimization of a hydrodynamic brake empowered with MRF [3]. The article tries to draft general guidelines for that class of devices. The second one deals with the application of MRF on industrial drive system clutches in continuous operation [4]. In this case, a theoretical design approach is devoted to computing electrical consumption, copper losses, and temperature distribution, parameterized with respect to the number of discs comprising the clutch device. The implementation of MRF systems to enhance the performance of an aircraft landing gear is then approached [5]. Therefore, this paper devotes attention towards aeronautical systems, specifically those that should be effective just once in a flight (or at most two, if take-off is also considered). The focus of the work addresses the pressure losses due to the square of the flow rate, which can be significant for high-speed mechanisms, such as the landing gears at the impact, differently from automotive suspensions, which instead work at relatively low velocities. Finally, an MRF-based semi-active air suspension is proposed to reduce vehicle vibrations and noise fields during motion [6]. It is interesting to remark that the manuscript presents an experimental characterization of the proposed MRF damper, which shows excellent performance with respect to ordinary devices, from both the energy dissipation and interior comfort points of view.

The three other papers of the collection deviate from the traditional concepts expressed above. One of them proposes an integrated concrete mixture charged with ferrous nanoparticles for better controlling the concrete mixture during casting [7]. The introduced methodology proved to be effective in the established aim by implementing an active stiffening control during the manufacturing process. An original issue was considered by the second paper, investigating the induced pressure change within an MRF system following the application of a magnetic field [8]. The correlation between control and observed variables is not linear, increasing for higher values of the input; at the same time, larger currents cause increased heat fluxes, which results in degradation of the coil performance and the magnetic field strength itself. The latest work provides readers with a comprehensive survey on particular applications of MRF, such as flow control devices or valves, [9]. Specifically, after a detailed and systematic review, the paper proposes classification criteria and categorized MRF valves into groups based on architecture similarities and performance envelopes.

The collection of papers in this Special Issue are well-representative of the most common MRF-based systems available in the scientific and technical literature. In turn, based on the wide bibliography represented herein, it can be considered as a starting base from which one can explore the fascinating world of MRF-based devices, which currently continue to exhibit untapped potential and, therefore, are expected to excel in the near future. This step is intimately correlated to the development of technologies, mainly addressed to limit the weights and sizes of their components.

Author Contributions: Conceptualization, A.C.; methodology, R.P.; software, S.A.; validation, I.D.; formal analysis, A.C.; investigation, R.P.; resources, All; data curation, I.D.; writing—original draft preparation, A.C.; writing—review and editing, R.P., S.A. and I.D.; visualization, All; supervision, A.C. All authors have read and agreed to the published version of the manuscript.

Conflicts of Interest: The author declares there is no conflict of interest.

References

1. Lenggana, B.W.; Ubaidillah, U.; Imaduddin, F.; Choi, S.-B.; Purwana, Y.M.; Harjana, H. Review of Magnetorheological Damping Systems on a Seismic Building. *Appl. Sci.* **2021**, *11*, 9339. [CrossRef]
2. Pecora, R. A Practical Approach for the Mitigation of Seismic-Induced Vibrations in Slender Metallic Structures through Magnetorheological Fluid Dampers. *Appl. Sci.* **2022**, *12*, 4155. [CrossRef]
3. Kesy, Z.; Musiałek, I.; Choi, S.-B. Design Optimization of a Hydrodynamic Brake with an Electrorheological Fluid. *Appl. Sci.* **2023**, *13*, 1089. [CrossRef]
4. Kluszczynski, K.; Pilch, Z. A Comparison Study on Magnetorheological Multi-Disc Clutches in Steady Continuous-Duty States from the Viewpoint of Electrical Energy Consumption and Spatial Temperature Distribution. *Appl. Sci.* **2022**, *12*, 7895. [CrossRef]
5. Kang, B.-H.; Hwang, J.-H.; Choi, S.-B. A New Design Model of an MR Shock Absorber for Aircraft Landing Gear Systems Considering Major and Minor Pressure Losses: Experimental Validation. *Appl. Sci.* **2021**, *11*, 7895. [CrossRef]
6. Li, G.; Ruan, Z.; Gu, R.; Hu, G. Fuzzy Sliding Mode Control of Vehicle Magnetorheological Semi-Active Air Suspension. *Appl. Sci.* **2021**, *11*, 10925. [CrossRef]
7. Chibulu, C.; Yardimci, M.Y.; De Schutter, G. Effect of Volume Fraction of Fine Sand on Magnetorheological Response and Blocking Mechanisms of Cementitious Mixtures Containing Fe_3O_4 Nanoparticles. *Appl. Sci.* **2022**, *12*, 10104. [CrossRef]
8. Widodo, P.J.; Budiana, E.P.; Ubaidillah, U.; Imaduddin, F. Magnetically-Induced Pressure Generation in Magnetorheological Fluids under the Influence of Magnetic Fields. *Appl. Sci.* **2021**, *11*, 9807. [CrossRef]
9. Gołdasz, J.; Sapinski, B.; Kubík, M.; Macháček, O.; Bańkosz, W.; Sattel, T.; Tan, A.S. Review: A Survey on Configurations and Performance of Flow-Mode MR Valves. *Appl. Sci.* **2022**, *12*, 6260. [CrossRef]

applied sciences

MDPI

Review

Review of Magnetorheological Damping Systems on a Seismic Building

Bhre Wangsa Lenggana [1], Ubaidillah Ubaidillah [1,*], Fitrian Imaduddin [1], Seung-Bok Choi [2,3,*], Yusep Muslih Purwana [4] and Harjana Harjana [5]

1 Department of Mechanical Engineering, Faculty of Engineering, Universitas Sebelas Maret, Surakarta 57126, Indonesia; bhrewangsa20@gmail.com (B.W.L.); fitrian@ft.uns.ac.id (F.I.)
2 Department of Mechanical Engineering, The State University of New York, Korea (SUNY Korea), Incheon 21985, Korea
3 Department of Mechanical Engineering, Industrial University of Ho Chi Minh City, 12 Nguyen Van Bao Street, Vap District, Ho Chi Minh City 700000, Vietnam
4 Department of Civil Engineering, Faculty of Engineering, Universitas Sebelas Maret, Surakarta 57126, Indonesia; ymuslih@yahoo.com
5 Department of Physics, Faculty of Mathematics and Physics Sciences, Universitas Sebelas Maret, Surakarta 57126, Indonesia; harjana@yahoo.com
* Correspondence: ubaidillah_ft@staff.uns.ac.id (U.U.); seungbok.choi@sunykorea.ac.kr (S.-B.C.); Tel.: +62-857-9952-7552 (U.U.)

Featured Application: The discussion in this article could be used as a literature study on the application of semi-active damping to building structures in general and the application of MR dampers to building structures specifically related to earthquake damage mitigation.

Abstract: Building structures are vulnerable to the shocks caused by earthquakes. Buildings that have been destroyed by an earthquake are very detrimental in terms of material loss and mental trauma. However, technological developments now enable us to anticipate shocks from earthquakes and minimize losses. One of the technologies that has been used, and is currently being further developed, is a damping device that is fitted to the building structure. There are various types of damping devices, each with different characteristics and systems. Multiple studies on damping devices have resulted in the development of various types, such as friction dampers (FDs), tuned mass dampers (TMDs), and viscous dampers (VDs). However, studies on attenuation devices are mostly based on the type of system and can be divided into three categories, namely passive, active, and semi-active. As such, each type and system have their own advantages and disadvantages. This study investigated the efficacy of a magnetorheological (MR) damper, a viscous-type damping device with a semi-active system, in a simulation that applied the damper to the side of a building structure. Although MR dampers have been extensively used and developed as inter-story damping devices, very few studies have analyzed their models and controls even though both are equally important in controlled dampers for semi-active systems. Of the various types of models, the Bingham model is the most popular as indicated by the large number of publications available on the subject. Most models adapt the Bingham model because it is the most straightforward of all the models. Fuzzy controls are often used for MR dampers in both simulations and experiments. This review provides benefits for further investigation of building damping devices, especially semi-active damping devices that use magnetorheological fluids as working fluids. In particular, this paper provides fundamental material on modeling and control systems used in magnetorheological dampers for buildings. In fact, magnetorheological dampers are no less attractive than other damping devices, such as tuned mass dampers and other viscous dampers. Their reliability is related to the damping control, which could be turned into an interesting discussion for further investigation.

Keywords: magnetorheological; MR damper; seismic; structure; earthquake; modeling; control

Citation: Lenggana, B.W.; Ubaidillah, U.; Imaduddin, F.; Choi, S.-B.; Purwana, Y.M.; Harjana, H. Review of Magnetorheological Damping Systems on a Seismic Building. *Appl. Sci.* **2021**, *11*, 9339. https://doi.org/10.3390/app11199339

Academic Editors: Antonio Concilio, Salvatore Ameduri, Ignazio Dimino, Vikram G. Kamble and Rosario Pecora

Received: 9 August 2021
Accepted: 30 September 2021
Published: 8 October 2021

1. Introduction

Earthquakes are responsible for the loss of thousands of lives and billions of dollars in damages to properties. Therefore, designing a structural building capable of withstanding seismic forces has become a primary concern in the field of engineering [1]. The most common conventional design approaches largely depend on a structure's ability to absorb an earthquake's input energy by deforming inelastic sections that were exclusively designed into a structural system while preserving the structural stability and integrity. Engineers have attempted to design and develop technologies that prevent and mitigate seismic hazards, such as earthquakes [2].

The development of seismic engineering technology goes hand in hand with discussions on earthquake disasters. Although humans cannot escape natural disasters, they can be mitigated to minimize damage [2,3]. To date, seismic engineering specialists have gained only a limited understanding of crustal behavior. As such, predicting the magnitude, time, and distance of an earthquake is exceptionally complicated and beyond the ability of scientific knowledge at times as earthquakes occur spontaneously [4–6].

Earthquakes are vibrations generated within the Earth by sudden movements when sheared rocks are stretched beyond their elasticity [7]. They are a unique phenomenon produced by tectonic activities within the Earth's crust. Earthquakes that occur in a region can, to a certain extent, be felt in the surrounding areas [8]. The impact of an earthquake is random and is measured by its magnitude. Large-magnitude earthquakes result in property damage in the earthquake region. Although the ground motion amplitude of an earthquake is determined by its magnitude, distance plays a crucial role in the mitigation of seismic hazards. Therefore, the characteristics and potential impact of an earthquake must be understood during a seismic hazard analysis [9–12]. Earthquakes occur when tectonic plate movements accumulate energy. When released, this energy radiates in all directions in the form of earthquake waves [13,14]. Earthquakes commonly produce two types of seismic waves: (1) body waves and (2) surface waves. These seismic waves often occur only after an earthquake or an explosion [13,15–17].

A body wave spreads throughout the inside of the Earth and causes minimal damage to building structures. These waves can be divided into two types: (1) P waves (longitudinal waves); and (2) S waves (transversal waves). A P wave, or a longitudinal wave, has a very swift time propagation compared with other waves. The velocity of a P wave is between 1.5 km/s and 8 km/s in the Earth's crust and can spread through solid, liquid, and gas media. An S wave, or a transversal wave, has a slower spread than a P wave. The velocity of an S wave is typically 60% to 70% slower than a P wave. The direction of particles in an S wave also differs from that in a P wave in that it is perpendicular to the direction of the propagation wave. A surface wave can be likened to water, as a surface wave spreads on the Earth's surface [18–23]. It not only has a slower propagation time than a body wave but a lower frequency as well. As such, surface waves are more damaging to building structures than body waves [24,25]. The amplitude of a surface wave declines rapidly as the depth increases. This is due to the dispersion of the surface wave, with the wave factorization based on the wavelength of the propagation wave [26,27].

Potential damage can be estimated and understood using the peak ground velocity (PGV), which relates to intensity. Trinufac and Brady (1975) established an empirical correlation between PGV and the modified Mercalli intensity scale (MMI); however, their regressions focused on intensity only [28,29]. As such, a PGV estimate makes it possible to approximate the MMI. Other studies have also conducted observational regressions of PGV intensity to estimate the correlation between PGV and MMI [30–38]. Wald et al. investigated the correlation between PGV and intensity; however, they used peak ground acceleration (PGA) in association with PGV to derive a correlation to intensity. This correlation is used when making earthquake maps [39]. Kaka and Atkinson created a new correlation to map earthquakes in eastern North America. They found the equation used by Wald et al. to be incompatible with the application of their study. This occurred as a result of different ground motion properties, where the frequency is much higher in stable continental regions such as eastern North America. As such, they modified the equation to use PGV to estimate high and low intensities [34,39]. Gerstenberg et al. (2005) then used the equations created by Kaka and Atkinson as well as Wald et al. to investigate a similar case [40].

The current design methods employed for seismic activities allow structures to undergo plastic deformation during large earthquakes while remaining elastic during small and medium earthquakes. The design of a structural damper must generally fulfil requirements for both wind and earthquake actions [2,41–44]. However, the calculations behind designing for the effects of wind forces and earthquakes differ. For instance, structural designs for wind loads use force as a consistent basis as the pressure on the exposed surface area tends to hit a building directly. This is called force-type loading. In structural designs for earthquake loads, building structures are more likely to experience random movements and vibrations from the ground floor. This causes an inertial force within the building and results in stress. This is called displacement-type loading [45–48]. Another way of expressing this difference is through the load–deformation curve of the building; i.e., the demand on a building is the type of force imposed by the displacement-type wind pressure imposed by random earthquake shocks.

Another distinguishing feature between force-type loading and displacement-type loading is the load–deformation curve of a building. The demand on a building structure is the type of force imposed by the pressure displacement-type loading due to random shocks [44]. Shakes from an earthquake are time-variant and have very random waves and accelerations. On the other hand, most of the designs represent the inertial force due to the earthquake as a definite effect of random vibrations [2,49,50].

The ground motion that occurs during the earthquake process is cyclic to the neutral position of the structure. Therefore, in a seismic action, there are many complete references and the time duration of the earthquake tends to be small, thereby causing stress on the building [45,49]. Several studies on the statistical equations of seismicity have been carried out without separating the use of ground motion data sets in terms of distances to the nearest fault line [51–55]. The engineering field has extensively investigated and developed algorithms and control systems. In this case, they are algorithms and control systems for mitigating natural disasters, particularly earthquakes [54–56].

Over the last few decades, the field of structural engineering has extensively investigated the development of smart structure technologies for structural seismic response controls. This is because smart structures are considered to be effective at handling earthquakes. Rather than rebuilding a structure, some studies have used a large number of innovative devices and systems to protect buildings from earthquakes [56].

However, as earthquakes have several kinds of waves, they may still damage a building. Therefore, several technological devices, such as friction dampers (FDs), tuned mass dampers (TMDs), and viscous dampers (VDs), that purport to adapt to seismic activity are considered capable of adequately handling earthquakes. Recently, smart materials, such as shape memory alloys and magnetorheological (MR) fluids, have been combined with devices to absorb seismic energy. However, semi-active systems have received more attention because they offer impressive adaptability and a very low power requirement [57–61].

As most flexible structures, such as heavy and tall buildings and bridges, have minimal structural damping and are prone to dynamic excitation from wind or seismic vibrations, different control mechanisms are used to prevent structural failures. There are three types of control systems for building structures: passive, active, and semi-active or hybrid systems. Although passive control systems have been extensively studied [62], active and semi-active systems have also been investigated as a control for structures [63].

2. Types of Dampers on Structural Buildings

2.1. Friction Dampers

Friction dampers (FDs) generate the desired energy discharges via the friction produced by two solid bodies sliding relative to each other. This is a common process used in the engineering field. It can also be applied to seismic building structures. This friction can also be used, on a smaller scale, to absorb kinetic motion energy [63–65]. As such, Pall et al. (1980) developed passive FDs to improve a structure's seismic responses. This was based on the resistance developed between two interfaces to remove a number of different input energies. During seismic stimulation, the device was found to provide the desired amount of energy dissipation under a predetermined load. It was also found to be immune to thermal effects and have reliable performance and stable hysterical behavior [66].

The device proposed by Pall et al. (1980) for the seismic control of large-paneled structures joined brake pads between steel plates to provide a consistent response to force shifting [66]. Some building structures in Canada have implemented a modern version of this device model [67]. Another similar study applied a copper alloy friction bearing to a Sumitomo friction damping device that slides along the inner surface of a cylindrical steel casing. The action of the spring on the resulting inner and outer sections provided the required normal force [68,69]. The device was able to reduce the displacement in comparison with the original configuration of the structure. The small modifications that were made in the new configuration at the base increased the shear forces and acceleration in some cases.

The reduction in seismic energy received by the device was due to the values of the forces acting on the FD. The load at which slip begins to occur and the damper begins to separate the seismic energy is called the FD slip load. This slip load determines the energy dissipation capacity of the FD. In this case, the FD had an optimal slip load. This corresponds to the lowest number of responses from the structure. A 10% to 15% variation in the mean slip load was sufficient to influence many structural responses. Multiple studies have arrived at the same conclusion [68].

As FDs have nonlinear behavior, nonlinear time history analysis is required to analyze the friction of damped structures. Large, rectangular, and almost sturdy FD hysteresis loops indicate a higher energy dissipation capacity. Therefore, a damped brace that generates the FD slip load can model the FD provided that it is in an elastic clamp.

Several studies have investigated the performance of FDs, with some focusing on FDs installed in reinforced concrete (RC) frame structures. The Structural Analysis Program2000® (SAP2000®) software was used in these cases to investigate samples from seismic zones in India. The skid load determination was adjusted to find the response of the frame without damping and the frame with a FD. The obtained results were then compared and the decrease in displacement and mean force were estimated [70,71].

The design of the slip load significantly affects the efficiency of the FD. The responses of a seismic structure can be described as the application of energy to the structure and the energy dissipation of the structure. As such, the difference between these two energies can be minimized through an optimal design [67,72,73]. However, the energy lost due to friction within the structure must also be maximized [74]. Several studies have investigated multiple optimal viscous and viscoelastic reducing designs [75–77]. However, only a handful of studies have investigated the optimization of FDs for seismic loads. Filiatrault and Cherry investigated the slip load spectrum of a design to maximize the energy dissipation while taking into account ground motion and structural properties. This study was performed to determine the optimal slip load distribution for FDs [72,78]. The authors also investigated the optimal slip load value for peak ground acceleration. To that end, it was assumed that the FD would not slip when subjected to strong wind loads. The FD was assumed to only slip when exposed to a strong earthquake load. Moreschi and Singh used genetic algorithms (GAs) to determine the optimal placement of friction absorbers on structures and metals in a reinforced steel frame. Fallah and Honarparast also investigated the use of GAs to optimize the slip load distribution and placement of Pall

FDs. This was accomplished by developing a nondominant genetic sorting algorithm for multilevel shear wires. Patro and Sinha found that a shear frame structure with dry friction produced a uniform slip load distribution for the optimal seismic response with various ground motion characteristics to determine the appropriate slip load [72,79,80].

Some researchers have concluded that retrofitting existing buildings with a passive control system for seismic activities is the most convenient solution. One study suggested using a FD and a bracing system in a steel frame. This not only reduced the seismic demand but also increased the ductility to generally improve the seismic performance [81]. The advantages of this method are the development of a stable, rectangular hysteresis loop and independence under environmental conditions such as temperature and loading rate. In addition to FDs, masonry filler panels were found to increase the seismic resistance of steel structures by increasing their lateral strength and stiffness as well as reducing the drift in the storyline. Zahrai et al. investigated the effect of brick-filling panels on the seismic performance in a four-story, three-span steel frame with Pall® FDs. They found that FDs in the steel frame increased the ductility and decreased the drift to less than 1%. However, the infill panels not only fulfilled their function during the imposition of drift but also increased the structural strength. Therefore, the use of infill panels in conjunction with FDs was found to reduce the number of dynamic structural responses as the infill panels dissipated the input energy of the earthquake by 4% to 10% depending on the thickness [80].

Borislav and Mualla proposed a new type of FD device for a steel frame story with seismic loading. At the time, the novelty was that the proposed material provided a very stable performance over a long cycle, was resistant to adhesive wear, and did not damage the surface of the steel plate. Their experimental study assessed the friction-bearing material, the performance of the damping unit, and the response of the scaled model's frame to harmonic excitation. This study was almost identical to the study by Zahrai et al., which added infill panels to a brick building. In any case, Borislav and Mualla found an increase in the dynamic response in comparison with conventional methods [80,82].

Various innovations have been used to improve the performance and demonstrate the efficacy of FDs in building structures. Software has also been used as an evaluation tool in several case studies. Baratia et al. studied the seismic behavior of a building that had been retrofitted with FDs. The seismic performance of steel structures that were either 6, 9, or 12 storys tall was evaluated with and without the use of dampers. A finite element modeling technique, namely SAP2000®, was used to analyze the performance of friction absorbers in these asymmetrical structures. The study found a significant increase in seismic behavior, which demonstrated the efficacy of FDs as tools for seismic reinforcement in these buildings [83]. Chandra et al. used the Extended Three-Dimensional Analysis of Building Systems® (ETABS®) software to dynamically analyze the nonlinear dimension time history of a new structural system of FD frames in an 18-story apartment building. The critical 5% viscous damping value was assumed to be in the initial elastic stage in order to account for the presence of nonstructural components [84].

In contrast, Shao et al. developed a cross-braced FD to seismically enhance a 36-foot tall concrete shear wall building with soft floors that had been built in the 1970s in the greater Seattle area. The first two floors of this building consisted of a relatively rigid concrete podium, while the subsequent two floors had concrete columns to support the sliding concrete walls of the upper floors. Therefore, the two floors above the podium were soft and considered to be earthquake-prone during seismic events (10%/50 years). The floor's reinforcement involved long clamps and tension bars at the junction of each clamp, which was where the FDs were placed. Two 890 kN capacity cross-supporting FDs were installed within the 12 perimeters of the soft floor, bringing the total number of friction absorbers used to 24. Two silencer prototypes were designed, built, and tested by Pall Dynamics Inc. to ensure the performance. The installation of these dampers was completed in 2005. The dampers were found to increase the structural integrity, improve the building's safety, and help minimize structural and nonstructural damage after an earthquake, thereby reducing the potential downtime and repair costs after a seismic event [85].

An effective mechanism that uses a similar principle in the case of seismic damping is the triangular-plate-added damping stiffness (TADAS) damper. The TADAS damper is an effective mechanism for reducing the energy input to structures during an earthquake via metal deformation that is not elastic. In general, the damper device consists of several triangular plates welded to the same base plate. Each triangular plate is inserted into the perforated base plate before welding. During an earthquake, interstellar drift causes movement of the upper end of the TADAS damper relative to the lower end. This causes the metal plate to be removed from the damper and, as a result, its energy is dissipated [86,87]. In addition, there are other damper models, such as shear panels or plates. Shear walls are innovative lateral force retaining systems that are able to effectively help buildings resist wind and earthquake forces [88–92]. Model shear panels or plates have properties that are fundamentally useful to resisting seismic forces and increasing the initial rigidity of a tall building structure. This model also has inherent redundancy and a substantial energy dissipation capacity, reducing the erection time and structure costs. Most importantly, compared with other shear walls, such as equivalent reinforced concrete, the low weight of the shear plates reduces, by about 30%, the gravity-related mass and seismic force (inertia) transmitted to the foundation [93–97]. Steel plate shear walls were in use from the 1970s to the mid-1980s. The failure mode of the plate shear wall is considered to be out-of-plane buckling of the inner plate. This resulted in experts building structures that are very rigid, which offers an economic advantage over concrete shear walls. In Japan, the trigger has been designed to be a rigid trigger or a thick plate to prevent local buckling [98–102]. Apart from the use of steel, pure aluminum is also used as a base isolation for the manufacture of innovative passive energy dissipation devices. Base isolation is the most effective method for designing seismic buildings. The pure aluminum is also intended to provide seismic protection in framed buildings. The use of this metal can be said to be a novelty in the field of civil engineering that is quite interesting [103].

2.2. *Tuned Mass Dampers*

Tuned mass dampers (TMDs) are a type of passive device often used to control the response of buildings and bridges. A TMD consists of a mass–spring system attached to the main structure of a building or bridge and is usually installed on the roof of the structure to counteract the ground's motion to reduce the dynamic response of the structure.

The dissipation of energy is achieved by the inertial damper forces acting on the structure [104,105]. These systems are primarily efficient at controlling wind-generated vibrations in lean structures, such as towers and tall buildings. TMDs can be classified into three categories: tuned liquid dampers; TMDs; and tuned liquid column dampers.

A TMD is the classic configuration of a tuned damper and consists of a properly adjusted damping element and a secondary mass with a spring to increase the attenuation of a primary structure resulting from frequency-dependent hysteresis. Therefore, TMDs can be credited with increasing the damping of a structural system. Although these devices effectively reduce wind-excited responses to stationary narrowband stimuli, they are less effective for broadband excitations, such as earthquakes [105].

The scheme of a TMD is shown in Figure 1 [106]. It consists of a mass, spring stiffness in the structure, and damping for earthquake excitations. Numerous numerical, analytical, experimental, and optimal structural studies have been carried out to determine the efficacy of TMDs in reducing the seismic response of structures. Lin et al. developed and implemented TMDs in building structures [106]. The study used real earthquake data in its numerical and statistical analysis of torsion in an installation in a five-story building. The results show that passive TMDs effectively reduced the response of the building during an earthquake. A similar experimental study conducted by Zuo et al. developed a TMD with multiple degrees of freedom and optimized the two degree of freedom system [107].

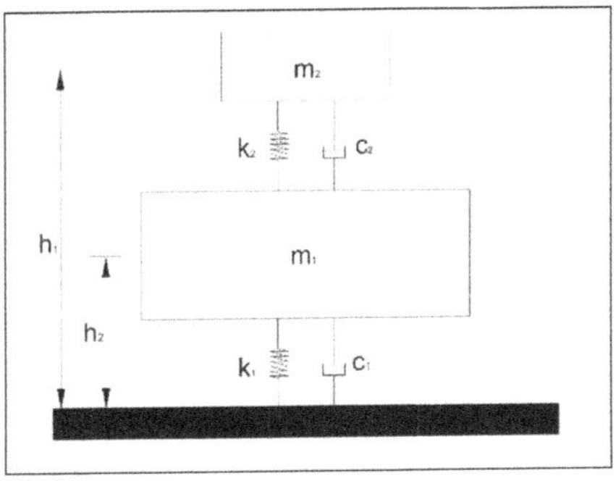

Figure 1. The primary structure of a TMD system.

During its development, the TMD was quite a popular device, and it has been applied to several building structures. Several theoretical and experimental studies have been conducted on TMDs. Peter (2006) theoretically and experimentally analyzed TMDs. Marano et al. (2007) used linear TMDs to control seismic excitation in buildings with limited reliability based on technical optimizations. While both Marano et al. (2007) and Peter (2006) agree that TMDs optimize a building's response to seismic loads [108,109], Pinkaew et al. found that TMDs were not effective at reducing vibrations [110].

Not all studies on TMDs have analyzed their performance. Other studies investigated TMDs in terms of cost optimizations [111–115]. Huang et al. compared the redistribution of structural materials and the application of TMDs to benchmarks for reinforced concrete structures with a 60-story steel belt frame with wind loading. The study concluded that a redesign of the building's standards, to meet inter-story deviations and the peak acceleration, would require a 14% increase in initial material costs, while using adjustable mass dampers would only require a 3.6% increase in the initial structural material costs [112]. However, TMDs can only reduce the free vibrations of a tall building structure. This also only happens when the earthquake ends. This is because TMDs have the task of mitigating the structural acceleration caused by wind on tall buildings. Various types of TMDs, including both passive and active TMDs, have been applied in several buildings in multiple countries. The most recent application of a TMD is shown in Figure 2.

The development of the TMD also gave rise to the new idea of using liquid as a mass to balance the structure of a building when it vibrates. A tuned liquid damper (TLD) is a type of TMD. In this case, the mass used is replaced with a liquid (usually water). The effects of external vibrations are balanced through the movement of water in the tube or tank. This concept is not new, and it has been used to stabilize ship structures for many years [114]. In TLDs, the horizontal tube movement controls the fluid flow, which is adjusted to a certain frequency [115–121]. However, the application of a TMD to a building structure, especially in order to overcome earthquake loads, requires further examination before it can become a reality.

Figure 2. Tuned mass damper applications: (**a**) 101-Taipei, Taipei—Taiwan; (**b**) Aspire Tower, Doha—Qatar; (**c**) Sanghai World Financial Center, Sanghai—China.

2.3. Viscous Dampers

One of the countermeasures for reducing seismic vibrations is to use an evaluation of the seismic design approach. In order to evaluate such a seismic design approach with nonlinear dynamic analysis, a robust model for the steel device and damper depending on the rating is required. Accurate models have been proposed for viscous, viscoelastic, and elastomeric fluids that are rate dependent [122,123]. Fully active control systems apply both dissipative and non-dissipative forces to a structure. A properly designed fully active control system is capable of significantly increasing the damping in comparison with a passive system. However, controlling these devices requires a large amount of power and suffers from instabilities due to time delays. Moreover, these active systems will probably malfunction in the event of an electrical failure or damage as the control system ceases to operate and the damping device cannot change the system to a passive system. In this case, an active system provides a wider range of technologies. However, the actuator controller relies on an external power source to apply force to the structure. Therefore, providing optimal responses in relation to demand can be accomplished by adding or removing the applied force [124].

In control engineering, an active system consists of four interconnected components: (1) the plant (i.e., the building); (2) the sensors; (3) the controller or control computer; and (4) the actuators. Each of these components functions as a subsystem and is integrated in a closed feedback control loop such that the output of one component is the input of another component [125]. Although full-scale active systems have been implemented in research structures, the lack of reliability and the lack of cost-effectiveness have limited their widespread use in non-research models, which prefer to adopt passive systems instead. Therefore, full-scale active systems are still under investigation by many users and researchers. However, many studies have developed damping devices that optimize the cost-effectiveness and the potential reliability to provide a solution for both passive and active systems. As such, researchers have shifted their focus to semi-active systems in recent times [126,127].

The development of damping devices that use semi-active systems progressed rapidly as they require significantly less energy to control than active systems. This is because semi-active systems do not require the addition of mechanical energy to the structural system. Therefore, the bounded input guarantees the stability of the bounded output. This type of device is also referred to as a controllable passive system [126,128]. Semi-active systems provide electrical safety as, if the electrical device malfunctions or fails, the system converts into a passive system. This differs greatly from an active system, which relies on a considerable amount of power to control the system and increase performance. Semi-active systems do not require tremendous amounts of power to increase performance as feedback from the external sensors in the internal control mechanism dictates its dissipative or resistive power. This enables the system to inversely combine elements from active to passive or vice versa [128].

Semi-active systems effectively minimize structural damage under large environmental loads and provide three main advantages over active and passive systems. As they are very dissipative, they do not add energy to the system, thereby guaranteeing stability. Unlike passive systems, semi-active systems are also able to change their structural behavior due to nonlinearity, degradation, or damage over time. A recent study on semi-active systems demonstrated their efficacy for seismic control using viscous fluid dampers. Magnetorheological (MR) dampers are relatively inexpensive semi-active devices that are not only reliable but also provide electrical failsafes.

3. Magnetorheological Dampers for Structural Buildings

3.1. Magnetorheological Fluids

When subjected to a magnetic field, magnetorheological fluids change their rheological behavior in response to growing yield stresses. As such, MR fluids have great potential in the development of electromechanical devices as they provide a simple, responsive, quiet, and quick interface between mechanical and electronic control systems [129–131].

MR fluids were first discovered by Jacob Rabinow and have only grown in popularity since then. MR fluid is considered to be a multifunctional intelligent fluid as it can be rapidly modified and reversed in a short period of time (milliseconds) when a magnetic field is applied. In the absence of a magnetic field, MR fluids behave like Newtonian fluids. The magnetic field applied to MR fluids changes the arrangement of particles to form a chain-like shape. This chain-like shape modifies the fluid's rheological properties by drastically changing the value of the viscosity. This change in viscosity results in yield stress changes depending on the magnitude and direction of the applied magnetic field. The characterization of the rheological behavior of these fluids occurs at two stages: pre-yield and post-yield [132,133].

Due to the unique characteristics of MR fluids, the application of magnetic fields can dramatically change their rheological properties. MR fluids have been used to successfully develop various brake systems, dampers, and other devices. In most of these applications, the surface of the device is in contact with the MR fluid because it requires relative motion, such as linear motion in the case of shock absorbers, to operate. However, the surface of the device wears out more rapidly due to the abrasive nature of the iron particles in the MR fluid. As such, selecting the right surface material is essential to ensure high resistance to wear as well as durability [133,134].

The main components in a MR fluid formation are the carrier fluid, magnetic particles, and the additives. Carbonyl iron, with a purity of 99%, is often used to provide the magnetic particles due to its high magnetic permeability and magnetization saturation [135]. Moreover, the chemical deposition of pentacarbonyl iron vapor produces carbonyl iron particles, which usually form into spheres, thereby reducing wear and tear on the walls of devices that use MR fluids [136,137]. However, the findings of one study suggest that fiber particles provide better yield stresses with lower viscosities. In some applications, carbonyl iron particles range between 3 μm and 5 μm in size with particle concentrations between 20% and 40% depending on the volume. As such, the likelihood of wear from erosion and friction is nominal due to the particles' minute size [138].

Oils with low viscosity, such as silicone oil, mineral oil, and other synthetic oils, are used as carrier fluids as they are excellent at forming MR fluids due to their wide range of viscosity-changing objectives. However, the carrier liquid must be non-reactive with the iron particles. Silicone oil is often used as a carrier fluid in vibration control applications due to its high viscosity index, low vibration, high shear strength, and high flash point [135].

The additives used are usually surfactants that prevent agglomeration and reduce the deposition rate of the magnetic particles [139]. This is essential as high-density particles tend to settle, which can render the device ineffective if left untreated [138]. Additives such as oils, thixotropic agents [140], and Span® 80 and TWEEN® 80 emulsifiers are often used to improve the sedimentation stability [141], while organic acid and stearic acid are often used to increase the density of the carrier liquid [135] and stabilize the sedimentation [138].

3.2. Application of MR Dampers in Building Structures

Due to their superior quality, MR dampers are in high demand for use in scientific case studies and in several real-world industrial applications. Carlson and Weiss concluded that MR dampers provide good operating reliability as temperature fluctuations and impurities in the fluid did not affect the performance and functionality. However, a significant drawback was their nonlinear characteristics, which involve hysteresis (force vs. velocity and force vs. displacement). This makes accurately modeling MR dampers and developing an efficient algorithm for improved performance a challenge [142,143].

Over the past 15 years, MR dampers have been widely used in a variety of fields for vibration control. These include [144–147], but are not limited to, building structures [148–151], bridges [149,150], suspension systems in automotive and high-speed trains [152–154], advanced artificial limb systems [155], large washing machines [156], landing gears for airplanes [157,158], commercial vehicle seats [159], complex mechanical systems [160], and rotor systems of helicopters [161]. However, despite their many advantages, MR dampers are difficult to commercialize due to their complex structures and user-dependent configurations [136,162].

Cesar and Baros (2010) discussed the use of semi-active structural control techniques within the framework of a civil engineering experimental model equipped with MR dampers. The study aimed to develop a semi-active damping device in the Comparison of Vibration Control in Civil Engineering Using Passive and Active Dampers (COVICO-CEPAD) project in the Eurocores S3T program framework. An experimental frame was set up at the Faculty of Engineering of the University of Porto (FEUP) to explore the calibration of MR dampers as well as identify the dynamic properties of small-scale metal frames with and without the use of specific MR devices (Figure 3). Some of the results from the simulation of the controlled frame under earthquake conditions were compared with the experimental results from a frame installed on a Quanser® Shake Table II [163,164].

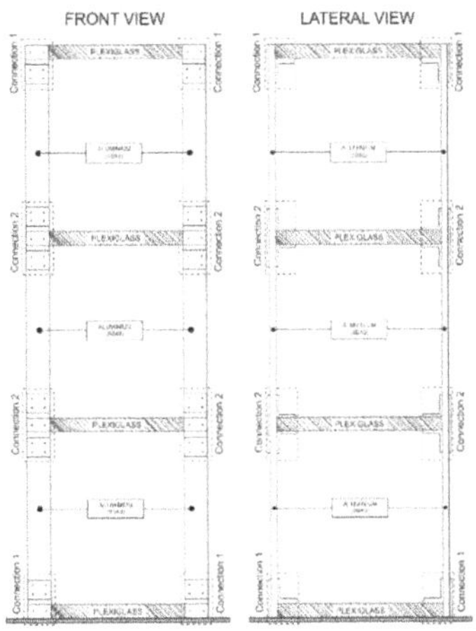

Figure 3. Building structure with a metal frame and an installation [163].

Data and characteristics that had been obtained from several earthquakes were used in the study to evaluate the performance of structures fitted with semi-active controllers. This information was fed into and calibrated by a Quanser® Shake Table II to experimentally and numerically compare the various control strategies. However, the study only yielded results when certain earthquake inputs, such as the El Centro earthquake, were used. In the numerical example, a three-storey structure was controlled using a MR damper on the first floor. The simulation showed that a clipped optimal control algorithm produced improvements in uncontrolled systems [163,164].

Another study investigated the performance of two variations of passive and semi-active damping devices in a 12-story bearing frame (Figure 4). A segmented and insulated floor constituted the upper floor. The passive and semi-active devices were installed to reduce the seismic response of the structure [165].

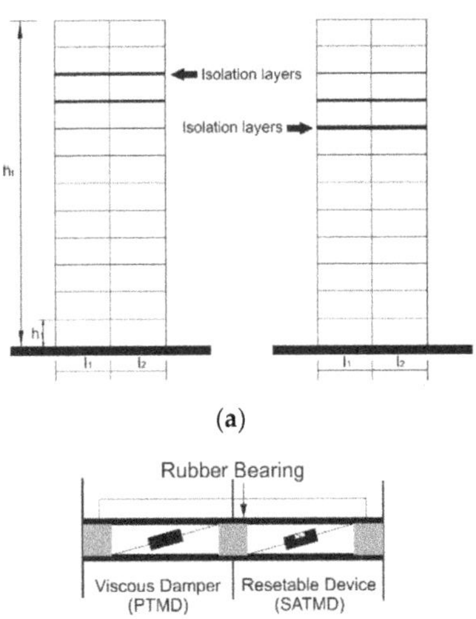

Figure 4. Segmentation of the 12-story building: (**a**) Sketch of the 12-story building structure with '10 + 2' and '8 + 4' models; (**b**) Schematic of isolated sections [165].

The study used a semi-active control approach. To manipulate reaction forces, feedback controls that reset the system according to structural responses were applied to the device. A simplified two degrees of freedom system was adopted as the control parameter of the semi-active damping device. Response reduction factors of various seismic intensities were analyzed using time history analysis. The study found that the semi-active damping device effectively reduced the movement of the structure under a given seismic load in comparison with the passive system. The variability in the study was more stringent for the semi-active system for all ground motions used and indicated a more robust control system [165].

Mehmet Eren Uz compared the seismic application of MR dampers in two buildings of different sizes under optimal parameters. His study investigated the efficacy of a damping device with semi-active systems in optimizing the responses of two structures that are located close to each other and connected to MR dampers under seismic loading. The challenge of this study was to effectively enhance the control strategy for MR dampers in two separate structures (Figure 5) [166].

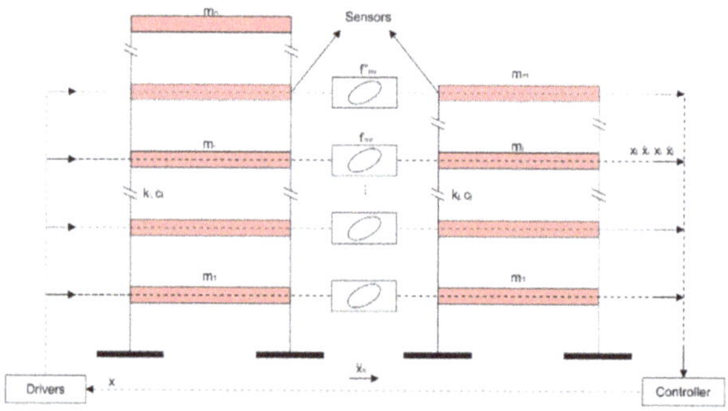

Figure 5. Schematic of an MR damper between two buildings [166].

Depending on the requirements, the control force and the number of damping devices were calculated and analyzed using MathWorks® Simulink, a MATLAB-based graphical programming environment. The linear–quadratic regulator (LQR) algorithm and linear–quadratic–Gaussian (LQG) control were used to achieve the desired control force while the law of truncated voltage was used to synthesize the voltage that was used. As a result, it was possible to control the displacement response of the MR dampers at lower stresses for shorter buildings. Improvements in the performance of a damping device that is attached to two adjacent buildings were intended in order to design an optimal nonlinear hysteretic damping device. The stochastic response of the two adjacent buildings linked to a nonlinear damping device was efficient. The stochastic linearization method was used in the optimal design to avoid the need for multiple nonlinear time history analyses. The results showed that high-voltage applications were not needed as they were not effective for MR dampers. The proposed optimal design method achieved improved seismic performance that enhanced the productivity on the economic side [166].

Constantinou et al. conducted an experimental study on a three-story steel structure and a bridge structure. The three-story model was tested with and without fluid viscous dampers (FVDs) installed at an approximately 35° angle on braces. Four dampers were installed on the first floor and six dampers were installed in pairs. Figure 6 shows the schematic design of the building structure used in this study. The bridge model was tested using 15 different isolated system configurations. Four FVDs, similar to those used in the three-story building, were added to the insulation system. The study found that the FVDs effectively reduced the seismic response of the tested structures [167,168].

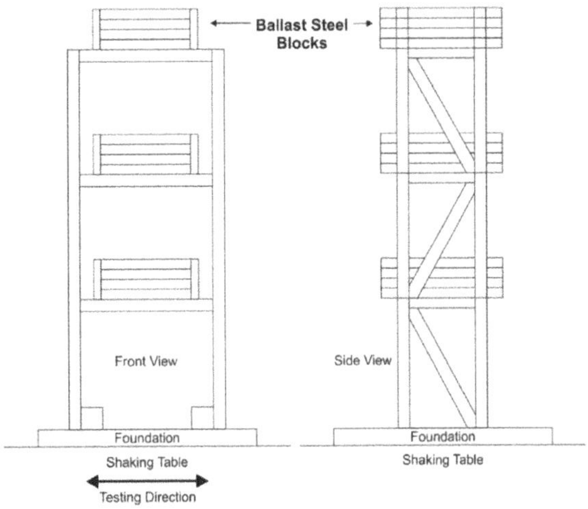

Figure 6. Model of the three-story building test structure [168].

Guo et al. designed and applied a seismic enhancement procedure to an existing building using FVDs and discussed some of the critical issues of using FVDs for seismic enhancement. This study included models of analytic dampers in major earthquakes as well as the layout of the dampers. The case study investigated the application of FVDs to a 21-story hotel that had been built in 1991 (Figure 7). The hotel was structurally modified and seismically upgraded by placing 56 damping devices on six selected floors due to limited space. A damping ratio of 5.3% was used in all the damping devices. The study found that the FVDs were able to minimize damage to the interior decorations of the building above the sixth floor as well as significantly reduce earthquake responses on the upper floors. Apart from that, the application of the damping devices made it possible for the hotel to execute short-term and economical construction projects [169].

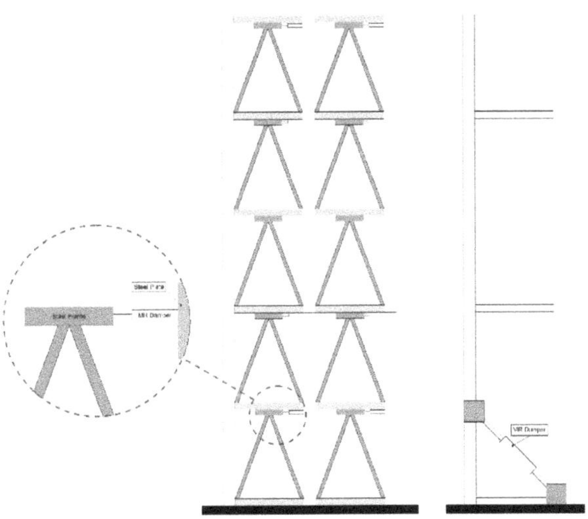

(**a**) (**b**)

Figure 7. Installation of a damper in a building structure: (**a**) Chevron-brace damper; (**b**) Diagonal-brace damper [169].

Wang and Muhin investigated the efficacy and financial feasibility of using damping devices to improve the seismic performance of a 35-story steel building. Three different types of additional damping devices were also used to prevent the building from collapsing. One of the types of damping devices investigated was a FVD. The factors taken into account in the study included architectural problems, programmatic issues, and construction capabilities. A simplified approach was used to achieve the same effective damping ratio and to characterize the mechanics of the case study. The approach was also used to mechanically characterize the floor displacement to make it consistent with optimal performance objectives. Based on a nonlinear dynamic analysis, the FVD was the most efficient of the three damping devices tested in this study. In terms of cost, the FVD was also the most cost-effective option for reducing economic losses arising from earthquakes [170].

In another study, Weber describes a new MR damper-style tracking control scheme. Installation balancing was performed by steady-state primary nonlinear compensation combined with the modeling mapping approach of the MR damper device. To that end, Weber reduced force tracing errors that were generated due to model imperfections. This was also due to parameter uncertainty, which was then reduced by the proportional feedback gain and the parallel integral, which is based on the absolute value of the actual device's damping force. In this case, Weber performed experimental validation of the force tracking control scheme on rotational and long-stroke MR dampers, demonstrating their robustness and efficacy. The results show that a more accurate force is obtained using the combined feedback approach compared with the other two methods used in the study [171].

Several applications of MR dampers in building structures were briefly described above. However, when MR dampers are used in a controlled damping device with semi-active systems in building structures, the models and controls of the MR dampers warrant closer examination than their application. Unfortunately, this is an area of study that has been grossly overlooked. Therefore, we examine and discuss the various types of models and controls that are used in MR dampers.

4. Modeling and Control of MR Dampers for Structural Buildings

4.1. Modeling of MR Dampers

The structure of a building is exceptionally vulnerable to seismic loads, especially seismic loads at higher frequencies [172,173]. Additionally, the control systems used are widely discussed [174,175]. This section discusses models and controls that have been used in MR dampers in previous studies [176,177]. Small-scale MR dampers are mostly described using the Bingham model due to its simplicity. This model consists of dashpot and friction elements connected in parallel, while the damper force is formulated as shown in Table 1.

The control scheme based on the Bouc–Wen model presented by Weber allows for real-time tracking of the desired control style. This control scheme is intended for MR damping devices without feedback from the force sensor. Weber used several Bouc–Wen models to estimate the MR damper force by parallel computation with a constant current. The MR damper's current is determined by a piecemeal linear interpolation scheme. Numerical and experimental validations were carried out. Tests show that the real-time control scheme is numerically stable and a force tracking error of not more than 0.078 indicates acceptable accuracy. One of the model studies conducted by Weber is shown in Figure 8 [178].

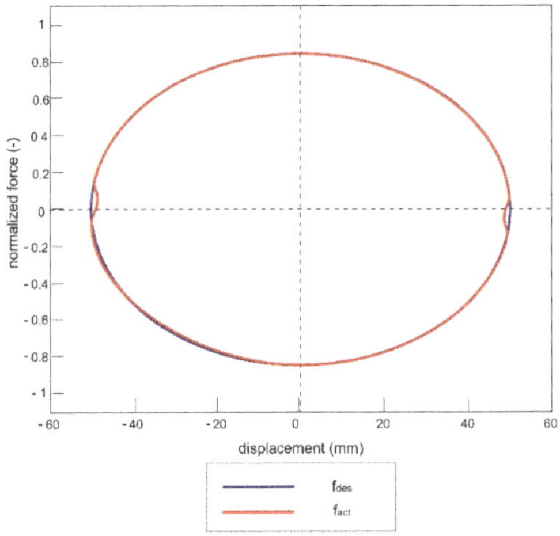

Figure 8. Simulated tracking of viscous damping: force displacement trajectories.

Table 1. Basic models applied for MR dampers.

Model	Equation	Figure
Bingham Model	$f = c_1\dot{x} + f_0 sgn(\dot{x})$ where c_1 and f_0 are equal to the damping coefficient and the shear friction force, respectively; and *sgn* () is a function of the signum. The damping force is linearly dependent on the damper speed, while the friction force depends on the velocity [177,179].	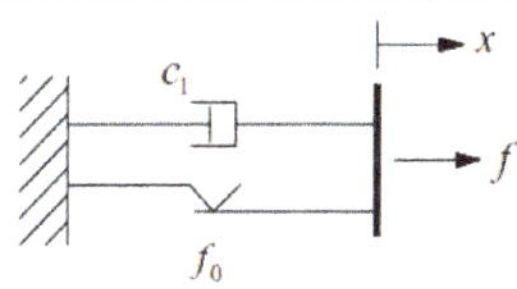
Gamota and Filisko Model	$\left.\begin{aligned} f &= k_1(x_2 - x_1) + c_1(\dot{x}_2 - \dot{x}_1) + f_0 \\ &= c_0\dot{x}_1 + f_c sgn(\dot{x}_1) + f_0 \\ &= k_2(x_3 - x_2) + f_0 \end{aligned}\right\}, \text{ if } \lvert f \rvert > f_c$ $\left.\begin{aligned} f &= k_1(x_2 - x_1) + c_1\dot{x}_2 + f_0 \\ &= k_2(x_3 - x_2) + f_0 \end{aligned}\right\}, \text{ if } \lvert f \rvert \le f_c$ where c_0 is the damping coefficient of the Bingham model, and k_1, k_2, and c_1 are the coefficients of stiffness and viscous damping, respectively, in relation to the linear solid material [179,180].	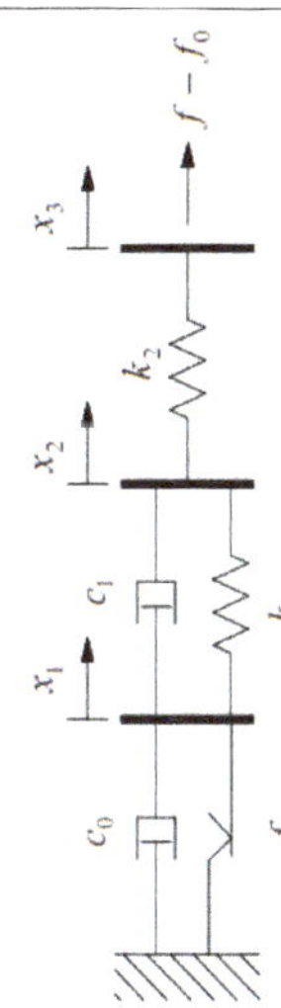

Table 1. *Cont.*

Model	Equation	Figure
Bouc-Wen Model	$f = c_1\dot{y} + k_1(x - x_0)$ $c_1\dot{y} = \alpha z + k_0(x - y) + c_0(\dot{x} - \dot{y})$ $\dot{z} = -\gamma\lvert\dot{x} - \dot{y}\rvert z\lvert z\rvert^{n-1} - \beta(\dot{x} - \dot{y})\lvert z\rvert^{n} + A(\dot{x} - \dot{y})$ k_1 represents the damper stiffness of the accumulator and c_0 is the dashpot coefficient associated with viscous damping at higher velocities. A dashpot c_1 was included in the model to produce the roll-off observed in the experimental data at a low velocity. k_0 is the stiffness control at a higher velocity, and x_0 is the initial displacement of the spring k_1 in relation to the nominal damper force due to the battery [180].	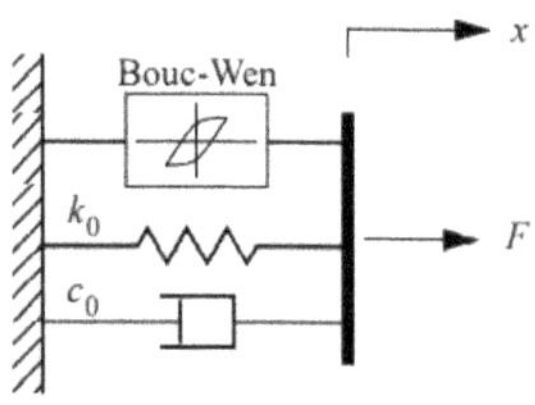
BingMax Model	$f(t) = k\int_0^t \exp\left(-\frac{t-\tau}{\lambda}\right)\dot{x}(\tau)d\tau + f_y sgn\left(\dot{x(t)}\right)$ where $a = c/k$ is the quotient of the dashpot c and the spring k, and f_y is the friction force on the slider [180].	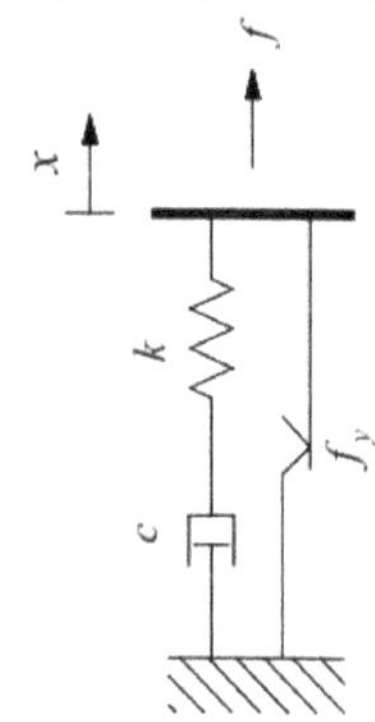
LuGre Model	$F_{ss}(v) = g(v)sgn(v) + f(v)$ where $g(v)$ is the Coulomb friction and the Stribeck effect, and $f(v)$ represents viscous friction.	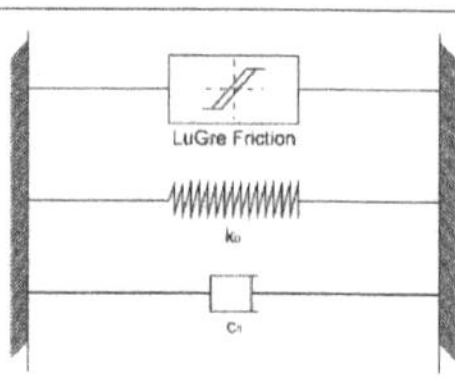

Ahn et al. conducted a MR damper modeling study to develop an alternative method for MR dampers. It used fuzzy self-tuning based on neural techniques as well as a control system that was validated using a number of simulations.

This case study used the Bingham model to describe the damper's behavior. The Bingham model's characteristic parameters were determined depending on the needs of the experiment. Figure 9 shows the predicted damping force generated in the Bingham model in comparison with the experimental responses.

When the acceleration had a negative value, the measured force had a positive value. Conversely, when the acceleration had a positive value, the measured force had a negative value. This occured when the velocity was zero. The self-tuning fuzzy model, a general model commonly used to improve MR dampers, was then utilized [177].

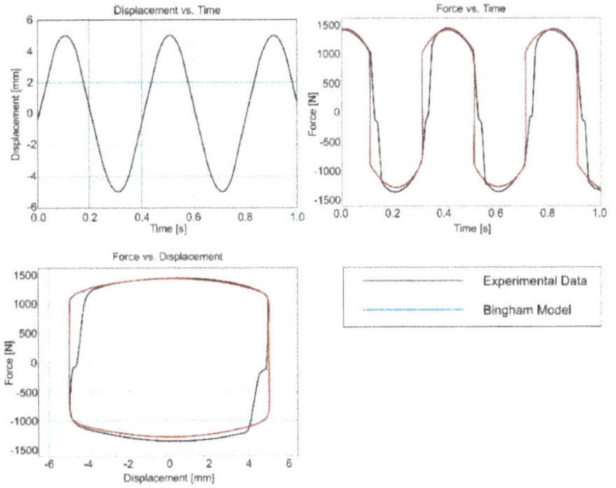

Figure 9. Comparison between experimental data and the predicted damping force [165].

Soltane et al. also used a self-tuning fuzzy model and a Bingham model in their study on MR dampers. The study proposed a new parametric dynamic model for MR fluid absorbers. It adapted the Bingham model and was set up to accurately reproduce the hysterical behavior of the damper. The optimal model parameters were then obtained by making the model's predictions as identical as possible to the experimental measurements. As one of the model's applications, the performance of MR dampers in reducing free and forced cable vibrations was investigated and numerically evaluated. The results of numerical simulations showed the accuracy and efficacy of the modified Bingham model in comparison with the standard Bingham model. A comparison of the restraining-wire-free vibration control of the standard Bingham model and the hysteretic-regularized Bingham model (HRB) is shown in Figure 10.

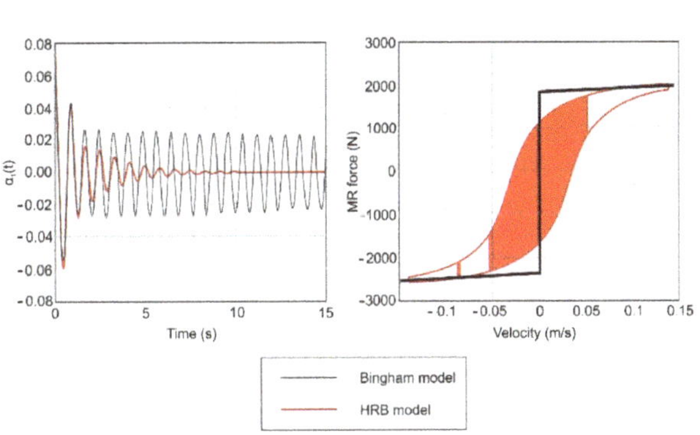

Figure 10. Control of restraining-cable-free vibrations: comparison between the proposed model and the Bingham model [179].

The results showed that the largest vibration response in the HRB model was approximately 70% of that obtained using the Bingham model. The speed of the vibration's decay was also much quicker. The equivalent viscous damping value of the Bingham model was 2.5%. However, the equivalent viscous attenuation value of the proposed HRB model was 9.2%. This indicated that the proposed HRB model was very effective at reducing structural responses [179].

Although the simplicity of the Bingham model helps to overcome several issues in MR damper case studies, some modifications are required when adapting the standard Bingham model to achieve better accuracy, such as those made by Ahn et al. and Soltane et al. [177,179].

Although Gamota and Filisko were the first to adapt the Bingham model to model electro-rheological dampers in 1991, Spencer et al. were the first to apply this model to MR damper modeling (in 1997). Their model consisted of a Bingham model in series with a standard linear solid model. The Gamota–Filisko models can be arranged with the equation shown in Table 1.

Spencer et al. determined that the parameters of the Gamota–Filisko model fit the 2.5 Hz data for cases in which the voltage to the driver was 1.5 V. The results of Spencer et al.'s study are shown in Figure 11 and provide a comparison between the predicted results and the experimental results. As expected, the force transfer behavior of the Gamota–Filisko damper models is well illustrated. Additionally, the predicted and experimental results on the velocity–force curve are almost identical. However, because the regulatory equation developed by Gamota and Filisko was still too rigid for this case, the numerical solution was difficult to handle. One of the main challenges of this model is deriving its numerical solution, which was also noted by Ehrgott and Masri (1994) [180].

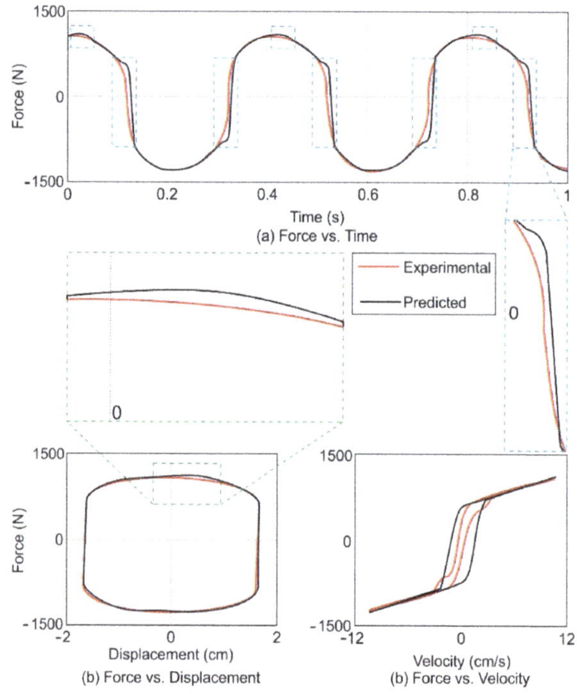

Figure 11. Comparison between predicted (black) and experimentally obtained (red) responses for the Gamota and Filisko model [180].

4.2. Semi-Active Controllers for MR Dampers

One of the advantages of MR dampers with semi-active systems is that they can be used in either passive or semi-active mode to control a building structure. When the device is not controlled, it becomes a passive system (Figure 12), where electrical current is constantly supplied to the MR damper. In this scenario, no feedback data are required. In this system, passive-on and passive-off controls correspond to when the maximum current and the minimum current, respectively, are applied to the damper [181].

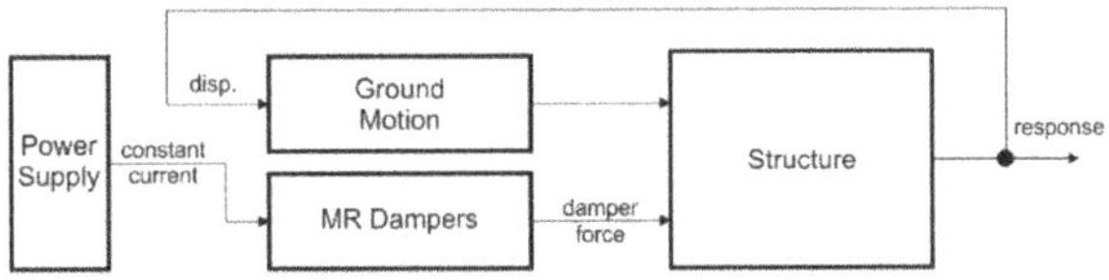

Figure 12. Block diagram of a passive system for a MR damper [181].

A semi-active control system is shown in Figure 13. As there is feedback, it requires controllers and sensors to provide feedback data according to their needs. In general, a semi-active control system can be defined as a device that does not require energy to be inputted into the controlled system. The current entering the MR damper is instead allocated to control the damping force by changing the magnetic flux intensity of the coil. However, under certain conditions, the current cannot change the direction of the damping force as this is performed by an active system damper. As such, a MR damper with a semi-active system requires several sensors, such as an accelerometer and load cell displacement transducers, to provide feedback so that it can optimally control the damping force. Therefore, several MR damper controllers, such as controllers based on a LQR, controllers based on a LQG, neural network controllers, and fuzzy controllers, have been developed.

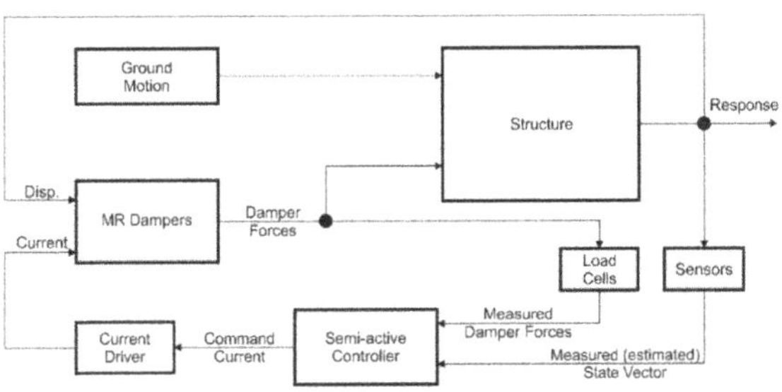

Figure 13. Block diagram of a semi-active controlled system for a MR damper [181].

A LQR is a control system that is often used to control structures via a feedback controller that aims to provide stable movement. Dyke et al. developed a similar type of control system—a H2/LQG-based clipped optimal control system—to counteract earthquake vibrations in three-story structures. Jansen and Dyke (2000) investigated the seismic responses of a six-story building that had been retrofitted with MR dampers on the first and second floors by using a variety of semi-active control algorithms, such as a truncated optimal controller, to control the building structure's response to ground motion. The study found that the algorithm that used the LQG method (the optimal truncated controller) was suitable for MR dampers in a manner similar to the Lyapunov control algorithm and the modulated homogeneous friction algorithm. The optimal truncated algorithm achieved a significant reduction in the seismic response (21.4% to 29.6%). These findings were corroborated by Yi et al. (2001), who also conducted a similar case study by applying a LQG controller to a small-scale six-story sliding building and compared the experimental responses obtained using a Lyapunov controller. The tests carried out in these two studies at various levels of excitation proved that MR dampers performed better than passive dampers in various situations. Additionally, semi-active systems were found to require only a small amount of power to outperform passive systems [182,183].

Neural network controllers are another type of controller investigated in the development of MR dampers. The use of these controllers is quite popular with researchers in the mechanical, electrical, and structural engineering fields. Unlike clipped optimal controllers, when modeling MR dampers, neural network controllers take into account the current input applied to the device and the piston velocity/displacement, which are often inaccurate. Therefore, the historical value of the input is used to improve the estimate. Wei et al. (2021) conducted a study in which the instantaneous variable was constructed via the Hilbert transform to represent the instantaneous characteristics of the excitation to further improve the accuracy of the MR damper model. The study suggested that, in the case of MR dampers, there is a correlation between the momentary variable and its nonlinear behavior. Therefore, studies related to this subject were investigated. In their study, Wei et al. repaired the MR damper neural network model for the first time using the instantaneous characteristics of the excitation. The results indicated that instantaneous characteristics play an important role in the model's accuracy. The study also found that the accuracy increased by more than 50% in comparison with the general model. Additionally, the target force could be tracked using the model's precise control strategy [184].

Fuzzy controllers are the most common type of controller used in MR damper applications for both structures and vehicles as the performance of traditional controllers, such as LQG and H2/LQG controllers, usually depends entirely on the accuracy of the system's dynamic modeling, while complex structural systems have nonlinearity and uncertainty in both the loading and magnitude of the structural properties. This makes it difficult to identify an accurate dynamic model for a traditional control design. As such, new control algorithms can solve these problems with very few optimizations. Several studies found that fuzzy controllers can solve this problem as well. Fuzzy logic control is based on set theory and consists of four components that simulate human reasoning: a fuzzy interface, a rule base, a decision-making interface, and defuzzification. Ndemanou and Nbendjo implemented a MR device with fuzzy control in seismic cases to investigate the nonstationary random responses of two adjacent tall buildings. Fuzzy control rules were used to model failure data by calculating the desired stress to produce reasonable control in each mode. As such, the process controls were optimized in conjunction with the first observation for the first vibration mode. This control strategy successfully increased the dynamic performance of the control devices, which could adjust the damping force by increasing the building's response without increasing the mechanical energy. Ndemanou and Nbendjo concluded that control algorithms are very important when optimizing the response of building structures to seismic loads [185].

Although this was corroborated by Mehrkian et al., unlike Ndemanou and Nbendjo, who applied MR dampers with fuzzy control in buildings, Mehrkian et al. used MR dampers as the base isolation of a building structure. The fuzzy control presented was also an intelligent multi-objective fuzzy–genetic controller. The controllers of their study aimed to improve the upper insulation characteristics and reduce the large bottom displacements under seismic loads. Unlike other controllers, the core of the fuzzy controller was conceptually constructed using factory-related control expertise. Moreover, a multi-genetic algorithm was applied to the fuzzy controller to optimize the controller's performance. This resulted in similar findings to those of Ndemanou and Nbendjo, where fuzzy controllers performed better than other controllers [186].

In addition, studies on control were also performed by Iemura et al., who argued about developing the law of autonomous semi-active control as a simplified semi-active control algorithm for seismic response reduction. The strategy of this method is to generate the desired hysteretic loop to absorb as much energy as possible with a semi-active device. The energy absorption capacity, which was similar to that of the friction damper in the study, is given by the hysteretic loop generated by the semi-active control. This minimizes the disadvantages of the friction damper, such as a large residual displacement and the generation of high frequencies in the damping force [187].

Several years earlier, Iemura et al. also conducted studies on controls in structural applications. The authors determined the effectiveness of the use of pseudo-negative control on a cable-stayed bridge. The combination of the pseudo-negative force hysteretic loop generated by the variable damper and the elastic strength of the tower–deck connection resulted in a hysterical loop that approximated the perfect rigid plastic force–deformation characteristic with a large damping ratio. The use of sensors was only required for damper connections for the purpose of measuring the relative displacement. The minimal use of sensors is one of the advantages of the study. Iemura et al. performed a comparison between passive, apparent negative power, and active control for a phase II comparison bridge. The pseudo-negative power control result was significantly better than the other results obtained in the study [188].

In contrast to previous studies, Weber et al. investigated the isolation performance of curved surface sliders (CSSs) with different damping mechanisms. This study considered two control strategies, namely amplitude proportional friction damping, which aims to straighten the friction damping during one cycle, and semi-actively controlled damping, and stiffness properties to improve the separation between the soil and the structure with no dynamic stiffness emulation. The considered CSS was assessed in terms of current error as a function of PGA, horizontal peak force and displacement, and peak structure acceleration. The results showed that the nonlinear characteristics can optimize the friction damping, the optimization of viscous damping is independent of PGA, the bowtie friction optimization improves the insulation at a low PGA while the insulation at a medium to high PGA, on the contrary, becomes worse, and the optimized amplitude proportional friction attenuation does not improve the insulation [189]. Other similar studies have been carried out as shown in Table 2.

Table 2. Recently published articles on controllers for MR dampers for structural buildings.

No.	Author and Year	Controller	Finding
1	Zafarani and Halabian (2020) [190]	Clipped optimal with LQG control	Control of the seismic inelastic torque response of multi-story buildings.
2	Mohebbi et. al. (2018) [191]	H2/LQG control	Modification of H2/LQG control to optimize the control system's performance
3	Zizouni et. al. (2019) [192]	Neural network control	Efficacy of neural network control on a three-story small-scale structure using the Tōhoku 2011 and Boumerdès 2003 earthquake data.
4	Bozorgvar and Zahrai (2019) [193]	Adaptive Neuro-Fuzzy inference system	Neuro-fuzzy optimization adapted to genetic algorithms.
5	Li and Liang (2018) [194]	Sliding mode control Fuzzy system	Developed a sliding mode control method based on a fuzzy system. Fuzzy logic control mitigates the chattering phenomenon.
6	Cesar and Barros (2017) [195]	Adaptive Neuro-Fuzzy inference system	Verified the efficacy of neuro-fuzzy controllers in reducing the responses of building structures equipped with MR dampers.
7	Al-Fahdawi and Barroso (2021) [196]	Adaptive Neuro-Fuzzy inference system and Simple adaptive control	Reduction of the seismic response of three-dimensional combined buildings under two-way seismic excitation with adaptive neuro-fuzzy inference system control and simple adaptive control.
8	Mousavi (2020) [197]	Fuzzy logic controller	Use of wavelet networks and fuzzy logic controllers to copy the inverse dynamics of MR dampers and nonlinear isolators.
9	Ndemanou and Nbendjo (2018) [185]	Fuzzy logic controller	Fuzzy logic controls are better than traditional controls and algorithmic controls and are critical when optimizing the response of a structure to seismic loads.
10	Mehrkian et. al. (2017) [186]	Fuzzy logic controller	Improving a fuzzy control system with a smart multi-objective fuzzy–genetic controller produced controls that were more effective than others.

5. Future Research

This study briefly described the various types of earthquake mitigation technologies. Technologies will continue to be developed in an effort to reduce losses. Various types of dampers with semi-active systems, such as friction dampers (FDs), tuned mass dampers (TMDs), and viscous dampers (VDs), that have been applied in building structures were investigated. However, the primary focus of this study was the models and controls of magnetorheological (MR) dampers, which are dampers with a semi-active system. Although the development of MR dampers for building structures continues to strive for optimal efficiency, setbacks and issues, such as difficulties with the manufacturing process, leakage of liquid from the device, and the wear rate of the piston and the interior of the cylinder, are not widely discussed. Therefore, discussion of these matters is expected to increase in the future. Oddly enough, although MR dampers use an iron powder fluid to operate, the wear and tear of these MR dampers are very rarely discussed. This is especially true as an increase in viscosity, caused by the magnetic flux, results in stronger friction. This will most definitely result in roughness appearing on the surfaces of the cylinders and pistons and reduce their performance.

Author Contributions: Conceptualization, U.U. and S.-B.C.; methodology, F.I. and H.H.; validation, S.-B.C.; data curation, F.I. and Y.M.P.; writing—original draft preparation, B.W.L.; writing—review and editing, B.W.L.; supervision, U.U. and S.-B.C. All authors have read and agreed to the published version of the manuscript.

Funding: This research was funded by Kemenristek/BRIN through Hibah WCR 2021, Republic of Indonesia.

Institutional Review Board Statement: Not applicable.

Informed Consent Statement: Not applicable.

Data Availability Statement: Not applicable.

Conflicts of Interest: The authors declare no conflict of interest.

References

1. Ozbulut, O.E.; Hurlebaus, S.; Desroches, R. Seismic Response Control Using Shape Memory Alloys: A Review. *J. Intell. Mater. Syst. Struct.* **2011**, *22*, 1531–1549. [CrossRef]
2. Takagi, J.; Wada, A. Recent earthquakes and the need for a new philosophy for earthquake-resistant design. *Soil Dyn. Earthq. Eng.* **2019**, *119*, 499–507. [CrossRef]
3. Tehseen, R.; Farooq, M.S.; Abid, A. Earthquake Prediction Using Expert Systems: A Systematic Mapping Study. *Sustainability* **2020**, *12*, 2420. [CrossRef]
4. Ahumada, A.; Altunkaynak, A.; Ayoub, A. Fuzzy logic-based attenuation relationships of strong motion earthquake records. *Expert Syst. Appl.* **2015**, *42*, 1287–1297. [CrossRef]
5. Bahrami, B.; Shafiee, M. Fuzzy Descriptor Models for Earthquake Time Prediction Using Seismic Time Series. *Int. J. Uncertain. Fuzziness Knowl.-Based Syst.* **2015**, *23*, 505–519. [CrossRef]
6. Nasrollahnejad, A. Prediction of Peak ground acceleration for earthquakes by using intelligent methods. In Proceedings of the 2017 5th Iranian Joint Congress on Fuzzy and Intelligent Systems, Tehran, Iran, 7–9 March 2017. [CrossRef]
7. Lamsal, M.R. Siesmic Activity and its Periphery. *Himal. Phys.* **2017**, *6*, 86–91. [CrossRef]
8. Komura, K.; Aiyama, K.; Nagata, T.; Sato, H.P.; Yamada, A.; Aoyagi, Y. Surface rupture and characteristics of a fault associated with the 2011 and 2016 earthquakes in the southern Abukuma Mountains, northeastern Japan, triggered by the Tohoku-Oki earthquake. *Earth Planets Space* **2019**, *71*, 1–23. [CrossRef]
9. Grünthal, G.; Wahlström, R.; Stromeyer, D. The unified catalogue of earthquakes in central, northern, and northwestern Europe (CENEC)—Updated and expanded to the last millennium. *J. Seism.* **2009**, *13*, 517–541. [CrossRef]
10. Dziewonski, A.M.; Chou, T.-A.; Woodhouse, J.H. Determination of earthquake source parameters from waveform data for studies of global and regional seismicity. *J. Geophys. Res. Space Phys.* **1981**, *86*, 2825–2852. [CrossRef]
11. Allen, T.I.; Wald, D.J.; Worden, C.B. Intensity attenuation for active crustal regions. *J. Seism.* **2012**, *16*, 409–433. [CrossRef]
12. Nievas, C.I.; Bommer, J.J.; Crowley, H.; Van Elk, J. *Global Occurrence and Impact of Small-to-Medium Magnitude Earthquakes: A Statistical Analysis*; Springer: Amsterdam, The Netherlands, 2020; Volume 18, pp. 1–35. [CrossRef]
13. Musson, R. Intensity and Intensity Scales, New Man. *Seismol. Obs. Pract.* **2012**, *2*, 1–41. [CrossRef]
14. Doglioni, C.; Barba, S.; Carminati, E.; Riguzzi, F. Fault on-off versus strain rate and earthquakes energy. *Geosci. Front.* **2015**, *6*, 265–276. [CrossRef]

15. Chung, T.-W. Attenuation of High-Frequency P and S Waves in the Crust of Southeastern South Korea. *Bull. Seism. Soc. Am.* **2001**, *91*, 1867–1874. [CrossRef]
16. Hamzehloo, H.; Doloei, G.J. Attenuation of high frequency Pand Swaves in the crust of the East-Central Iran. *Geophys. J. Int.* **2009**, *179*, 1669–1678. [CrossRef]
17. Kennett, B.; Furumura, T. Significant P wave conversions from upgoing S waves generated by very deep earthquakes around Japan. *Prog. Earth Planet. Sci.* **2019**, *6*, 49. [CrossRef]
18. Stern, T.; Lamb, S.; Moore, J.D.P.; Okaya, D.; Hochmuth, K. High mantle seismic P-wave speeds as a signature for gravitational spreading of superplumes. *Sci. Adv.* **2020**, *6*, eaba7118. [CrossRef] [PubMed]
19. Chen, W.; Sharma, R.; Rizzo, A.N.; Siegler, J.H.; Garcia, J.G.N.; Jacobson, J.R. Role of Claudin-5 in the Attenuation of Murine Acute Lung Injury by Simvastatin. *Am. J. Respir. Cell Mol. Biol.* **2013**, *50*, 1–33. [CrossRef]
20. Kern, H. Measuring and Modeling of P- and S-Wave Velocities on Crustal Rocks: A Key for the Interpretation of Seismic Reflection and Refraction Data. *Int. J. Geophys.* **2011**, *2011*, 1–9. [CrossRef]
21. Takemura, S.; Yoshimoto, K.; Tonegawa, T. Scattering of trapped P and S waves in the hydrated subducting crust of the Philippine Sea plate at shallow depths beneath the Kanto region, Japan. *Geophys. J. Int.* **2015**, *203*, 2261–2276. [CrossRef]
22. Ma, X.; Huang, Z. Attenuation of High-Frequency P and S Waves in the Crust of Central and Western Tien Shan. *Pure Appl. Geophys.* **2020**, *177*, 4127–4142. [CrossRef]
23. Asudeh, I. Seismic structure of Iran from surface and body wave data. *Geophys. J. Int.* **1982**, *71*, 715–730. [CrossRef]
24. Wu, X.; Lu, H.; Huang, K.; Wu, S.; Qiao, W. Frequency Spectrum Method-Based Stress Analysis for Oil Pipelines in Earthquake Disaster Areas. *PLoS ONE* **2015**, *10*, e0115299. [CrossRef] [PubMed]
25. Maiti, B.B. Ajayswarup Fan Footing Soil Foundation to Safeguard High and Low Rise Buildings from Seismic Waves. *Int. J. Adv. Eng. Res. Sci.* **2018**, *5*, 11–21. [CrossRef]
26. Heyburn, R.; Selby, N.D.; Fox, B. Estimating earthquake source depths by combining surface wave amplitude spectra and teleseismic depth phase observations. *Geophys. J. Int.* **2013**, *194*, 1000–1010. [CrossRef]
27. Maupin, V. Encyclopedia of Earthquake Engineering. *Encycl. Earthq. Eng.* **2014**, 1–15. [CrossRef]
28. Trifunac, M.D.; Brady, A.G. Correlations of peak acceleration, velocity and displacement with earthquake magnitude, distance and site conditions. *Earthq. Eng. Struct. Dyn.* **1976**, *4*, 455–471. [CrossRef]
29. Society, S. A study on the duration of strong earthquake ground motion. 14F, 1T, Refs. *Int. J. Rock Mech. Min. Sci. Géoméch. Abstr.* **1976**, *13*, 28. [CrossRef]
30. Elhout, E.A. The correlation between the ground motion intensity measure parameters of earthquakes. *Asian J. Civ. Eng.* **2020**, *21*, 829–840. [CrossRef]
31. Pejovic, J.; Jankovic, S. Selection of Ground Motion Intensity Measure for Reinforced Concrete Structure. *Procedia Eng.* **2015**, *117*, 588–595. [CrossRef]
32. Ye, L.; Ma, Q.; Miao, Z.; Guan, H.; Zhuge, Y. Numerical and comparative study of earthquake intensity indices in seismic analysis. *Struct. Des. Tall Spéc. Build.* **2013**, *22*, 362–381. [CrossRef]
33. Atkinson, G.M.; Kaka, S.I. Relationships between Felt Intensity and Instrumental Ground Motion in the Central United States and California. *Bull. Seism. Soc. Am.* **2007**, *97*, 497–510. [CrossRef]
34. Kaka, S.I.; Atkinson, G.M. Relationships between Instrumental Ground-Motion Parameters and Modified Mercalli Intensity in Eastern North America. *Bull. Seism. Soc. Am.* **2004**, *94*, 1728–1736. [CrossRef]
35. Pailoplee, S. Relationship between Modified Mercalli Intensity and peak ground acceleration in Myanmar. *Nat. Sci.* **2012**, *04*, 624–630. [CrossRef]
36. Nemati, M. Relationships between Modified Mercalli Intensity and Engineering Ground-Motion of the Earthquakes in Persia. *J. Earthq. Eng.* **2016**, *20*, 795–808. [CrossRef]
37. Atkinson, G.M.; Kaka, S.I. Relationships Between Felt Intensity and Instrumental Ground Motion for New Madrid Shake Maps. *Dep. Earth Sci. Carlet. Univ.* **2006**, 1–27.
38. Wu, Y.-M.; Chang, C.-H.; Hsiao, N.-C.; Wu, F.T. Relocation of the 1998 Rueyli, Taiwan, earthquake sequence using three-dimensions velocity structure with stations corrections. *Terr. Atmos. Ocean. Sci.* **2003**, *14*, 421. [CrossRef]
39. Wald, D.; Quitoriano, V.; Heaton, T.H.; Kanamori, H.; Scrivner, C.W.; Worden, C.B. TriNet "ShakeMaps": Rapid Generation of Peak Ground Motion and Intensity Maps for Earthquakes in Southern California. *Earthq. Spectra* **1999**, *15*, 537–555. [CrossRef]
40. Gerstenberger, M.C.; Wiemer, S.; Jones, L.M.; Reasenberg, P.A. Real-time forecasts of tomorrow's earthquakes in California. *Nat. Cell Biol.* **2005**, *435*, 328–331. [CrossRef] [PubMed]
41. Yang, T.Y.; Tung, D.P.; Li, Y. Equivalent Energy Design Procedure for Earthquake Resilient Fused Structures. *Earthq. Spectra* **2018**, *34*, 795–815. [CrossRef]
42. Sahin, C. Seismic Retrofitting of Existing Structures. *Int. Assoc. Erathquake Eng.* **2014**, *111*, 1–111. Available online: http://pdxscholar.library.pdx.edu/cengin_gradprojects%0Ahttp://pdxscholar.library.pdx.edu/cengin_gradprojects?utm_source=pdxscholar.library.pdx.edu/cengin_gradprojects/7&utm_medium=PDF&utm_campaign=PDFCoverPages (accessed on 28 August 2021).
43. Balendra, T. Earthquake-resistant Design of Buildings. *Vib. Build. Wind Earthq. Loads* **1993**, 115–144. [CrossRef]
44. Meheta, V.V.; Murty, C.V.R.; Goswami, R.; Vijayanarayanan, A.R. *Earthquake Behaviour of Buildings*; Gujarat State Disaster Management Authority: Gujarat, India, 2012; p. 268.

45. Thilakarathna, S.N.; Anwar, N.; Norachan, P.; Naja, F.A. The Effect of Wind Loads on the Seismic Performance of Tall Buildings. *Athens J. Technol. Eng.* **2018**, *5*, 251–276. [CrossRef]
46. Heiza, K.M.; Tayel, M.A. Comparative Study of The Effects of Wind and Earthquake Loads on High-rise Buildings. *Concr. Res. Lett.* **2012**, *3*, 386–405.
47. Cherry, S.; Ward, H.S.; Dalgliesh, W.A. *Earthquake and Wind Loads in Building Design*; National Research Canada: Ottawa, ON, Canada, 1963.
48. Shashidhar, K.; Balaji, K.V.G.D.; Poleswararao, K.; Malavika, P.G. Chakravarthy, Comparison of influence of wind and earthquake forces on low-rise and high-rise multi storey structures. *Int. J. Appl. Eng. Res.* **2015**, *10*, 35810–35816.
49. Tao, D.; Lin, J.; Lu, Z. Time-Frequency Energy Distribution of Ground Motion and Its Effect on the Dynamic Response of Nonlinear Structures. *Sustainability* **2019**, *11*, 702. [CrossRef]
50. Fajfar, P. Analysis in seismic provisions for buildings: Past, present and future. The fifth Prof. Nicholas Ambraseys lecture. *Bull. Earthq. Eng.* **2018**, *16*, 2567–2608. [CrossRef]
51. Kagawa, T.; Irikura, K.; Somerville, P.G. Differences in ground motion and fault rupture process between the surface and buried rupture earthquakes. *Earth Planets Space* **2004**, *56*, 3–14. [CrossRef]
52. Mavroeidis, G.P.; Papageorgiou, A.S. Effect of Fault Rupture Characteristics on Near-Fault Strong Ground Motions. *Bull. Seism. Soc. Am.* **2010**, *100*, 37–58. [CrossRef]
53. Yang, D.; Chen, G. Recent advances in engineering characteristics of near-fault ground motions and seismic effects of building structures. In Proceedings of the Second International Conference on Performance-based and Life-cycle Structural Engineering (PLSE 2015), Brisbane, QLD, Australia, 9–11 December 2015; pp. 1296–1305.
54. Zhao, Z.; Zhao, Z.; Xu, J.; Kubota, R.; Liu, L. Strong ground motion simulation for seismic hazard assessment in an urban area. *J. Geophys. Eng.* **2007**, *4*, 308–316. [CrossRef]
55. Athamnia, B.; Ounis, A.; Abdeddaim, M. Effect of a Near Fault on the Seismic Response of a Base-Isolated Structure with a Soft Storey. *Slovak J. Civ. Eng.* **2017**, *25*, 34–46. [CrossRef]
56. Rai, D.C. Future trends in earthquake-resistant design of structures. *Curr. Sci.* **2000**, *79*, 1291–1300.
57. Sarwar, W.; Sarwar, R. Vibration Control Devices for Building Structures and Installation Approach: A Review. *Civ. Environ. Eng. Rep.* **2019**, *29*, 74–100. [CrossRef]
58. Shekhar, S.; Shukla, S.P.; Zafar, S. Seismic Isolation Devices. *J. Civ. Eng. Environ. Technol.* **2017**, *4*, 336–340.
59. Gao, H.; Wang, C.; Huang, C.; Shi, W.; Huo, L. Development of a Frequency-Adjustable Tuned Mass Damper (FATMD) for Structural Vibration Control. *Shock. Vib.* **2020**, *2020*, 1–16. [CrossRef]
60. Alam, M.; Youssef, M.; Nehdi, M. Analytical prediction of the seismic behaviour of superelastic shape memory alloy reinforced concrete elements. *Eng. Struct.* **2008**, *30*, 3399–3411. [CrossRef]
61. El-Sinawi, A.H.; Alhamaydeh, M.H.; Jhemi, A.A. Optimal Control of Magnetorheological Fluid Dampers for Seismic Isolation of Structures. *Math. Probl. Eng.* **2013**, *2013*, 1–7. [CrossRef] [PubMed]
62. Frahm, H. Device for damping vibration of bodies. U.S. Patent US989958A, 18 April 1911.
63. Ontiveros-Pérez, S.P.; Miguel, L.F.F.; Miguel, L.F. Optimization of location and forces of friction dampers. *REM Int. Eng. J.* **2017**, *70*, 273–279. [CrossRef]
64. Ontiveros-Pérez, S.P.; Miguel, L.F.F.; Miguel, L.F.F. A New Assessment in the Simultaneous Optimization of Friction Dampers in Plane and Spatial Civil Structures. *Math. Probl. Eng.* **2017**, *2017*, 1–18. [CrossRef]
65. Constantinou, M.C.; Soong, T.T.; Dargush, G.F. *Passive Energy Dissipation Systems for Structural Design and Retrofit*; National Science Foundation: Arlington, VA, USA, 1998.
66. Pall, A.S.; Marsh, C.; Fazio, P. Friction Joints for Seismic Control of Large Panel Structures. *PCI J.* **1980**, *25*, 38–61. [CrossRef]
67. Desjardins, C.; Avtar, S.P.; Cedric, M.; T.S.N.C. Group. Response of friction damped braced frames. *ASCE J. Struct. Div.* **1982**, *108*, 1313–1323.
68. Aiken, I.D.; Kelly, J.M. *Earthquake Simulator Testing and Analytical Studies of Two Energy-Absorbing Systems for Multi-Story Structures*; Earthquake Engineering Research Center, National Technical Information Service: Berkeley, CA, USA, 1990; p. 276.
69. Ray, S.S.; Sahoo, S.; Das, S. Formulation and solutions of fractional continuously variable order mass-spring-damper systems controlled by viscoelastic and viscous-viscoelastic. *Adv. Mater. Eng.* **2016**, *8*, 1687814016646505. [CrossRef]
70. Raut, B.R.; Jangid, R.S. Seismic behavior of benchmark building with semi active variable friction damper. *Asian J. Civ. Eng.* **2015**, *16*, 417–436.
71. Pengyun, L.; Jiedong, L.; Ming, N.; Wanli, Z.; Anguo, H. Vibration Energy of Earthquake Excited Building Structures Based on Computer Simulation. *Energy Procedia* **2012**, *17*, 1116–1123. [CrossRef]
72. Armali, M.; Damerji, H.; Hallal, J.; Fakih, M. Effectiveness of friction dampers on the seismic behavior of high rise building vs. shear wall system. *Eng. Rep.* **2019**, *1*, 1–14. [CrossRef]
73. Apostolakis, G. Optimal Evolutionary Seismic Design of Three-Dimensional Multistory Structures with Damping Devices. *J. Struct. Eng.* **2020**, *146*, 04020205. [CrossRef]
74. Pall, A.S.; Pall, R.T. Performance-Based Design Using Pall Friction Dampers—An Economical Design Solution. In Proceedings of the 13th World Conference on Earthquake Engineering, Vancouver, BC, Canada, 1–6 August 2004.
75. Park, J.-H.; Kim, J.; Min, K.-W. Optimal design of added viscoelastic dampers and supporting braces. *Earthq. Eng. Struct. Dyn.* **2004**, *33*, 465–484. [CrossRef]

76. Levy, R.; LaVan, O. Fully stressed design of passive controllers in framed structures for seismic loadings. *Struct. Multidiscip. Optim.* **2006**, *32*, 485–498. [CrossRef]
77. Whittle, J.K.; Williams, M.S.; Karavasilis, T.; Blakeborough, A. A Comparison of Viscous Damper Placement Methods for Improving Seismic Building Design. *J. Earthq. Eng.* **2012**, *16*, 540–560. [CrossRef]
78. Filiatrault, B.A.; Member, A.; Cherry, S. Seismic design spectra for friction. *Damped Struct.* **1990**, *116*, 1334–1355.
79. Moreschi, L.M.; Singh, M.P. Design of yielding metallic and friction dampers for optimal seismic performance. *Earthq. Eng. Struct. Dyn.* **2003**, *32*, 1291–1311. [CrossRef]
80. Fallah, N.; Honarparast, S. NSGA-II based multi-objective optimization in design of Pall friction dampers. *J. Constr. Steel Res.* **2013**, *89*, 75–85. [CrossRef]
81. De Matteis, G.; Brando, G.; Caldoso, F.; D'Agostino, F. Seismic performance of dual steel frames with dissipative metal shear panels. *Ing. Sismica* **2018**, *35*, 124–141.
82. Zahrai, S.M.; Moradi, A.; Moradi, M. Using friction dampers in retrofitting a steel structure with masonry infill panels. *Steel Compos. Struct.* **2015**, *19*, 309–325. [CrossRef]
83. Mualla, I.H.; Belev, B. Performance of steel frames with a new friction damper device under earthquake excitation. *Eng. Struct.* **2002**, *24*, 365–371. [CrossRef]
84. Barati, F.; Esfandiari, A. Exploring the Efficiency of Dampers for Repair and Strengthening of Existing Buildings. *J. Struct. Eng. Geo-Tech.* **2012**, *2*, 67–73.
85. Chandra, R.; Masand, M.; Nandi, S.K.; Tripathi, C.P.; Pall, R.; Pall, A. Friction-Dampers for Seismic Control of La Gardenia Towers South City, Gurgaon, India. *Engineering* **2008**, 1–8.
86. Saeedi, F.; Shabakhty, N.; Mousavi, S.R. Seismic assessment of steel frames with triangular-plate added damping and stiffness devices. *J. Constr. Steel Res.* **2016**, *125*, 15–25. [CrossRef]
87. Mahmoudi, M.; Abdi, M.G. Evaluating response modification factors of TADAS frames. *J. Constr. Steel Res.* **2012**, *71*, 162–170. [CrossRef]
88. De Matteis, G.; Sarracco, G.; Brando, G. Experimental tests and optimization rules for steel perforated shear panels. *J. Constr. Steel Res.* **2016**, *123*, 41–52. [CrossRef]
89. Vian, D.; Bruneau, M.; Purba, R. Special Perforated Steel Plate Shear Walls with Reduced Beam Section Anchor Beams. II: Analysis and Design Recommendations. *J. Struct. Eng.* **2009**, *135*, 221–228. [CrossRef]
90. Valizadeh, H.; Sheidaii, M.; Showkati, H. Experimental investigation on cyclic behavior of perforated steel plate shear walls. *J. Constr. Steel Res.* **2012**, *70*, 308–316. [CrossRef]
91. Chan, R.; Albermani, F.; Kitipornchai, S. Experimental study of perforated yielding shear panel device for passive energy dissipation. *J. Constr. Steel Res.* **2013**, *91*, 14–25. [CrossRef]
92. Egorova, N.; Eatherton, M.R.; Maurya, A. Experimental study of ring-shaped steel plate shear walls. *J. Constr. Steel Res.* **2014**, *103*, 179–189. [CrossRef]
93. Brando, G.; De Matteis, G. Design of low strength-high hardening metal multi-stiffened shear plates. *Eng. Struct.* **2014**, *60*, 2–10. [CrossRef]
94. Brando, G.; De Matteis, G. Experimental and numerical analysis of a multi-stiffened pure aluminium shear panel. *Thin-Walled Struct.* **2011**, *49*, 1277–1287. [CrossRef]
95. De Matteis, G.; Brando, G.; Panico, S.; Mazzolani, F.M. Bracing type pure aluminium stiffened shear panels: An experimental study. *Adv. Steel Constr.* **2009**, *5*, 106–119.
96. Hamed, A.A.; Mofid, M. On the equivalent simple models of braced steel shear panels. *Proc. Inst. Civ. Eng. Struct. Build.* **2015**, *168*, 570–577. [CrossRef]
97. Hamed, A.A.; Mofid, M. On the experimental and numerical study of braced steel shear panels. *Struct. Des. Tall Spéc. Build.* **2015**, *24*, 853–872. [CrossRef]
98. Zhang, C.; Zhu, J.; Wu, M.; Yu, J.; Zhao, J. The Lightweight Design of a Seismic Low-Yield-Strength Steel Shear Panel Damper. *Materials* **2016**, *9*, 424. [CrossRef]
99. Kang, T.H.-K.; Martin, R.D.; Park, H.-G.; Wilkerson, R.; Youssef, N. Tall building with steel plate shear walls subject to load reversal. *Struct. Design Tall Spec. Build.* **2013**, *22*, 500–520. [CrossRef]
100. Jain, S.; Rai, D.C.; Sahoo, D.R. Postyield Cyclic Buckling Criteria for Aluminum Shear Panels. *J. Appl. Mech.* **2008**, *75*, 021015. [CrossRef]
101. Deng, K.; Pan, P.; Li, W.; Xue, Y. Development of a buckling restrained shear panel damper. *J. Constr. Steel Res.* **2015**, *106*, 311–321. [CrossRef]
102. De Matteis, G.; D'Agostino, F.; Brando, G. Experimental tests on steel buckling inhibited shear panels. *Open Constr. Build. Technol. J.* **2014**, *8*, 279–288.
103. De Matteis, G.; Brando, G.; Mazzolani, F.M. Pure aluminium: An innovative material for structural applications in seismic engineering. *Constr. Build. Mater.* **2012**, *26*, 677–686. [CrossRef]
104. Liu, L.; Tan, P.; Ma, H.; Yan, W.; Zhou, F. A Novel Energy Dissipation Outrigger System with Rotational Inertia Damper. *Adv. Struct. Eng.* **2018**, *21*, 1865–1878. [CrossRef]
105. Housner, G.W.; Bergman, L.A.; Caughey, T.K.; Chassiakos, A.G.; Claus, R.O.; Masri, S.F.; Skelton, R.E.; Soong, T.T.; Spencer, B.F.; Yao, J.T.P. Structural Control: Past, Present, and Future. *J. Eng. Mech.* **1997**, *123*, 897–971. [CrossRef]

106. Soong, T.; Spencer, B. Supplemental energy dissipation: State-of-the-art and state-of-the-practice. *Eng. Struct.* **2002**, *24*, 243–259. [CrossRef]
107. Chakraborty, S.; Roy, B.K. Reliability based optimum design of Tuned Mass Damper in seismic vibration control of structures with bounded uncertain parameters. *Probabilistic Eng. Mech.* **2011**, *26*, 215–221. [CrossRef]
108. Zuo, L.; Nayfeh, S.A. The Two-Degree-of-Freedom Tuned-Mass Damper for Suppression of Single-Mode Vibration Under Random and Harmonic Excitation. *J. Vib. Acoust.* **2005**, *128*, 56–65. [CrossRef]
109. James, K.R.; Haritos, N.; Ades, P.K. Mechanical stability of trees under dynamic loads. *Am. J. Bot.* **2006**, *93*, 1522–1530. [CrossRef]
110. Marano, G.C.; Greco, R.; Trentadue, F.; Chiaia, B. Constrained reliability-based optimization of linear tuned mass dampers for seismic control. *Int. J. Solids Struct.* **2007**, *44*, 7370–7388. [CrossRef]
111. Pinkaew, T.; Lukkunaprasit, P.; Chatupote, P. Seismic effectiveness of tuned mass dampers for damage reduction of structures. *Eng. Struct.* **2003**, *25*, 39–46. [CrossRef]
112. Matta, E. Lifecycle cost optimization of tuned mass dampers for the seismic improvement of inelastic structures. *Earthq. Eng. Struct. Dyn.* **2018**, *47*, 714–737. [CrossRef]
113. Huang, M.F.; Tse, K.T.; Chan, C.M.; Lou, W. Integrated Structural Optimization and Vibration Control for Improving Wind-Induced Dynamic Performance Of Tall Buildings. *Int. J. Struct. Stab. Dyn.* **2011**, *11*, 1139–1161. [CrossRef]
114. Wang, D.; Tse, T.K.; Zhou, Y.; Li, Q. Structural performance and cost analysis of wind-induced vibration control schemes for a real super-tall building. *Struct. Infrastruct. Eng.* **2014**, *11*, 990–1011. [CrossRef]
115. Ronagh, M. Plastic Hinge Length of RC Columns Subjected to Both Far-Fault and Near-Fault Ground Motions Having Forward Directivity. *Struct. Des. Tall Spec. Build.* **2011**, *24*, 421–439. [CrossRef]
116. Tse, K.T.; Kwok, K.; Tamura, Y. Performance and Cost Evaluation of a Smart Tuned Mass Damper for Suppressing Wind-Induced Lateral-Torsional Motion of Tall Structures. *J. Struct. Eng.* **2012**, *138*, 514–525. [CrossRef]
117. Bhattacharjee, E.; Halder, L.; Sharma, R.P. An experimental study on tuned liquid damper for mitigation of structural response. *Int. J. Adv. Struct. Eng.* **2013**, *5*, 3. [CrossRef]
118. Min, L. Studies on Tuned Liquid Damper (Tld) By Free-Oscillation. *Struct. Eng.* **1988**, *5*, 381–391.
119. Park, E.; Heo, J.S.; Lee, S.K.; Lee, S.H.; Kim, H.J.; Jo, J.S.; Cho, B.C.; Joo, S.J.; Min, K.W. Performance Test of a Tuned Liquid Mass Damper installed in a Real-Scaled Structure. *J. Comput. Struct. Eng. Inst. Korea* **2017**, *21*, 6.
120. Gur, S.; Roy, K.; Mishra, S.K. Tuned liquid column ball damper for seismic vibration control. *Struct. Control. Health Monit.* **2015**, *22*, 1325–1342. [CrossRef]
121. Fei, Z.; Jinting, W.; Feng, J.; Liqiao, L. Control performance comparison between tuned liquid damper and tuned liquid column damper using real-time hybrid simulation. *Earthq. Eng. Eng. Vib.* **2019**, *18*, 695–701. [CrossRef]
122. Karavasilis, T.L.; Kerawala, S.; Hale, E. Hysteretic model for steel energy dissipation devices and evaluation of a min-imal-damage seismic design approach for steel buildings. *J. Constr. Steel Res.* **2012**, *70*, 358–367. [CrossRef]
123. Brando, G.; D'Agostino, F.; De Matteis, G. Seismic performance of MR frames protected by viscous or hysteretic dampers. *Struct. Des. Tall Spéc. Build.* **2015**, *24*, 653–671. [CrossRef]
124. Ribakov, Y.; Gluck, J. Active viscous damping system for control of MDOF structures. In Proceedings of the 25th International Conference Noise Vibration Enginering ISMA, Leuven, Belgium, 13–15 September 2000; pp. 147–154.
125. Franklin, G.F.; Powell, J.D.; Emami-Naeini, A. *Feedback Control of Dynamic Systems*; Prentice Hall: Upper Saddle River, NJ, USA, 1995.
126. Oliveira, F.; Morais, P. Semi-active control of a fluid viscous damper for vibration mitigation. In Proceedings of the 15th World Conference on Earthquake Engineering, Lisbon, Portugal, 24–28 September 2012.
127. Waghmare, M.V.; Madhekar, S.N.; Matsagar, V.A. Semi-Active Fluid Viscous Dampers for Seismic Mitigation of RC Elevated Liquid Storage Tanks. *Int. J. Struct. Stab. Dyn.* **2019**, *19*. [CrossRef]
128. Bogdan, P.; Stefania, P.; Muresan, I. Active, Semi-active and Hibrid Control Systems Properties and Applications. *Acta Tech. Napoc. Civ. Eng. Archit.* **2011**, *54*, 153–161.
129. Ashour, O.; Rogers, C.A.; Kordonsky, W. Ashour. Magnetorheological Fluids: Materials, Characterization, and Devices. *J. Intell. Mater. Syst. Struct.* **1996**, *7*, 123–130. [CrossRef]
130. Carlson, J.D.; Weiss, K.D. Magnetorheological Materials Based on Alloy Particles. U.S. Patent No. 5,382,373, 17 January 1995.
131. Özsoy, K.; Usal, M.R. A mathematical model for the magnetorheological materials and magneto reheological devices. *Eng. Sci. Technol. Int. J.* **2018**, *21*, 1143–1151. [CrossRef]
132. Kordonski, W.; Gorodkin, S. The behavior of a magnetorheological (MR) fluid under compressive deformation. *J. Rheol.* **2016**, *60*, 129–139. [CrossRef]
133. Spaggiari, A. Properties and applications of Magnetorheological fluids. *Frat. Integrità Strutt.* **2012**, *7*, 48–61. [CrossRef]
134. Morillas, J.R.; De Vicente, J. Magnetorheology: A review. *Soft Matter* **2020**, *16*, 9614–9642. [CrossRef]
135. Ashtiani, M.; Hashemabadi, S.; Ghaffari, A. A review on the magnetorheological fluid preparation and stabilization. *J. Magn. Magn. Mater.* **2015**, *374*, 716–730. [CrossRef]
136. Wang, D.H.; Liao, W.-H. Magnetorheological fluid dampers: A review of parametric modelling. *Smart Mater. Struct.* **2011**, *20*. [CrossRef]
137. Robinson, J.; Xi, K.; Kumar, R.V.; Ferrari, A.C.; Au, H.; Titirici, M.-M.; Parra Puerto, A.; Kucernak, A.; Fitch, S.D.S.; Garcia-Araez, N. 2021 roadmap on lithium sulfur batteries. *J. Phys. Energy* **2020**, *2*, 1–31.

138. Kumar, J.S.; Paul, P.S.; Raghunathan, G.; Alex, D.G. A review of challenges and solutions in the preparation and use of magnetorheological fluids. *Int. J. Mech. Mater. Eng.* **2019**, *14*, 13. [CrossRef]
139. Bombard, A.J.F.; Antunes, L.S.; Gouvêa, D. Redispersibility in magnetorheological fluids: Surface interactions between iron powder and wetting additives. *J. Phys. Conf. Ser.* **2009**, *149*, 012038. [CrossRef]
140. Premalatha, S.E.; Chokkalingam, R.; Mahendran, M. Magneto Mechanical Properties of Iron Based MR Fluids. *Am. J. Polym. Sci.* **2012**, *2*, 50–55. [CrossRef]
141. Zhang, J.Q.; Jing, Q. Effect of seven different additives on the properties of MR fluids. *J. Physics: Conf. Ser.* **2009**, *149*, 012086. [CrossRef]
142. Carlson, J.D. What Makes a Good MR Fluid? *J. Intell. Mater. Syst. Struct.* **2002**, *13*, 431–435. [CrossRef]
143. Balbás, L.; Borstel, G.; Alonso, J.A. Nonlocal density functional calculation of the electron affinity of atoms. *Phys. Lett. A* **1986**, *114*, 236–240. [CrossRef]
144. Aly, A.M. Vibration Control of Buildings Using Magnetorheological Damper: A New Control Algorithm. *J. Eng.* **2013**, *2013*, 1–10. [CrossRef]
145. Khan, S.A.; Suresh, A.; Seetha, R.N. Principles, Characteristics and Applications of Magneto Rheological Fluid Damper in Flow and Shear Mode. *Procedia Mater. Sci.* **2014**, *6*, 1547–1556. [CrossRef]
146. Dobre, A.; Andreescu, C.N.; Stan, C. The influence of the current intensity on the damping characteristics for a magneto-rheological damper of passenger car. *IOP Conf. Ser. Mater. Sci. Eng.* **2016**, *147*, 012110. [CrossRef]
147. Paciello, V.; Pietrosanto, A. Magnetorheological Dampers: A New Approach of Characterization. *IEEE Trans. Instrum. Meas.* **2011**, *60*, 1718–1723. [CrossRef]
148. Gordaninejad, F.; Wang, X.; Hitchcock, G.; Bangrakulur, K.; Ruan, S.; Siino, M. Modular High-Force Seismic Magne-to-Rheological. *J. Struct. Eng.* **2010**, *136*, 135–143. [CrossRef]
149. Jung, H.; Spencer, B.J.; Ni, Y.; Lee, I. State-of-the-art of semiactive control systems using MR fluid dampers in civil engineering applications. *Struct. Eng. Mech.* **2004**, *17*, 493–526. [CrossRef]
150. Roesset, J.M.; Yao, J.T.P. State of the Art of Structural Engineering. *J. Struct. Eng.* **2002**, *128*, 965–975. [CrossRef]
151. Lenggana, B.W.; Ubaidillah, U.; Imaduddin, F.; Widyarso, W.; Purnomo, E.D.; Utami, D.; Mazlan, S.A. Performance prediction of a novel modular magnetorheological damper for seismic building. *J. Adv. Res. Fluid Mech. Therm. Sci.* **2019**, *58*, 275–286.
152. Kim, H.-C.; Shin, Y.-J.; You, W.; Jung, K.C.; Oh, J.-S.; Choi, S.-B. A ride quality evaluation of a semi-active railway vehicle suspension system with MR damper: Railway field tests. *Proc. Inst. Mech. Eng. Part F J. Rail Rapid Transit* **2016**, *231*, 306–316. [CrossRef]
153. Lau, Y.K.; Liao, W.-H. Design and Analysis of Magnetorheological Dampers for Train Suspension. *Proc. Inst. Mech. Eng. Part F J. Rail Rapid Transit* **2005**, *219*, 261–276. [CrossRef]
154. Wang, D.H.; Liao, W.-H. Semi-active suspension systems for railway vehicles using magnetorheological dampers. Part I: System integration and modelling. *Veh. Syst. Dyn.* **2009**, *47*, 1305–1325. [CrossRef]
155. Carlson, J.; Matthis, W. Structures and, undefined 2001, Smart prosthetics based on magnetorheological fluids, Spiedigitallibrary.Org. In Proceedings of the SPIE's 8th Annual International Symposium on Smart Structures and Materials, Newport Beach, CA, USA, 4–8 March 2001; 2001; Volume 4332, pp. 308–316. Available online: https://www.spiedigitallibrary.org/conference-proceedings-of-spie/4332/0000/Smart-prosthetics-based-on-magnetorheological-fluids/10.1117/12.429670.short (accessed on 28 August 2021).
156. Spelta, C.; Previdi, F.; Savaresi, S.M.; Fraternale, G.; Gaudiano, N. Control of magnetorheological dampers for vibration reduction in a washing machine. *Mechatronics* **2009**, *19*, 410–421. [CrossRef]
157. Lee, D.Y.; Nam, Y.J.; Yamane, R.; Park, M.K. Performance evaluation on vibration control of MR landing gear. *J. Phys. Conf. Ser.* **2009**, *149*, 012068. [CrossRef]
158. Ubaidillah, U.; Lenggana, B.W.; Son, L.; Imaduddin, F.; Widodo, P.J.; Harjana, H.; Doewes, R.I. A New Magnetorheological Fluids Damper for Unmanned Aerial Vehicles. *J. Adv. Res. Fluid Mech. Therm. Sci.* **2020**, *73*, 35–45. [CrossRef]
159. Choi, S.-B.; Nam, M.-H.; Lee, B.-K. Vibration Control of a MR Seat Damper for Commercial Vehicles. *J. Intell. Mater. Syst. Struct.* **2000**, *11*, 936–944. [CrossRef]
160. Milecki, A.; Hauke, M. Application of magnetorheological fluid in industrial shock absorbers. *Mech. Syst. Signal Process.* **2012**, *28*, 528–541. [CrossRef]
161. Wereley, N.M.; Hu, W.; Kothera, C.S.; Chen, P.C.; Ngatu, G.T. Magnetorheological Fluid Elastic Lag Damper for Helicopter Rotors. U.S. Patent No. 8,413,772, 9 April 2013.
162. Guo, C.; Gong, X.; Zong, L.; Peng, C.; Xuan, S. Twin-tube- and bypass-containing magneto-rheological damper for use in railway vehicles. *Proc. Inst. Mech. Eng. Part F J. Rail Rapid Transit* **2015**, *229*, 48–57. [CrossRef]
163. Braz-César, M.T.; Barros, R.C. Semi-active control of an experimental frame using mr dampers: Numerical results and experimental validation. In Proceedings of the 3rd International Conference on Computational Methods in Structural Dynamics and Earthquake Engineering, Corfu, Greece, 26–28 May 2011; pp. 25–28.
164. Barros, R.C. Experimental Behaviour and Numerical Analysis of MR Dampers dampers. In Proceedings of the 15th World Conference on Earthquake Engineering, Lisbon, Portugal, 24–28 September 2012.
165. Chey, M.-H.; Chase, J.G.; Mander, J.B.; Carr, A.J. Semi-active tuned mass damper building systems: Design. *Earthq. Eng. Struct. Dyn.* **2009**, *39*, 119–139. [CrossRef]

166. Uz, M.E. Seismic Design of Magnetorheological Dampers Between Different Sizes Buildings Under Optimum Parameters. *Celal Bayar Üniversitesi Fen Bilimleri Derg.* **2017**, *13*, 1–13. [CrossRef]
167. Makris, N.; Constantinou, M. *Viscous Dampers: Testing Modeling and Application in Vibration and Seismic Isolation*; National Center for Earthquake Engenering Research, State University of New York Buffalo: Buffalo, NY, USA, 1990. [CrossRef]
168. Constantinou, M.C.; Symans, M.D. Experimental study of seismic response of buildings with supplemental fluid dampers. *Struct. Des. Tall Build.* **1993**, *2*, 93–132. [CrossRef]
169. Guo, T.; Xu, J.; Xu, W.; Di, Z. Seismic Upgrade of Existing Buildings with Fluid Viscous Dampers: Design Methodologies and Case Study. *J. Perform. Constr. Facil.* **2015**, *29*, 04014175. [CrossRef]
170. Wang, S.; Mahin, S.A. Seismic Upgrade of an Existing Tall Building Using Different Supplemental Energy Dissipation Devices. *J. Struct. Eng.* **2018**, *144*, 04018091. [CrossRef]
171. Weber, F. Robust force tracking control scheme for MR dampers. *Struct. Control. Health Monit.* **2015**, *22*, 1373–1395. [CrossRef]
172. Guo, S.; Yang, S.; Pan, C. Dynamic Modeling of Magnetorheological Damper Behaviors. *J. Intell. Mater. Syst. Struct.* **2006**, *17*, 3–14. [CrossRef]
173. Singru, P.; Raizada, A.; Krishnakumar, V.; Garg, A.; Tai, K.; Raj, V. Modeling of a magneto-rheological (MR) damper using genetic programming. *J. Vibroengineering* **2017**, *19*, 3169–3177. [CrossRef]
174. Guan, X.; Guo, P.; Ou, J. Modeling and Analyzing of Hysteresis Behavior of Magneto Rheological Dampers. *Procedia Eng.* **2011**, *14*, 2756–2764. [CrossRef]
175. Rossi, A.; Orsini, F.; Scorza, A.; Botta, F.; Belfiore, N.P.; Sciuto, S.A. A Review on Parametric Dynamic Models of Magnetorheological Dampers and Their Characterization Methods. *Actuators* **2018**, *7*, 16. [CrossRef]
176. Zapateiro, M.; Luo, N.; Rodellar, J. Modeling and Identification of Hysteretic Dynamics of Mr Dampers and Application to Semiactive Vibration. In Proceedings of the 14th World Conference on Earthquake Engineering, Beijing, China, 12–17 October 2008.
177. Ahn, K.K.; Truong, D.Q.; Islam, M.A. Modeling of a magneto-rheological (MR) fluid damper using a self tuning fuzzy mechanism. *J. Mech. Sci. Technol.* **2009**, *23*, 1485–1499. [CrossRef]
178. Weber, F. Bouc–Wen model-based real-time force tracking scheme for MR dampers. *Smart Mater. Struct.* **2013**, *22*, 045012. [CrossRef]
179. Soltane, S.; Montassar, S.; Ben Mekki, O.; El Fatmi, R. A hysteretic Bingham model for MR dampers to control cable vibrations. *J. Mech. Mater. Struct.* **2015**, *10*, 195–206. [CrossRef]
180. Spencer, B.F., Jr.; Dyke, S.; Sain, M.K.; Carlson, J.D. Phenomenological Model for Magnetorheological Dampers. *J. Eng. Mech.* **1997**, *123*, 230–238. [CrossRef]
181. Chae, Y. Seismic Hazard Mitigation of Building Structures Using Magneto-Rheological Dampers. PhD Thesis, Lehigh University, Bethlehem, PA, USA, 2011; p. 449.
182. Jansen, L.M.; Dyke, S.J. Semiactive Control Strategies for MR Dampers: Comparative Study. *J. Eng. Mech.* **2000**, *126*, 795–803. [CrossRef]
183. Rodriguez, R.; Stathis, J.H.; Linder, B.P. Modeling and Experimental Verification of the Effect of Gate Oxide Breakdown on CMOS Inverters. In Proceedings of the 2003 IEEE International Reliability Physics Symposium Proceedings, Dallas, TX, USA, 30 March–4 April 2003; pp. 11–16.
184. Wei, S.; Wang, J.; Ou, J. Method for improving the neural network model of the magnetorheological damper. *Mech. Syst. Signal Process.* **2021**, *149*, 107316. [CrossRef]
185. Ndemanou, B.P.; Nbendjo, B.R.N. Fuzzy magnetorheological device vibration control of the two Timoshenko cantilever beams interconnected under earthquake excitation. *Struct. Des. Tall Spéc. Build.* **2018**, *27*, e1541. [CrossRef]
186. Mehrkian, B.; Bahar, A.; Chaibakhsh, A. Semiactive conceptual fuzzy control of magnetorheological dampers in an irregular base-isolated benchmark building optimized by multi-objective genetic algorithm. *Struct. Control. Health Monit.* **2019**, *26*, e2302. [CrossRef]
187. Iemura, H.; Igarashi, A.; Pradono, M.H.; Kalantari, A. Negative stiffness friction damping for seismically isolated structures. *Struct. Control. Health Monit.* **2006**, *13*, 775–791. [CrossRef]
188. Iemura, H.; Pradono, M.H. Application of pseudo-negative stiffness control to the benchmark cable-stayed bridge. *J. Struct. Control.* **2003**, *10*, 187–203. [CrossRef]
189. Weber, F.; Distl, J.; Meier, L.; Braun, C. Curved surface sliders with friction damping, linear viscous damping, bow tie friction damping, and semiactively controlled properties. *Struct. Control. Health Monit.* **2018**, *25*, e2257. [CrossRef]
190. Zafarani, M.M.; Halabian, A.M. A new supervisory adaptive strategy for the control of hysteretic multi-story irregular buildings equipped with MR-dampers. *Eng. Struct.* **2020**, *217*, 110786. [CrossRef]
191. Mohebbi, M.; Dadkhah, H.; Dabbagh, H.R. A genetic algorithm–based design approach for smart base isolation systems. *J. Intell. Mater. Syst. Struct.* **2017**, *29*, 1315–1332. [CrossRef]
192. Zizouni, K.; Fali, L.; Sadek, Y.; Bousserhane, I.K. Neural network control for earthquake structural vibration reduction using MRD. *Front. Struct. Civ. Eng.* **2019**, *13*, 1171–1182. [CrossRef]
193. Bozorgvar, M.; Zahrai, S.M. Semi-active seismic control of buildings using MR damper and adaptive neural-fuzzy intelligent controller optimized with genetic algorithm. *J. Vib. Control.* **2018**, *25*, 273–285. [CrossRef]

194. Li, L.; Liang, H. Semiactive Control of Structural Nonlinear Vibration Considering the MR Damper Model. *J. Aerosp. Eng.* **2018**, *31*, 04018095. [CrossRef]
195. Braz-César, M.; Barros, R. Optimization of a Fuzzy Logic Controller for MR Dampers Using an Adaptive Neuro-Fuzzy Procedure. *Int. J. Struct. Stab. Dyn.* **2016**, *17*, 1740007. [CrossRef]
196. Al-Fahdawi, O.A.; Barroso, L.R. Adaptive neuro-fuzzy and simple adaptive control methods for full three-dimensional coupled buildings subjected to bi-directional seismic excitations. *Eng. Struct.* **2021**, *232*, 111798. [CrossRef]
197. Mousavi, S.H. Modeling and controlling a semi-active nonlinear single-stage vibration isolator using intelligent inverse model of an MR damper. *J. Mech. Sci. Technol.* **2020**, *34*, 1–8. [CrossRef]

applied sciences

MDPI

Article

A Practical Approach for the Mitigation of Seismic-Induced Vibrations in Slender Metallic Structures through Magnetorheological Fluid Dampers

Rosario Pecora

Industrial Engineering Department, Aerospace Division, University of Naples "Federico II", 80125 Naples, Italy; rosario.pecora@unina.it

Abstract: The mitigation of seismic-induced vibrations is essential for the effective protection of buildings and occupants during earthquakes. This especially applies to slender buildings with metallic frames; in this case, the structure's geometrical layout and relatively low damping properties favor an excessive and potentially catastrophic oscillatory response to a seismic event. Semiactive systems for energy dissipation are among the most commonly used strategies to control this oscillatory response. They offer the right balance between the reliability of passive devices and the versatility and adaptability of fully active systems. In this work, a vibration-suppression system based on dissipative bracings that integrate commercial magnetorheological fluid dampers (MRDs) was designed and validated through experimental tests on a true-scale structural model that was representative of a five-story slender building with a metallic frame. A practical and robust approach was proposed for: (1) The definition of the MRD type in compliance with a predefined mitigation target for seismic-induced accelerations on each floor of the structure; (2) The modeling of the MRDs, contribute to the dynamic response of the structural system. The approach involves a linearized formulation of the characteristic damping curves of the MRDs at different values of the activating current. By relying upon this linearization, a rapidly converging iterative process was set up to simulate the seismic response of the structure in the case of activated or deactivated dampers. The reference structure and the vibration-suppression system were then manufactured and tested on a sliding table, which provided realistic seismic excitation. The good correlation levels between the numerical predictions and the experimental measurements proved the effectiveness of the conceived system and of the approaches that were used for its design and simulation.

Keywords: slender metallic structures; dynamic response; vibration suppression; seismic protection devices; magnetorheological fluid dampers; damping braces

Citation: Pecora, R. A Practical Approach for the Mitigation of Seismic-Induced Vibrations in Slender Metallic Structures through Magnetorheological Fluid Dampers. *Appl. Sci.* **2022**, *12*, 4155. https://doi.org/10.3390/app12094155

Academic Editor: Andrea Paglietti

Received: 18 March 2022
Accepted: 15 April 2022
Published: 20 April 2022

1. Introduction

The mitigation of the oscillatory responses of buildings to seismic events has gained ever increasing attention in the last three decades. Although seismic protection has always been a topic of relevant importance for the scientific community, it is only with the recent design trends in constructions that it has become a key factor in the assessment of new proposals of civil structures.

Slender buildings that rise to ever more competitive heights, or bridges that extend along previously unimaginable spans, are just a few examples of how modern constructions are becoming extremely prone to damage and collapse in the case of earthquakes. The augmented safety risk is due to the challenging geometrical configurations and the specific structural arrangements that are adopted, which primarily involve light metallic trusses with intrinsic low damping properties.

The integration of antiseismic devices is a way to preserve the integrity of these structures and their contents during seismic events, even if they are of moderate intensity.

According to the definition that is provided by the applicable European standard [1], antiseismic devices are tools that are capable of modifying the structure's response to seismic action. Such modifications can generally be achieved in three different ways:

a. By isolating the structure from the ground;
b. By creating permanent or temporary restraints to the motion via rigid connections;
c. By dissipating energy.

Seismic isolation indeed represents the most effective solution to protect the structure, thanks to the relevant reduction in the acceleration, even at the highest floors [2].

Conversely, isolating devices must be designed together with the building that they are called upon to protect and installed during its construction. Moreover, the manufacturing and installation processes are much more expensive than those of the devices that are used to implement Seismic Mitigation Strategies b and c.

Therefore, the latter two strategies become preferable from a cost-saving standpoint, and they are the only viable solutions when dealing with the seismic retrofit of already built constructions. Rigid connections that prevent the relative motion between different parts of the structure are convenient for their simplicity. However, since their working principle is based on adding extra stiffness to the structure, their effectiveness is limited to a specific seismic excitation of a given intensity-and-frequency content [3]. This drawback is not present in the broad set of devices that mitigate the dynamic response of the structure by properly dissipating the energy that is released by the earthquake. In such a case, additional damping is provided to the system, which thereby lowers its vibration levels more robustly against a variety of seismic excitation characteristics.

Of these devices, magnetorheological dampers (MRDs) are among the most effective for vibration control since their properties can be adjusted in real time with relatively low power requirements, and, unlike active devices, they do not inject energy into the system that is being controlled [4]. MRDs exploit the unique characteristics of magnetorheological fluids to provide multiple levels of damping when subjected to magnetic fields with different intensities [5].

A magnetic field is generated by a continuous current that has a tunable intensity and that runs along coils that are wrapped around the chamber that hosts the fluid; the current is provided by the electric grid and/or by high-performances batteries [6,7]. In the case of an earthquake, a dedicated control system can be triggered to set the current intensity into the dampers on the basis of a predefined control algorithm. A semiactive control strategy of the dynamic response is implemented in this way since the control system does not directly apply force to the structure but, instead, is used to control the properties of the tunable passive damper [8].

Several control algorithms have been developed for the use of MRDs for the mitigation of the seismic responses of buildings [9]. Dyke et al. [10,11] propose clipped-optimal control, which is the most popular algorithm in the literature thanks to its relative simplicity and effectiveness. More sophisticated approaches have also been adopted, which rely upon Lyapunov functions [12,13], stochastic control [14], linear quadratic Gaussian with loop-transfer-recovery control [15], sliding mode control [16,17], and intelligent control, such as neural-network-based control [18,19] and fuzzy logic control [20,21].

Even though all the algorithms have been proven to be successful on paper or in experiments conducted on simplified structural models, the intrinsic level of complexity makes their practical application to actual buildings difficult; thus, simpler and more robust strategies are needed in order to preserve the integrity of the structure and the safety of the occupants in cases of seismic events.

This work proposes a practical approach to define a robust seismic protection system that is based on dissipative bracings that integrate magnetorheological fluid dampers (MRDs). The system is suitable for retrofitting civil constructions in areas with moderate and intense seismic activity, and it adopts commercially available dampers. The activation current of the MRDs is assumed to be constant during the seismic event, and its optimal intensity (i_{opt}) is set on the basis of the most probable seismic accelerogram and peak ground acceleration (PGA) that is expected in the geographical area of the structure. During the actual seismic event, the system is triggered by a sensor that measures the acceleration (a_b) at the structure's base. The intensity of the current that is sent to all MRDs is kept equal to the i_{opt} as long as the measured acceleration is lower than the PGA, and it is set to $i_{opt} \cdot a_b / PGA$ in the (remote) case that the a_b exceeds the expected PGA for the area of interest. This approach, therefore, combines the low level of sophistication of a passive system with the tuneability of the MRDs in order to implement a "multi-level passive control" of the seismic response, rather than a semiactive one. The low level of sophistication of the control law plays in favor of the higher safety and robustness of the protection system that is designed to work optimally with reference to the most probable seismic event(/intensity) for the area of interest, and conservatively in the remote case of more intense seismic activity. The key element for the design of such a system is the identification of the optimal MRDs and the activation current intensity in correspondence to a given seismic accelerogram(/PGA) for the area of interest and a mitigation target for the seismic response of the structure. To address this task, the dynamic response of a slender metallic structure to a typical seismic accelerogram at its base is first investigated to define the amount of modal damping to be introduced into the system to mitigate its vibrations, according to a predefined target.

The modal-damping requirement is then converted in terms of the damping levels to be produced by each MRD; for such a purpose, a linearization method is proposed to simulate the dynamic behavior of the MRD through a linear viscous damper that dissipates the same amount of energy during the seismic event. Thanks to this methodology and the underlying adopted formulation, fast parametric analyses of the reference structure dynamics are carried out in order to individuate a damper that is suitable for the specific application.

The characteristic damping curves of the selected device (damping force versus piston speed), which are experimentally determined by the manufacturer, are used to feed an iterative process for predicting the seismic response of the structure in correspondence with different activation levels of the MRDs. At the end of the process, the minimum activation current is determined in compliance with the vibration-suppression targets that are assumed for the seismic protection system.

Finally, dynamic tests of the true-scale structure are carried out, with and without the mitigation system installed, in order to prove its effectiveness, along with the effectiveness of the modeling approaches that were followed for its design.

2. Reference Structure and Seismic Acceleration

A slender metallic truss was selected as a reference structure for the numerical and experimental activities that are addressed in this study. The truss is dynamically representative of a five-story ultra-high-rise building, according to a geometrical scale factor of l = 4. The planform of the truss is square, with a side length of 640 cm, while each story is 900 cm in height (Figure 1). The pillars and floor beams are made of FE340 and are characterized by rectangular box sections with a 2 mm thickness and width x height dimensions of 40 mm × 40 mm and 40 mm × 20 mm, respectively. Each floor consists of a corrugated steel sheet with a reinforced concrete slab surmounted by steel blocks to achieve the global inertial characteristics, which are recapped in Table 1.

The structure considered was subjected to a monodirectional seismic acceleration (a_Y) along the Y-axis of the reference system that is reported in Figure 1. The time history of the a_y was taken from [22], which refers to the earthquake accelerograms that are compliant with Eurocode 8 [23], and which are related to Zone 3, Ground Type A, and a peak ground acceleration (PGA) equal to 1.23 m/s^2 (Figure 2, left side). The zone's category, ground type, and PGA were selected by considering the recent earthquake history and the typical geological characteristics of the seismic regions in Southern Italy. To coherently apply the real earthquake accelerogram to the scaled structure, the acceleration time history was adjusted by scaling the time vector by the factor ($\sqrt{\lambda}$); the scaled time history of the seismic acceleration is reported on the right side of Figure 2 and is hereafter referred to as the "seismic excitation" ($a_y(t)$).

Table 1. Inertial properties of the floors.

	Mass (kg)	Polar Inertia * (kgm^2)
First four floors	385.85	23.15
Covering roof	266.57	15.99

* The polar inertia relates to the center of gravity of the floor.

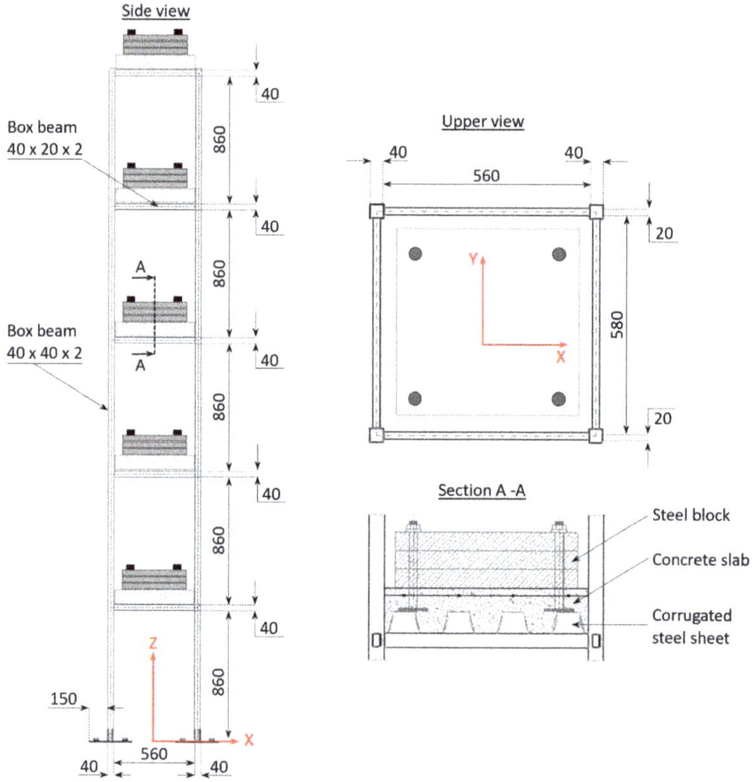

Figure 1. Reference structure, layout, and main dimensions (mm).

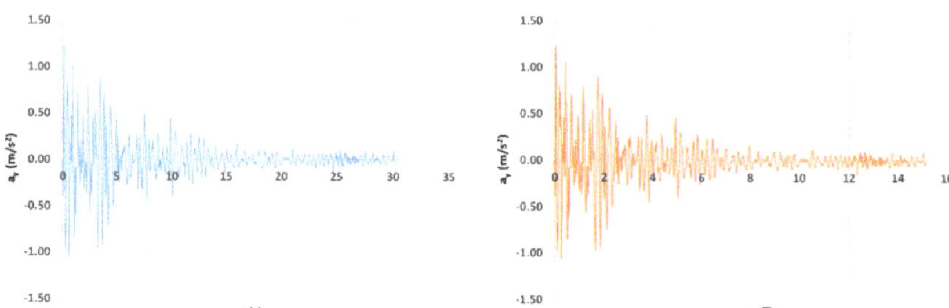

Figure 2. Seismic acceleration vs. time: real earthquake accelerogram (**left**); scaled accelerogram (**right**).

3. Seismic Response of the Reference Structure and Mitigation Target

A finite element model of the reference structure was generated in the UGS-FEMAP® environment [24]. The mesh geometry and the relevant data are reported in Figure 3.

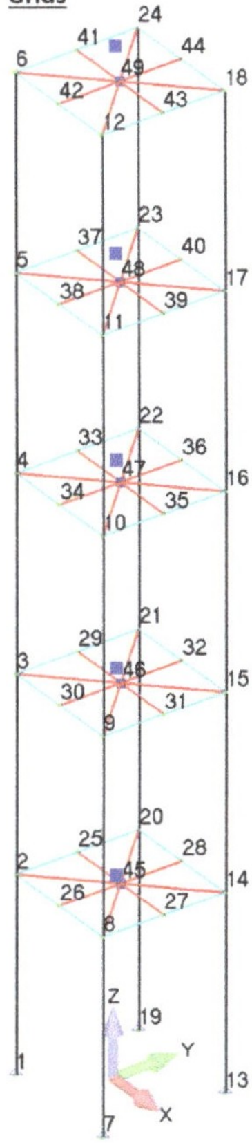

NODE ID	X (m)	Y (m)	Z (m)
1	−0.3	−0.3	0.0
2	−0.3	−0.3	0.9
3	−0.3	−0.3	1.8
4	−0.3	−0.3	2.7
5	−0.3	−0.3	3.6
6	−0.3	−0.3	4.5
7	0.3	−0.3	0.0
8	0.3	−0.3	0.9
9	0.3	−0.3	1.8
10	0.3	−0.3	2.7
11	0.3	−0.3	3.6
12	0.3	−0.3	4.5
13	0.3	0.3	0.0
14	0.3	0.3	0.9
15	0.3	0.3	1.8
16	0.3	0.3	2.7
17	0.3	0.3	3.6
18	0.3	0.3	4.5
19	−0.3	0.3	0.0
20	−0.3	0.3	0.9
21	−0.3	0.3	1.8
22	−0.3	0.3	2.7
23	−0.3	0.3	3.6
24	−0.3	0.3	4.5
25	−0.3	0.0	0.9

NODE ID	X (m)	Y (m)	Z (m)
26	0.0	−0.3	0.9
27	0.3	0.0	0.9
28	0.0	0.3	0.9
29	−0.3	0.0	1.8
30	0.0	−0.3	1.8
31	0.3	0.0	1.8
32	0.0	0.3	1.8
33	−0.3	0.0	2.7
34	0.0	−0.3	2.7
35	0.3	0.0	2.7
36	0.0	0.3	2.7
37	−0.3	0.0	3.6
38	0.0	−0.3	3.6
39	0.3	0.0	3.6
40	0.0	0.3	3.6
41	−0.3	0.0	4.5
42	0.0	−0.3	4.5
43	0.3	0.0	4.5
44	0.0	0.3	4.5
45	0.0	0.0	0.9
46	0.0	0.0	1.8
47	0.0	0.0	2.7
48	0.0	0.0	3.6
49	0.0	0.0	4.5

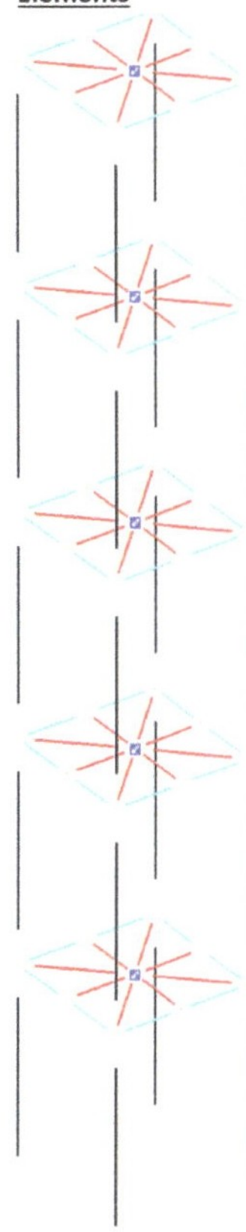

Pillar
(mesh size: 1 element)

Floor beam
(mesh size: 2 elements)

Rigid body element
(1 element per floor)

Lumped mass/inertia
(1 element per floor)

Figure 3. Finite element model of the reference structure, grids, and elements.

Linear bar elements were used for the mesh of the pillars (one single element per structural item) and horizontal beams (two elements per structural item), while each floor was modeled as a rigid body. To implement the rigid-body condition, auxiliary nodes were generated at the center of each floor and were then rigidly connected to all the nodes of the horizontal beams that support the floor; in these rigid connections, the degrees of freedom of the floor-beam grids were slaved to those of the auxiliary nodes at the center of the floors.

The central floor nodes were also used to define the inertial properties of each floor by means of lumped mass elements.

The grids at the base of the model (Grids 1, 7, 13, and 19) were constrained in all six degrees of freedom, and a modal analysis was launched in the NX-Nastran environment to evaluate the first 10 normal modes of the structure, together with their natural frequencies and generalized masses.

The Lanczos method [25] was selected for the calculation of the first twenty modes and related parameters (frequency and generalized mass); the obtained results are recapped in Table 2 and Figure 4 with reference to the first ten modes only.

Table 2. Modal parameters of the reference structure (first 10 modes only).

Mode Number and Name		Frequency (Hz)	Generalized Mass (kgm^2)
1	1st bending in the XZ plane	1.41	902.43
2	1st bending in the YZ plane	1.41	902.43
3	1st torsion	3.17	61.47
4	2nd bending in the XZ plane	4.43	926.30
5	2nd bending in the YZ plane	4.43	926.30
6	3rd bending in the XZ plane	7.94	1271.39
7	3rd bending in the YZ plane	7.94	1271.39
8	2nd torsion	9.63	60.92
9	4th bending in the XZ plane	11.51	773.40
10	4th bending in the YZ plane	11.51	773.40

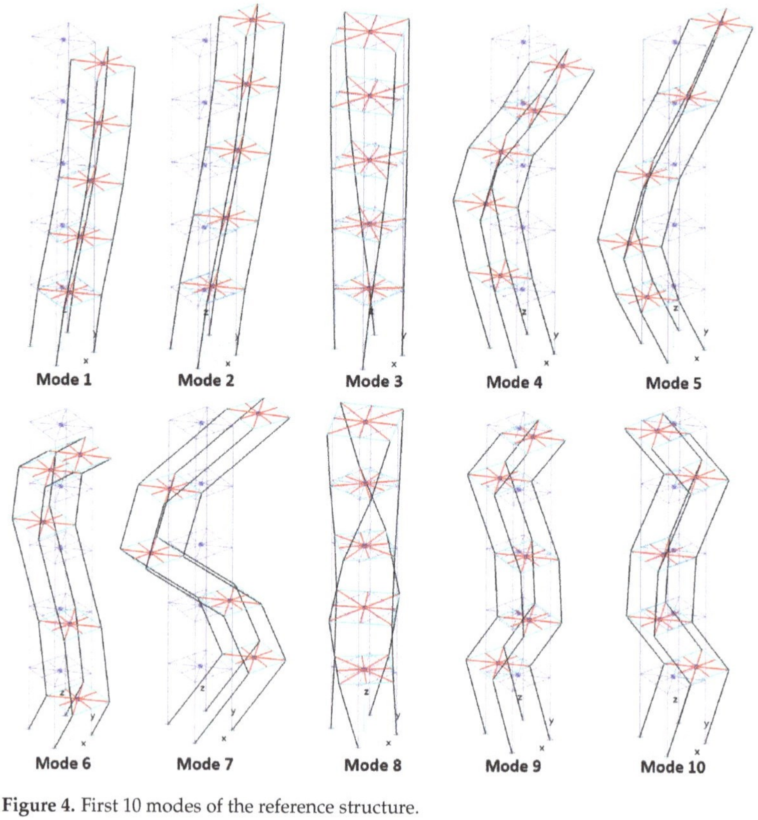

Figure 4. First 10 modes of the reference structure.

The modal database of the reference structure and the earthquake accelerogram ($a_y(t)$) were given as the input to a set of in-house developed MATLAB® [26] routines that implement the formulation that is described in Appendix A for the evaluation of the seismic response of the structure. A viscous modal damping of 0.015 (typical value for metallic structures, [27]) was considered for all modes.

Only the first six modes that were characterized by displacements along the direction of excitation (Modes 2,3,5,7,8,10) were taken into account since the residual inertia associated with all the remaining modes was equal to only 2% of the overall inertia that was exhibited by the structure when moving along the *y*-axis. The definition of "residual inertia" and the explanation of its link to the number of selected modes are given in Appendix A.

The response of the structure is synthesized here in terms of the time histories of the following parameters:

- The floor acceleration, which is intended as the acceleration (a_{yj}) of the j-th floor (j = 1,2, ... , 5) with respect to the ground and along the direction of the excitation (y-axis);
- The interstory acceleration/speed, which is intended as the difference ($\Delta a_{yj}(/\Delta v_{yj})$) between the acceleration (speed) of the j-th and (j − 1)-th floor (j = 1,2, ... ,5). When j = 1, the interstory acceleration/speed coincides with the first floor's acceleration (speed) with respect to the ground.

The obtained time histories are plotted in Figures 5–19, while the peak values of the relevant parameters (a_{yj}, Δa_{yj}, Δv_{yj} (j = 1,2, ... , 5)) are recapped in Table 3.

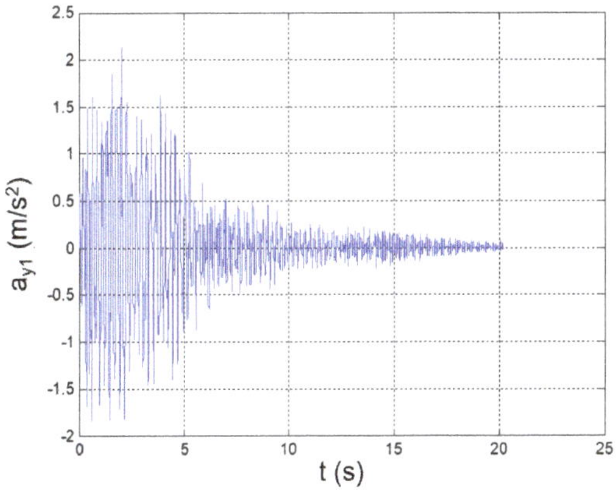

Figure 5. Seismic response, 1st-floor acceleration (Node 45) with respect to the ground.

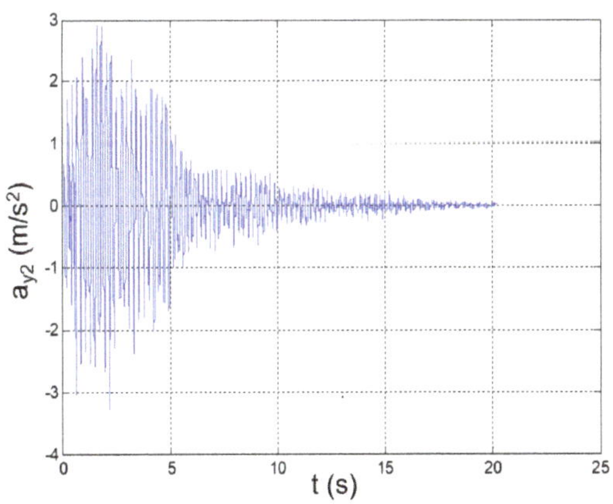

Figure 6. Seismic response, 2nd-floor acceleration (Node 46) with respect to the ground.

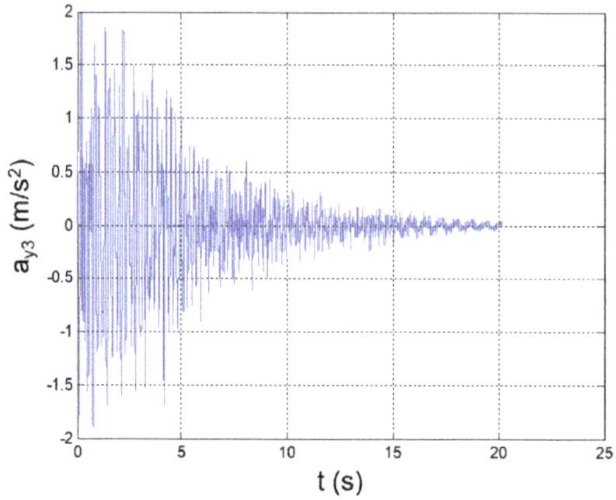

Figure 7. Seismic response, 3rd-floor acceleration (Node 47) with respect to the ground.

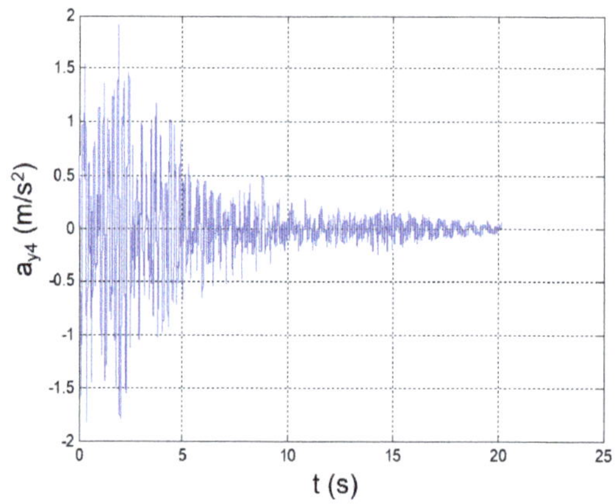

Figure 8. Seismic response, 4th-floor acceleration (Node 48) with respect to the ground.

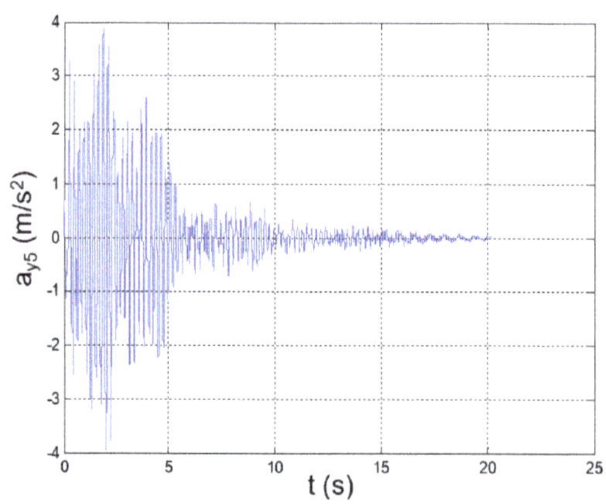

Figure 9. Seismic response, 5th-floor acceleration (Node 49) with respect to the ground.

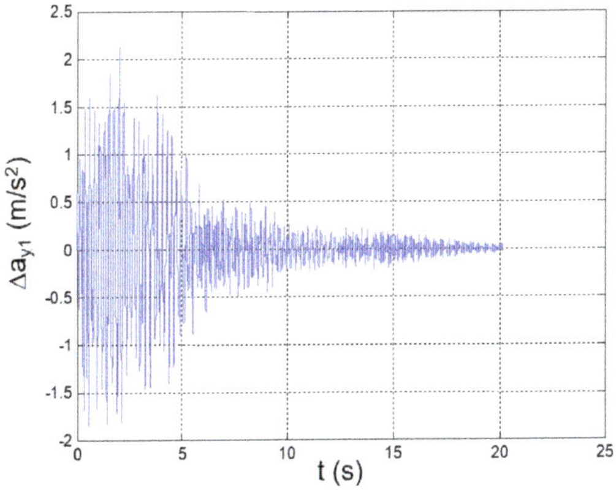

Figure 10. Seismic response, interstory acceleration at the 1st floor ($\Delta a_{y1} = a_{y1}$).

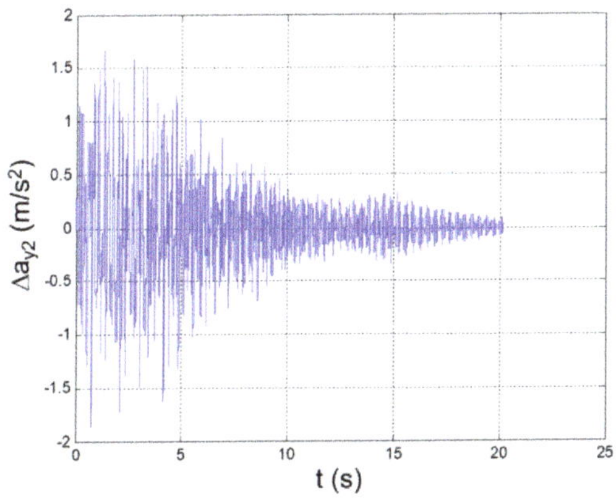

Figure 11. Seismic response, interstory acceleration at the 2nd floor ($\Delta a_{y2} = a_{y2} - a_{y1}$).

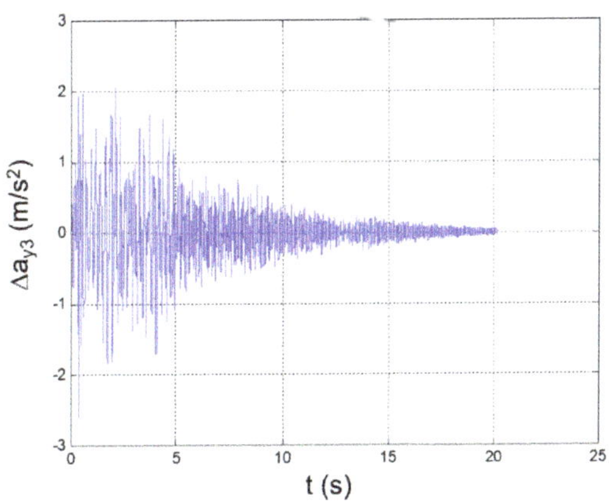

Figure 12. Seismic response, interstory acceleration at the 3rd floor ($\Delta a_{y3} = a_{y3} - a_{y2}$).

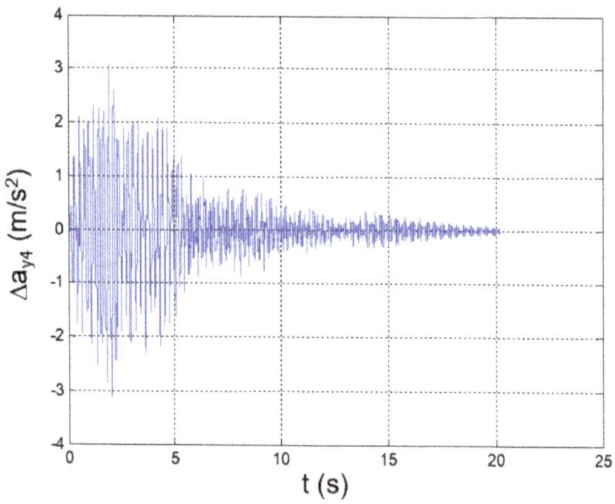

Figure 13. Seismic response, interstory acceleration at the 4th floor ($\Delta a_{y4} = a_{y4} - a_{y3}$).

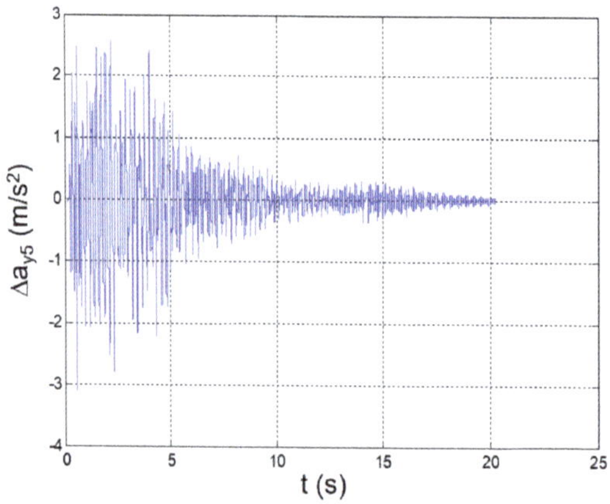

Figure 14. Seismic response, interstory acceleration at the 5th floor ($\Delta a_{y5} = a_{y5} - a_{y4}$).

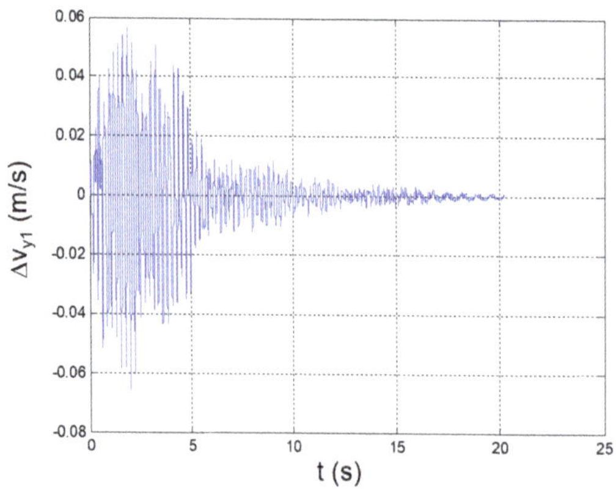

Figure 15. Seismic response, interstory speed at the 1st floor ($\Delta v_{y1} = v_{y1}$: speed of Node 45 with respect to the ground).

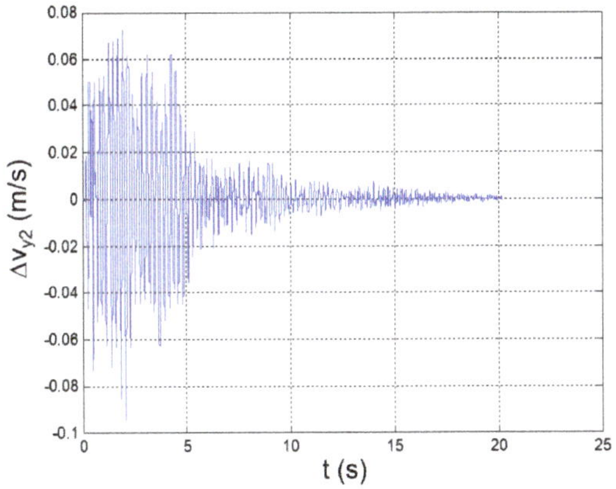

Figure 16. Seismic response, interstory speed at the 2nd floor ($\Delta v_{y2} = v_{y2} - v_{y1}$).

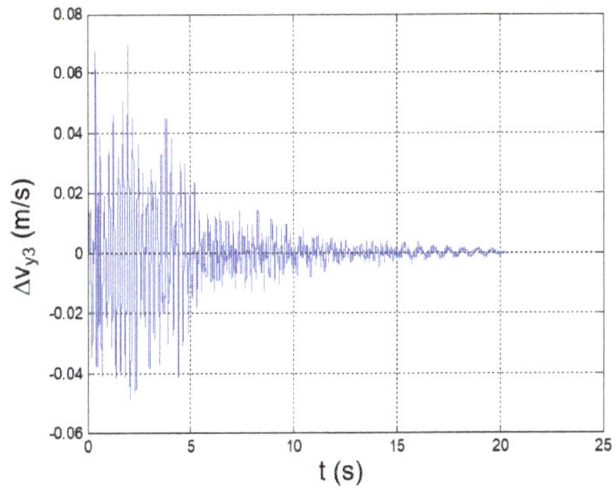

Figure 17. Seismic response, interstory speed at the 3rd floor ($\Delta v_{y3} = v_{y3} - v_{y2}$).

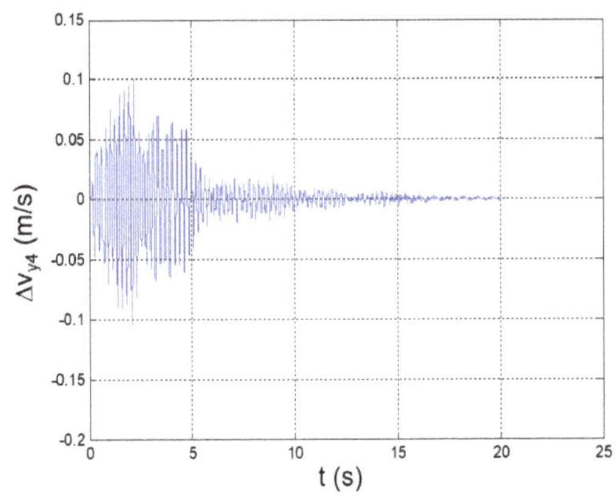

Figure 18. Seismic response, interstory speed at the 4th floor ($\Delta v_{y4} = v_{y4} - v_{y3}$).

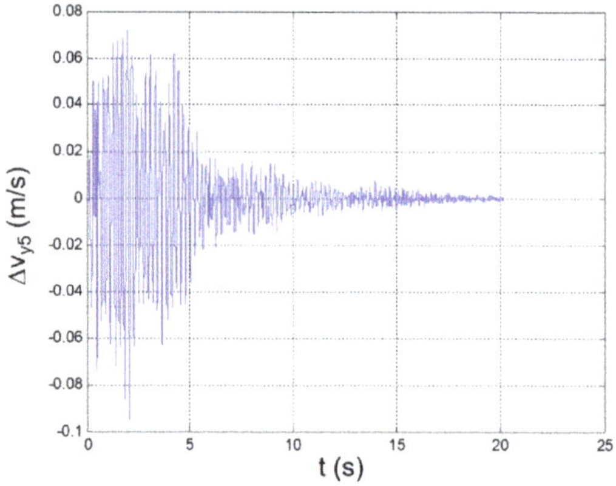

Figure 19. Seismic response, interstory speed at the 5th floor ($\Delta v_{y5} = v_{y5} - v_{y4}$).

Table 3. Seismic response of the reference structure; peak values of relevant parameters (maximum and minimum values in bold).

Story (j) →		1	2	3	4	5
$a_{yj,min}$	(m/s^2)	−1.8519	−3.3017	−1.8979	−1.8335	**−3.9358**
$a_{yj,max}$	(m/s^2)	2.1315	2.9111	1.9864	1.9148	**3.8772**
$\Delta a_{yj,min}$	(m/s^2)	−1.8519	−1.865	−2.6038	**−3.1147**	−3.1007
$\Delta a_{yj,max}$	(m/s^2)	2.1315	1.6767	2.0622	**3.0768**	2.5508
ΔV_{min}	(m/s)	−0.0661	−0.0484	−0.0491	**−0.105**	−0.0947
ΔV_{max}	(m/s)	0.0564	0.0375	0.0695	**0.1005**	0.0724

The maximum and minimum values of the floor accelerations and the interstory accelerations (speeds) occurred at the fifth and fourth floors, respectively. This fully agrees with the typical dynamics of slender buildings, where the highest earthquake-induced accelerations and speeds are generally registered at the top floors [28]. A relevant acceleration peak also occurs at the second floor; this circumstance is justifiable on the basis of the high participation of the second bending mode in the YZ plane in the dynamic response of the structure.

In addition, the maximum floor acceleration is 3.4-fold greater than the peak ground acceleration, which thus shows how relevant the amplification factor of the seismic response could be in the case of predominantly vertical constructions that are sustained by metallic trusses.

For this type of construction, even a moderate earthquake may produce devastating effects; therefore, the adoption of vibration dampers is recommended in order to efficiently mitigate the seismic response.

In this study, a 50% reduction in the acceleration-response peaks at each floor was set as a challenging target for designing a vibration-suppression system (VSS) based on dissipative structural bracings that integrate magnetorheological fluids dampers (MRD) that are available on the market.

In the next section, the theoretical approach that is adopted for the definition of the VSS is discussed.

4. Vibration-Suppression-System Layout and Individuation of the Appropriate MRD

The adoption of dissipative bracings that integrate magnetorheological fluid dampers (MRDs) was considered as a simple and effective solution to mitigate the seismic response of the reference structure.

The bracings were positioned according to the layout in Figure 20, which is in compliance with three practical design requirements:

a. To use the lowest number of bracings per floor;
b. To preserve the symmetry of the structure with respect to the Y–Z plane (thus preventing relevant changes in its seismic response after the installation of the bracings);
c. To dampen the vibrations of the structure along the direction of excitation (*Y*-axis).

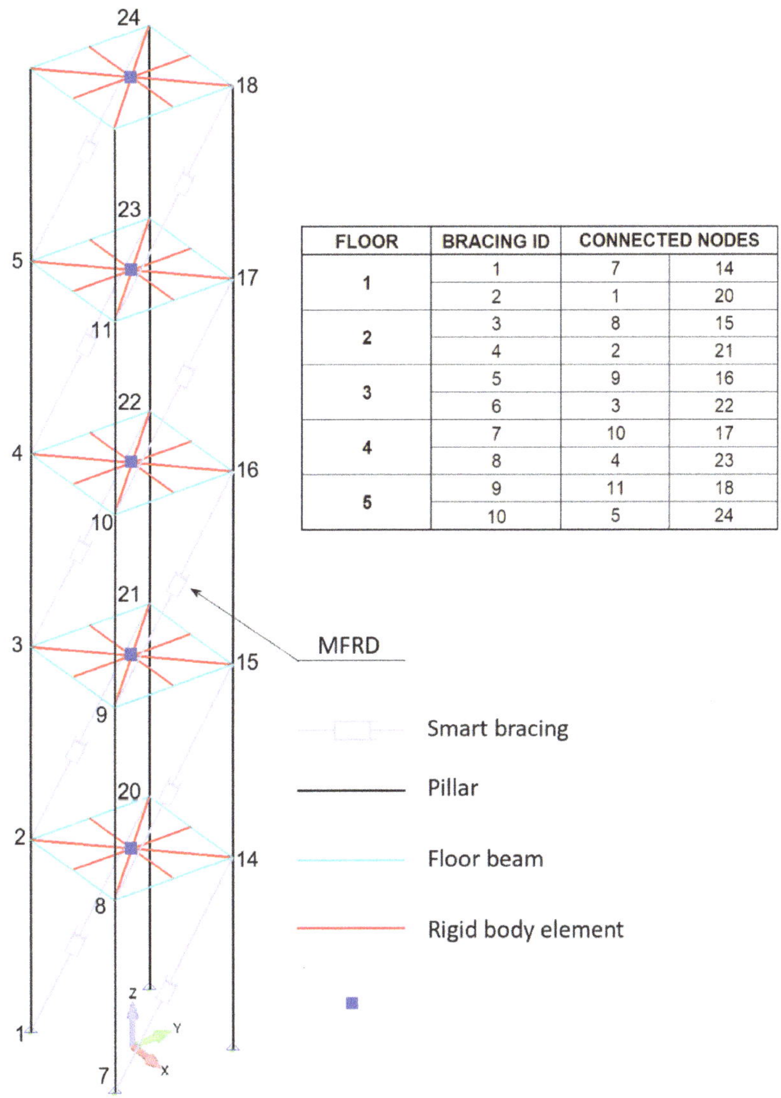

FLOOR	BRACING ID	CONNECTED NODES	
1	1	7	14
	2	1	20
2	3	8	15
	4	2	21
3	5	9	16
	6	3	22
4	7	10	17
	8	4	23
5	9	11	18
	10	5	24

Figure 20. FEM of the reference structure equipped with the vibration-suppression system.

The selection of the best candidate among the MRD models that are available on the market resulted from the qualitative analysis of the acceleration-response spectrum to the seismic signal (Figure 21).

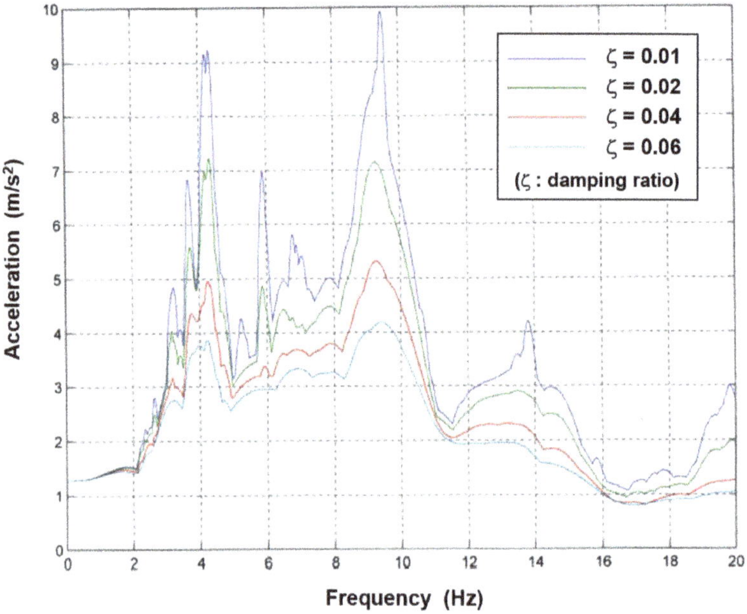

Figure 21. Acceleration-response spectrum to the seismic signal.

The spectrum is a graphical representation of the transient acceleration input, in terms of how a single-degree-of-freedom (SDOF) system (such as a mass on a spring and a damper) would respond to that input. The horizontal axis shows the natural frequency of a hypothetical SDOF. The vertical axis shows the peak acceleration that this SDOF would undergo as a consequence of the input signal and in correspondence with different damping levels [29].

From the curves in Figure 21, it was found that, to mitigate the acceleration response by 50%, it would have been necessary to reach at least a value of 0.04 for the modal damping of all the modes participating in the response. A generalized damping matrix characterized by 4% modal damping for all modes was then assumed for the structure and was converted into physical coordinates. The result was found to be consistent with the expression of the damping matrix that was obtained by assembling the contributions of the linear viscous dampers installed on each bracing, with each one characterized by a damping coefficient in the order of 10^5 Ns/m. The MRD-1005-3 from Lord Co.® (Cary, NC, USA) proved to ensure equivalent damping right in the order of 10^5 Ns/m, and it was therefore selected as a suitable device for the specific application (Figure 22; see datasheet in Appendix B). The equivalent physical damping is intended here as the damping (C_{eq}) that is exhibited by a linear viscous damper that dissipates the same energy as the MRD during the seismic event. The evaluation of this parameter can be carried out by means of the following approach.

The damping force (D) produced by an MRD is a nonlinear function (f) of the piston speed ($\dot{\delta}$: $D = f\left(\dot{\delta}\right)$). If we linearize this function by approximating it to $D_{eq}\dot{\delta}$, then the quadratic error that arises is:

$$\varepsilon^2\left(D_{eq}, t\right) = [f\left(\dot{\delta}\right) - D_{eq}\dot{\delta}]^2 = f^2\left(\dot{\delta}\right) - 2f\left(\dot{\delta}\right)D_{eq}\dot{\delta} + D_{eq}^2\dot{\delta}^2 \quad (1)$$

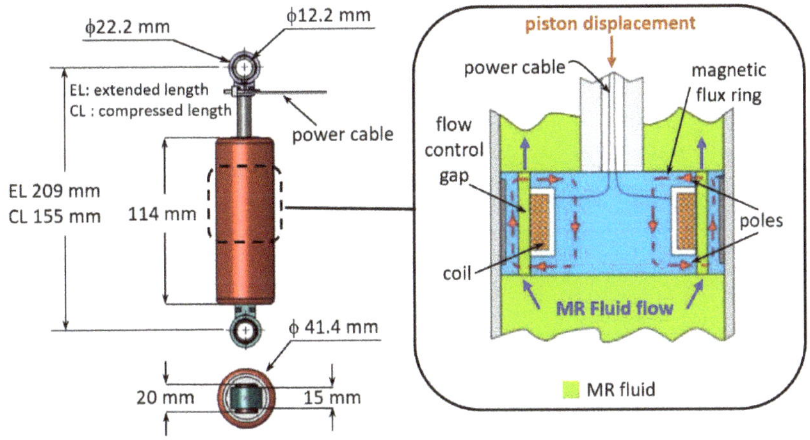

Figure 22. LORD MRD-1005-3: main dimensions and sketch of the working principle.

The quadratic error that is due to the linearization is minimal along the entire duration (T) of the seismic event if Equation (2) is satisfied:

$$\frac{1}{T}\int_0^T \frac{\partial \varepsilon^2(D_{eq},t)}{\partial D_{eq}} dt = 0 \tag{2}$$

By substituting the expression of ε^2 from Equation (1) into Equation (2), we obtain:

$$\frac{1}{T}\int_0^T \left[-2f\left(\dot{\delta}\right)\dot{\delta} + 2D_{eq}\dot{\delta}^2\right] dt = 0$$

and, hence:

$$D_{eq} = \frac{\int_0^T f\left(\dot{\delta}\right)\dot{\delta}dt}{\int_0^T \dot{\delta}^2 dt} \tag{3}$$

For the generic MRD-connecting nodes (h and k) of the structure, the order of magnitude of the equivalent damping was calculated by means of Equation (3), and by considering: $\dot{\delta}(t) = V_k(t) - V_h(t)$, where $V_k(t)(/V_h(t))$ is the speed response of the node ($k(/h)$) along the direction (kh) to the seismic signal exciting the reference (clean) structure and under the hypothesis of 4% modal damping for all the modes taken into account.

5. MRD Activation Current and Evaluation of the Mitigated Seismic Response of the Structure

As in the case of the oleopneumatic dampers, the energy that is dissipated by an MRD is due to the work that is performed by the forces that act on a piston that slides into a viscous fluid. These forces mainly depend on the piston speed, and the graphical representation of this dependence is generally referred to as the "operative curve of the damper". The substantial difference between oleopneumatic dampers and MRDs is not only in terms of the different trends of the operative curves but is also in terms of the fact that MRDs can operate along several curves.

The operative curve of an oleopneumatic damper is unique and it depends on the device's geometry and the mechanical properties of the damping fluid.

The same also applies to an MRD, but, in this case, the mechanical properties of the magnetorheological (MR) fluid can be modified by the magnetic field that is generated by an embedded electromagnet.

In more detail, during the motion of the piston, the MR fluid is forced to flow into tight gaps that are hosted by the piston's head (Figure 22). When the MRD is powered on, the electric current flows into a solenoid that is installed in the piston's head, and a magnetic field arises within the gaps. The ferromagnetic particles that are suspended in the MR fluid align themselves to the magnetic field and generate a system of microscopic structures (fibrils) that hamper the MR fluid's flow through the gaps [30].

At a macroscopic level, this increases the force that is exhibited by the damper, which depends on the intensity of the magnetic field and, consequently, on the intensity of the current into the solenoid. Therefore, different activation currents are associated with different working modes of the MRD and with different operative curves.

The operative curves of the selected MRD are reported in Figure 23 [31] and they correspond with three relevant intensities of the activation current: 0 A (no activation), 1 A (maximum intensity), and 0.5 A (medium intensity).

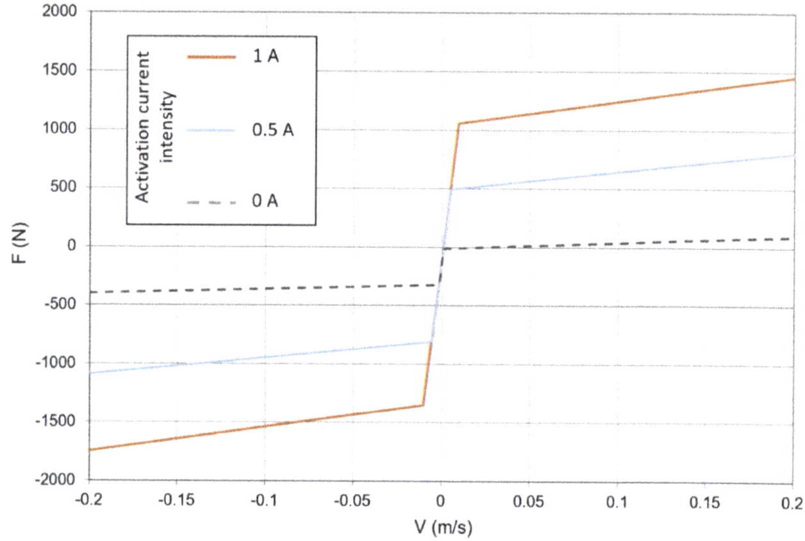

Figure 23. LORD MRD-1005-3, relevant operative curves.

The seismic response of the structure was evaluated in conjunction with different activation levels of the MRDs, starting from 0 A, and increasing by 0.25 A until the mitigation target of the VSS was reached (Figure 24). At each step, the activation current was equal for all the MRDs in order to preserve the structural symmetry with respect to the Y–Z plane.

Each MRD was treated as a linear damper that was characterized by a specific (equivalent) damping coefficient. Under this assumption, the differential equation that governs the seismic response of the system is linear and takes the form (see Appendix A):

$$\underline{\underline{M}}\,\underline{\Delta\ddot{d}} + \underline{\underline{D}}\,\underline{\Delta\dot{d}} + \underline{\underline{K}}\,\underline{\Delta d} = -\underline{\underline{M}}\,\underline{e}a(t), \tag{4}$$

where:

$\underline{\underline{M}}$ is the mass matrix of the structure;
$\underline{\underline{K}}$ is the stiffness matrix of the structure that is constrained at its base;
$\underline{\underline{D}}$ is the damping matrix of the structure, the elements of which are related to the equivalent damping coefficients that are assumed for the MRDs;
$\underline{\Delta d}$ is the displacement vector of the nodes of the structure with respect to the base of the structure;
$a(t)$ is the seismic acceleration at the base of the structure;

$\underline{e}$ is the vector that defines the direction of the seismic acceleration into the same reference frame that is used for the definition of $\underline{\underline{M}}$, $\underline{\underline{K}}$, $\underline{\underline{D}}$, and $\underline{\Delta d}$.

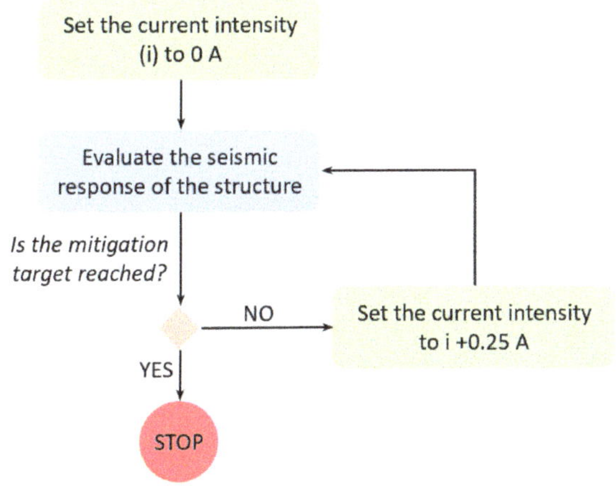

Figure 24. Calculation loop for the individuation of the activation current compliant with the vibration-mitigation target (defined in Section 3).

Equation (4) was solved by referring to the same methodology that was adopted for the evaluation of the seismic response of the structure without dampers: by relying upon the formulation reported in Appendix A, the response was first evaluated in the generalized coordinates domain, and then in terms of the acceleration and speed of the nodes with respect to the base of the structure.

It is worth noting that, in the generalized coordinates domain, the equation that was used to evaluate the seismic response of the structure that was equipped with dampers is formally the same as that used in the case where no dampers were installed. The only difference is in the elements of the generalized damping matrix.

In the case of dampers not being installed on the structure, the generalized damping matrix was built coherently to a modal damping equal to 0.015 for all the relevant modes involved in the seismic response. In the presence of dampers, this matrix is instead given by: $\underline{\underline{\Phi}}_V^T \, \underline{\underline{D}} \, \underline{\underline{\Phi}}_V$, where $\underline{\underline{\Phi}}_V$ is the modal matrix.

The damping matrix ($\underline{\underline{D}}$) was assembled on the basis of the equivalent damping that was exhibited by each MRD, and was calculated by means of the following iterative method (Figure 25):

Step 1: The seismic response of the structure is evaluated in the case of no dampers installed (clean structure). For the generic MRD to be placed on the i-th bracing that connects the nodes (h and k), the equivalent damping ($D_{eq,i}$) was found according to Equation (3) and considering:

$\dot{\delta}(t) = \Delta V_i(t) = V_h(t) - V_k(t)$, where $V_h(t)$ and $V_k(t)$ are the speed responses of the nodes, h and k, respectively, along the h–k direction; the function (f) is coincident with the operative curve of the damper that corresponds to the selected activation current; T is equal to the duration of the seismic event.

Step 2: The seismic response of the structure is evaluated in the case of an MRD that is installed and that exhibits the values of the equivalent damping that was found in the previous step.

Step 3: The equivalent damping of the MRDs is recalculated on the basis of the speed responses ($\Delta V_i(t)$) that were evaluated in the previous step.

For each MRD, the new value of the equivalent damping, $D'_{eq,i}$, is compared to that assumed at the previous step by evaluating the error margin, $\mu = \frac{D'_{eq,i} - D_{eq,i}}{D_{eq,i}}$. If m is below a predefined threshold (t), the process stops, as well as the mitigated seismic response of the structure that is evaluated in the current step.

If not, Steps 2 and 3 are repeated until m < t for each MRD.

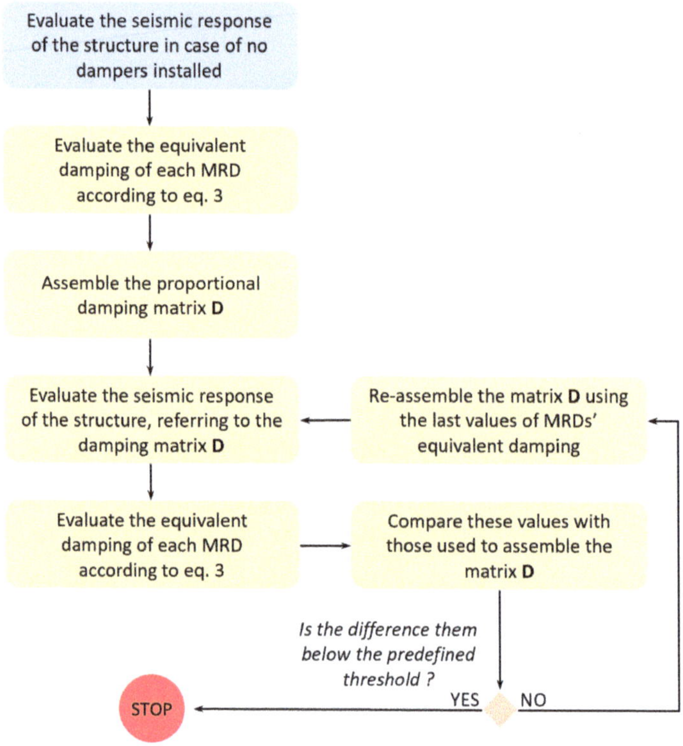

Figure 25. Iterative method for the calculation of MRDs' equivalent damping.

The above-described iteration was executed at the second step of the loop in Figure 24 in order to evaluate the mitigated seismic response of the structure and the equivalent damping of the MRDs that corresponded to the input value of the activation current.

The mitigation target for the seismic-induced vibrations was reached in the case of all MRDs activated with a current of 0.5 A; the convergence on the equivalent damping was reached after one iteration only, with a precision threshold of t = 0.001.

The values of the equivalent damping obtained at each step are reported in Table 4. The mitigated seismic response of the structure is plotted in Figures 26–40, for the time histories of the floor accelerations with respect to the base, the interstory accelerations, and the speeds. The unmitigated seismic response is also plotted for comparison.

Table 4. Equivalent damping values for the MRD activated with a current of 0.5 A.

Floor	Bracings ID (Figure 20)	Equivalent Damping of the MRD (Ns/m) Step 1	Steps 2, 3 (1st Iteration)	Steps 2, 3 (2nd Iteration)
1	1,2	3.9733×10^4	1.0479×10^5	1.1321×10^5
2	3,4	5.5285×10^4	1.0986×10^5	1.1598×10^5
3	5,6	4.9329×10^4	1.1510×10^5	1.2042×10^5
4	7,8	2.8692×10^4	1.1754×10^5	1.2053×10^5
5	9,10	3.4315×10^4	1.2053×10^5	1.2053×10^5

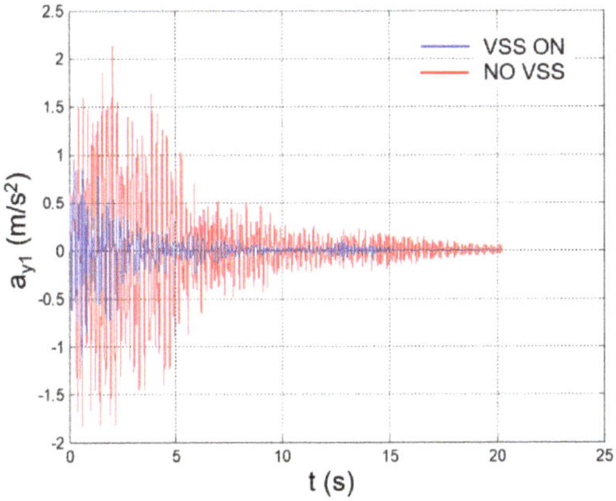

Figure 26. Seismic response, 1st-floor acceleration (Node 45) with respect to the ground.

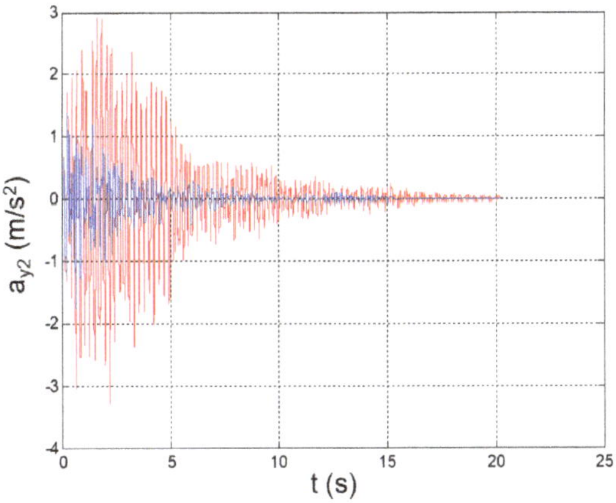

Figure 27. Seismic response, 2nd-floor acceleration (Node 46) with respect to the ground.

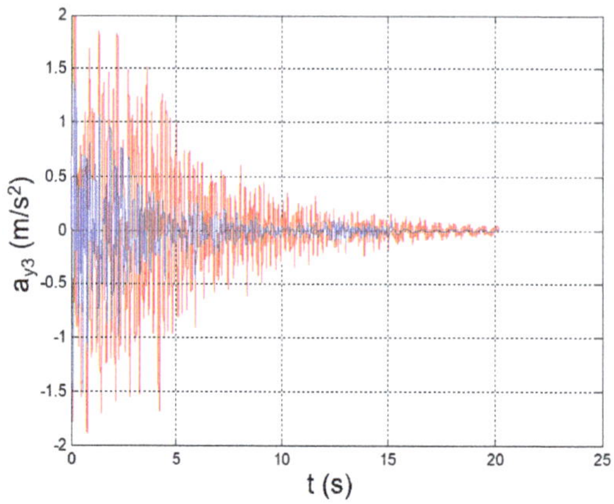

Figure 28. Seismic response, 3rd-floor acceleration (Node 47) with respect to the ground.

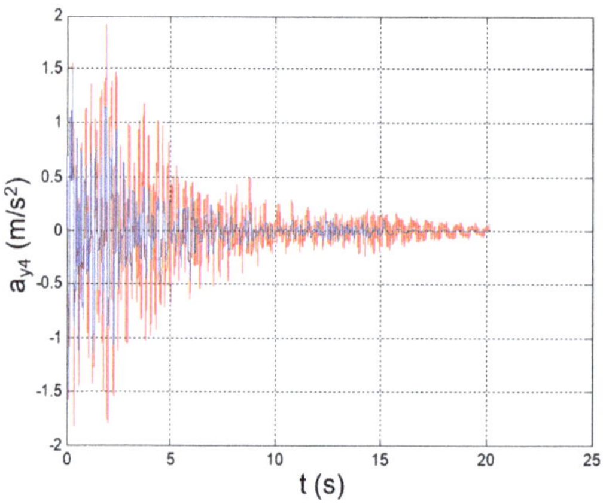

Figure 29. Seismic response, 4th-floor acceleration (Node 48) with respect to the ground.

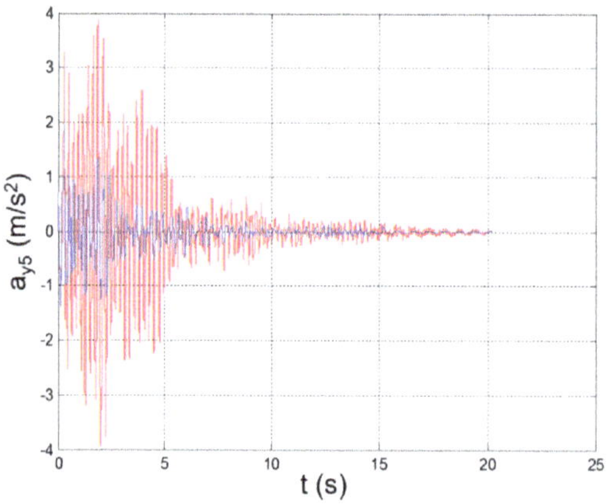

Figure 30. Seismic response, 5th-floor acceleration (Node 49) with respect to the ground.

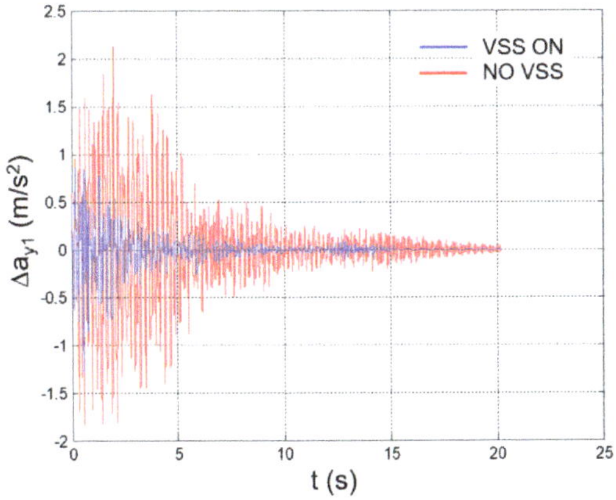

Figure 31. Seismic response, interstory acceleration at the 1st floor ($\Delta a_{y1} = a_{y1}$).

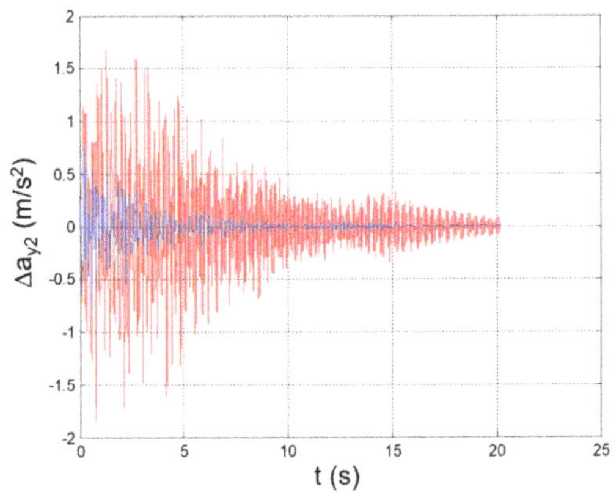

Figure 32. Seismic response, interstory acceleration at the 2nd floor ($\Delta a_{y2} = a_{y2} - a_{y1}$).

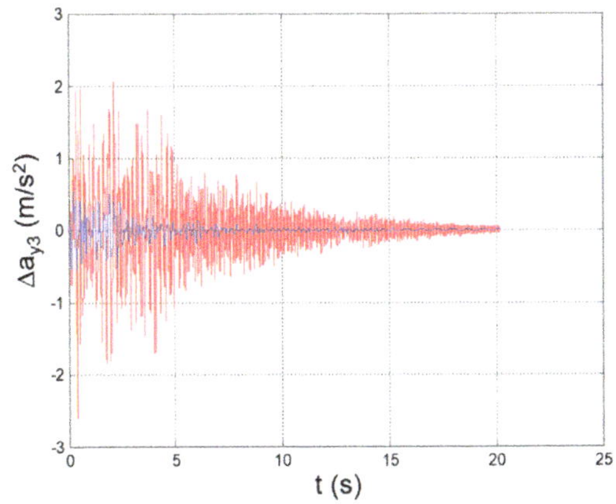

Figure 33. Seismic response, interstory acceleration at the 3rd floor ($\Delta a_{y3} = a_{y3} - a_{y2}$).

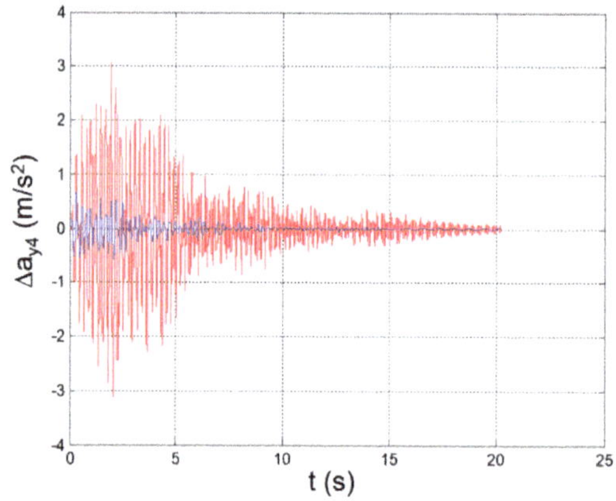

Figure 34. Seismic response, interstory acceleration at the 4th floor ($\Delta a_{y4} = a_{y4} - a_{y3}$).

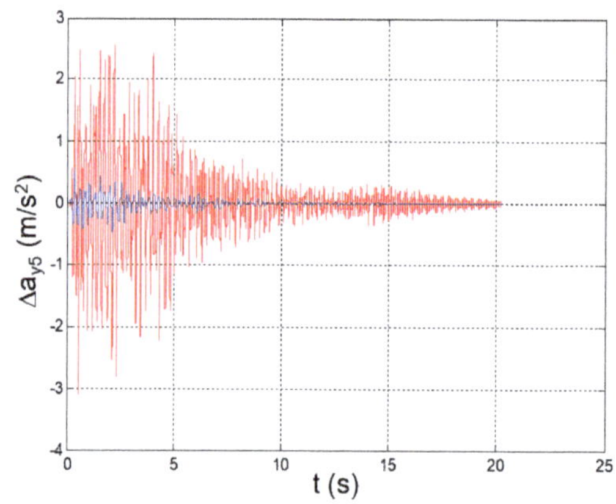

Figure 35. Seismic response, interstory acceleration at the 5th floor ($\Delta a_{y5} = a_{y5} - a_{y4}$).

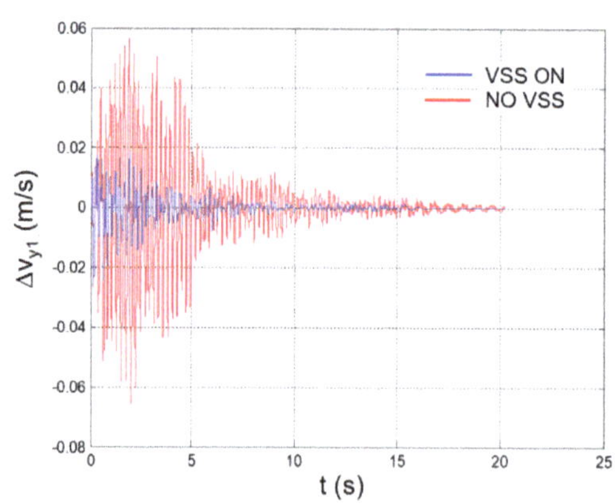

Figure 36. Seismic response, interstory speed at the 1st floor ($\Delta v_{y1} = v_{y1}$: speed of Node 45 with respect to the ground).

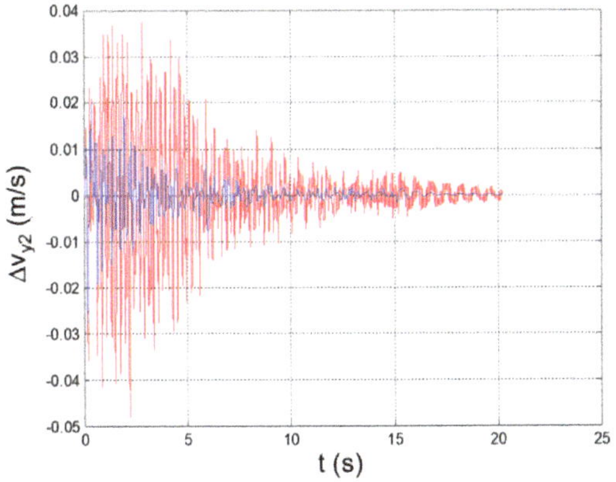

Figure 37. Seismic response, interstory speed at the 2nd floor ($\Delta v_{y2} = v_{y2} - v_{y1}$).

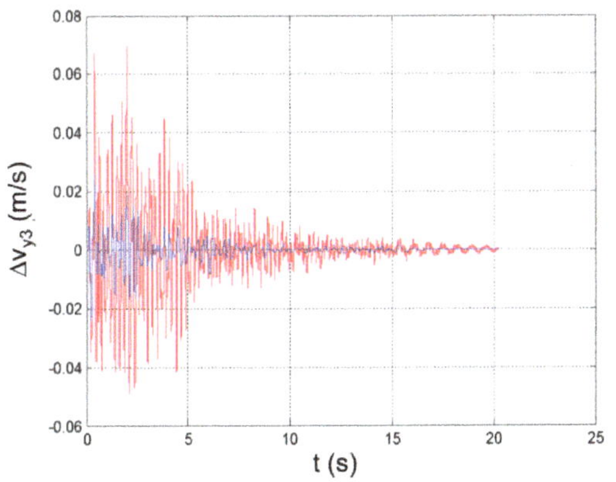

Figure 38. Seismic response, interstory speed at the 3rd floor ($\Delta v_{y3} = v_{y3} - v_{y2}$).

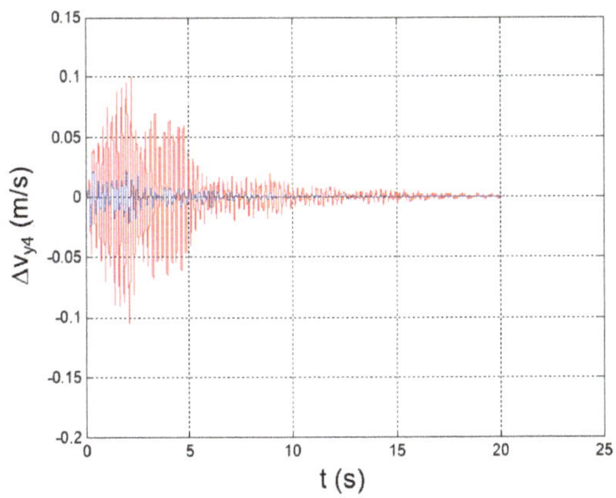

Figure 39. Seismic response, interstory speed at the 4th floor ($\Delta v_{y4} = v_{y4} - v_{y3}$).

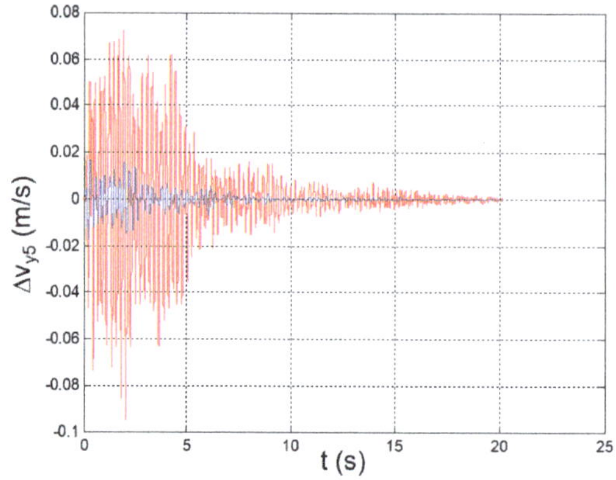

Figure 40. Seismic response, interstory speed at the 5th floor ($\Delta v_{y5} = v_{y5} - v_{y4}$).

Peak values of the plotted parameters are recapped in Table 5, together with the achieved mitigation coefficients.

Table 5. Seismic response of the reference structure, peak values of the relevant parameters, and mitigation coefficients.

	Story (j) →		1	2	3	4	5
VSS ON (MRD Current 0.5 A)	$a_{yj,min}$	(m/s²)	−0.6466	−0.9931	−0.9402	−0.9203	−1.3862
	$a_{yj,max}$	(m/s²)	0.5085	0.8558	0.9923	0.9652	1.3080
	$\Delta a_{yj,min}$	(m/s²)	−0.6466	−0.4042	−0.2608	−0.1716	−0.0954
	$\Delta a_{yj,max}$	(m/s²)	0.5085	0.3706	0.2573	0.1756	0.0989
	ΔV_{min}	(m/s)	−0.0169	−0.0147	−0.0105	−0.0072	−0.0041
	ΔV_{max}	(m/s)	0.0155	0.0136	0.0096	0.0066	0.0037
NO VSS (Table 3 values)	$a'_{yj,min}$	(m/s²)	−1.8519	−3.3017	−1.8979	−1.8335	−3.9358
	$a'_{yj,max}$	(m/s²)	2.1315	2.9111	1.9864	1.9148	3.8772
	$\Delta a'_{yj,min}$	(m/s²)	−1.8519	−1.865	−2.6038	−3.1147	−3.1007
	$\Delta a'_{yj,max}$	(m/s²)	2.1315	1.6767	2.0622	3.0768	2.5508
	$\Delta V'_{min}$	(m/s)	−0.0661	−0.0484	−0.0491	−0.105	−0.0947
	$\Delta V'_{max}$	(m/s)	0.0564	0.0375	0.0695	0.1005	0.0724
Mitigation coefficients	$\frac{a'_{yi,min} - a_{yi,min}}{a'_{yi,min}}$		65.08%	69.92%	50.46%	49.81%	64.78%
	$\frac{a'_{yi,max} - a_{yi,max}}{a'_{yi,max}}$		76.14%	70.60%	50.05%	49.59%	66.26%
	$\frac{\Delta a'_{yi,min} - \Delta a_{yi,min}}{\Delta a'_{yi,min}}$		65.08%	78.33%	89.98%	94.49%	96.92%
	$\frac{\Delta a'_{yi,max} - \Delta a_{yi,max}}{\Delta a'_{yi,max}}$		76.14%	77.90%	87.52%	94.29%	96.12%
	$\frac{\Delta V'_{min} - \Delta V_{min}}{\Delta V'_{min}}$		74.43%	69.63%	78.62%	93.14%	95.67%
	$\frac{\Delta V'_{max} - \Delta V_{max}}{\Delta V'_{max}}$		72.52%	63.73%	86.19%	93.43%	94.89%

6. Experimental Validation of the Vibration-Suppression System (VSS)

The efficacy of the vibration-suppression system was proven by means of true-scale experimental tests that were carried out at the structural dynamics laboratories of the ENEA Casaccia Research Center (Rome, Italy) [32].

The reference structure was installed on a shaking table, which reproduced the accelerogram of the earthquake that was considered for the VSS design.

Three test cases were addressed:

Test 1: Earthquake striking the clean structure (intended as the reference structure without MRDs installed);
Test 2: Earthquake striking the structure with selected MRDs (LORD MRD-1005-3) that were installed and not activated;
Test 3: Earthquake striking the structure with selected MRDs that were installed and activated by a current of 0.5 A.

For each test case, the seismic response of the structure was acquired through five monaxial accelerometers that were installed at the center of each floor and oriented along the direction of the excitation (Figure 41).

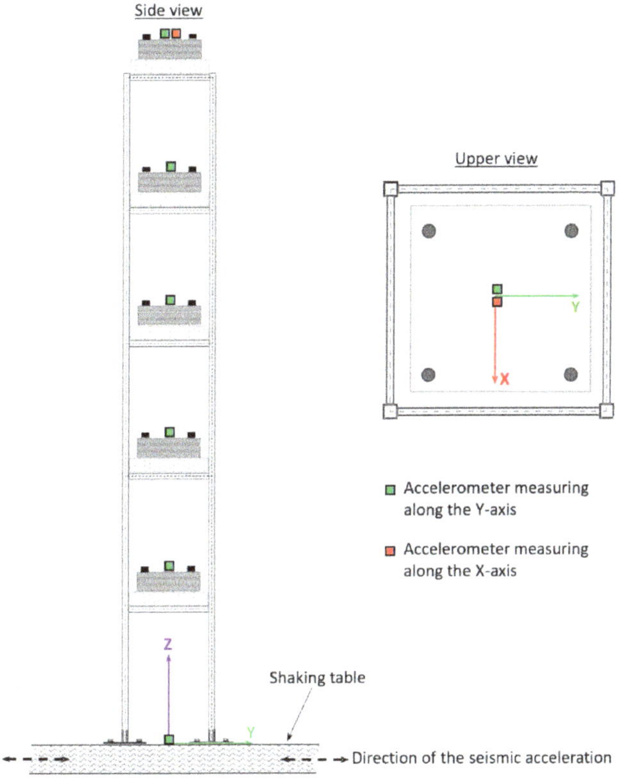

Figure 41. Accelerometers' location and type.

Two extra-accelerometers were placed at the base and on the top of the structure. The accelerometer at the base measured along the direction of excitation and was used to check the coherence of the ground acceleration with the reference seismic accelerogram. The accelerometer at the top of the structure was oriented perpendicularly to the symmetry plane of the structure and was used to verify the absence of significant out-of-plane accelerations.

Siemens LMS SCADAS mobile® was used to drive the shaking table and to record the accelerometers' signals. The input signal generation (shaking-table acceleration) and acquisition processes were controlled and synchronized via LMS Test-Lab software [33].

Each dissipative bracing was realized by joining two squared steel pipes at the ends of the MRD. Hinged joints, which were implemented by means of pin and bolts, were used to connect the MRD to the pipes and the pipes to the structure (Figure 42).

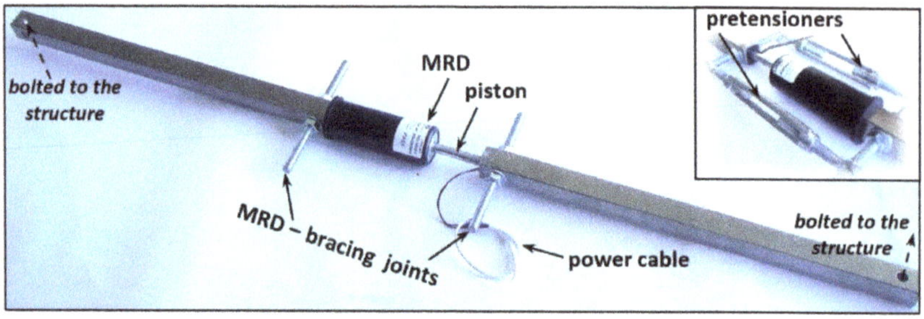

Figure 42. Dissipative bracing that integrates the MRD.

Manual pretensioners were used to set the initial stroke of the MRD's piston at half of its maximum value, which thus allowed for equal maximum excursions of the piston in both compression and tension. The pretensioners were removed after the installation of the bracings on the structure.

In Figure 43, the layout of the reference structure with all dissipative bracing installed is reported.

Figure 43. Reference structure and dissipative bracings (VSS).

The intensity of the electric current into each MRD was regulated by using the potentiometers that were provided by the MRDs supplier (see Appendix B for details).

In Figures 44–48, the time histories of the floor accelerations that were recorded during the three test cases are reported. The interstory accelerations were obtained by the elaboration of these measurements and they are plotted in Figures 49–53. In all the test cases, the extra-accelerometer placed at the top of the structure showed the absence of accelerations out of the symmetry plane of the structure.

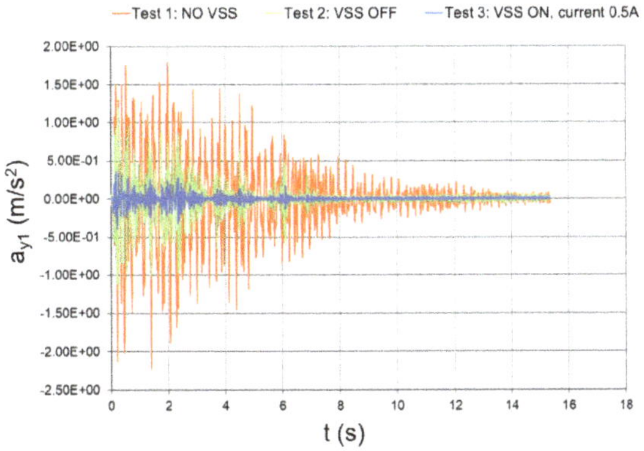

Figure 44. Seismic response, 1st-floor acceleration with respect to the ground (experimental data).

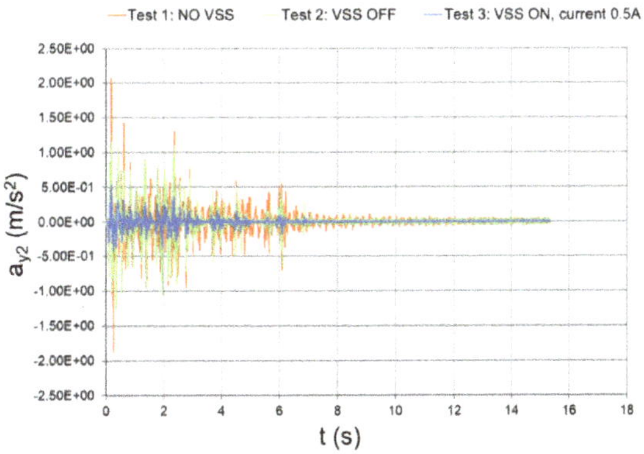

Figure 45. Seismic response, 2nd-floor acceleration with respect to the ground (experimental data).

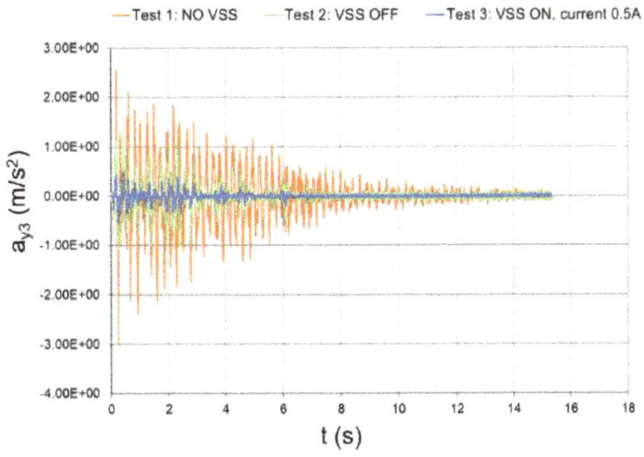

Figure 46. Seismic response, 3rd-floor acceleration with respect to the ground (experimental data).

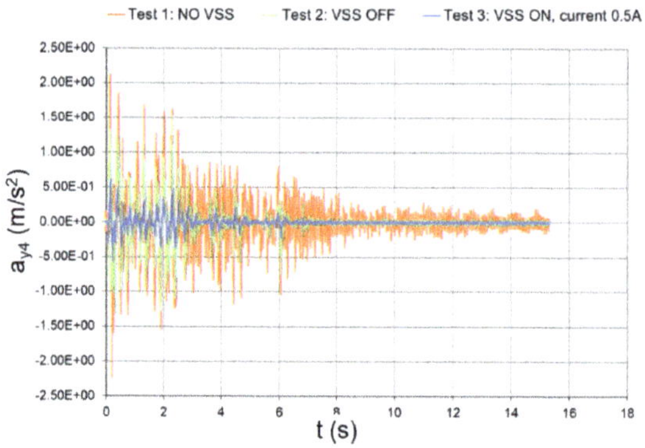

Figure 47. Seismic response, 4th-floor acceleration with respect to the ground (experimental data).

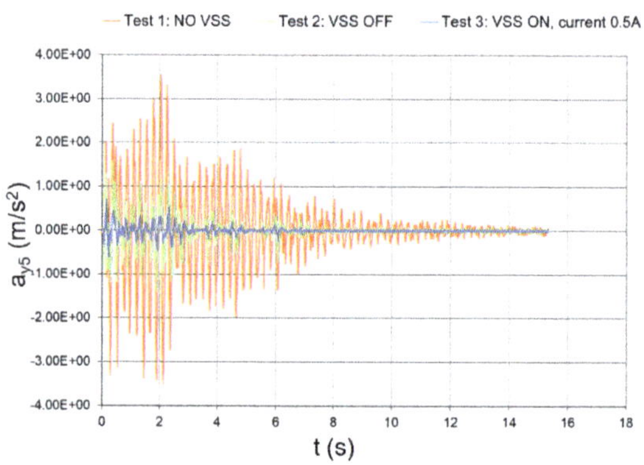

Figure 48. Seismic response, 5th-floor acceleration with respect to the ground (experimental data).

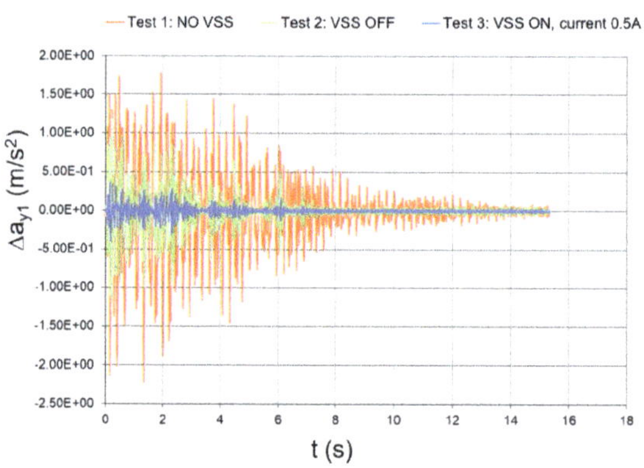

Figure 49. Seismic response (experimental data), interstory acceleration at the 1st floor ($\Delta a_{y1} = a_{y1}$).

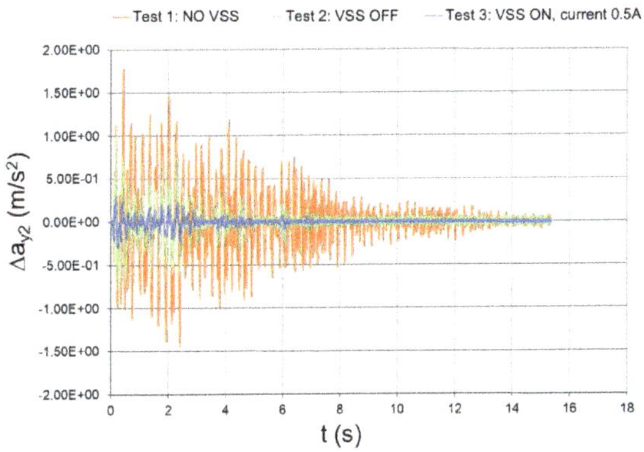

Figure 50. Seismic response (experimental data), interstory acceleration at the 2nd floor ($\Delta a_{y2} = a_{y2} - a_{y1}$).

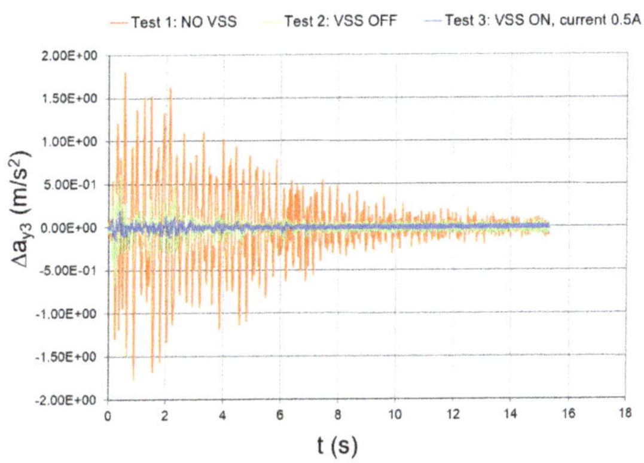

Figure 51. Seismic response (experimental data), interstory acceleration at the 3rd floor ($\Delta a_{y3} = a_{y3}a_{y2} - a_{y2}$).

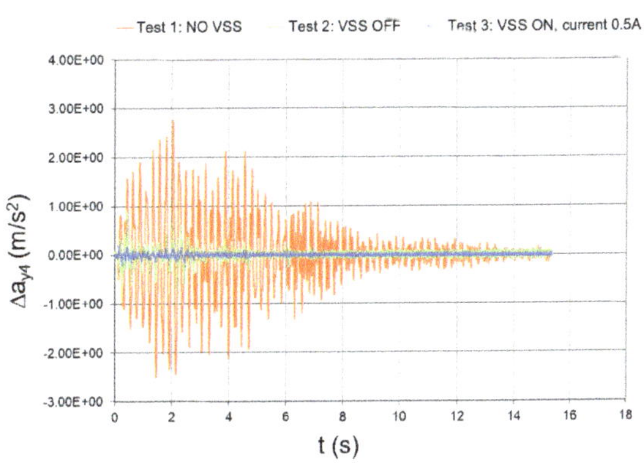

Figure 52. Seismic response (experimental data), interstory acceleration at the 4th floor ($\Delta a_{y4} = a_{y4}a_{y2} - a_{y3}$).

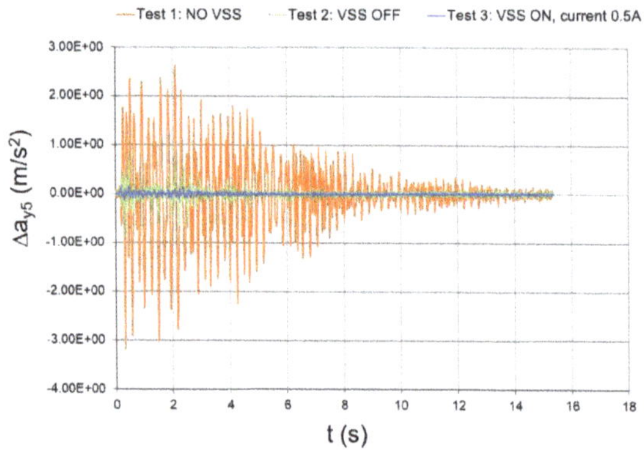

Figure 53. Seismic response (experimental data), interstory acceleration at the 5th floor ($\Delta a_{y5} = a_{y5}a_{y2} - a_{y4}$).

The peak values of the floor and interstory accelerations are reported in Table 6, together with the mitigation coefficients that were obtained with inactivated and activated dampers. In the case of activated dampers, the mitigation coefficients were even higher than those numerically predicted, which thus proves the effectiveness of the conceived vibration-suppression system.

Table 6. Seismic response of the reference structure, peak values of the relevant parameters, and mitigation coefficients (experimental data).

		Story (j) →		1	2	3	4	5
	NO VSS	$a'_{yj,min}$	(m/s²)	−2.2255	−1.8682	−3.0143	−2.2346	−3.506
		$a'_{yj,max}$	(m/s²)	1.7844	2.0731	2.5607	2.1352	3.5508
		$\Delta a'_{yj,min}$	(m/s²)	−2.2255	−1.4388	−1.7743	−2.497	−3.1873
		$\Delta a'_{yj,max}$	(m/s²)	1.7844	1.7855	1.8097	2.7786	2.6375
	VSS OFF	$a''_{yj,min}$	(m/s²)	−1.181	−1.2661	−1.1193	−1.3145	−1.3973
		$a''_{yj,max}$	(m/s²)	1.1894	1.1604	1.5001	1.5167	1.5635
		$\Delta a''_{yj,min}$	(m/s²)	−1.181	−0.7767	−0.6741	−0.9475	−0.8877
		$\Delta a''_{yj,max}$	(m/s²)	1.1894	0.9953	0.6177	0.9255	0.7862
	VSS ON (0.5 A)	$a_{yj,min}$	(m/s²)	−0.3999	−0.4198	−0.5552	−0.5613	−0.6203
		$a_{yj,max}$	(m/s²)	0.3589	0.5208	0.4761	0.6056	0.6997
		$\Delta a_{yj,min}$	(m/s²)	−0.3999	−0.2981	−0.1461	−0.2125	−0.1313
		$\Delta a_{yj,max}$	(m/s²)	0.3589	0.2932	0.1954	0.2086	0.1451
Mitigation coefficients	VSS OFF/ NO VSS	$\frac{a'_{yi,min} - a''_{yi,min}}{a'_{yi,min}}$		46.93%	32.23%	62.87%	41.18%	60.15%
		$\frac{a'_{yi,max} - a''_{yi,max}}{a'_{yi,max}}$		33.34%	44.03%	41.42%	28.97%	55.97%
		$\frac{\Delta a'_{yi,min} - \Delta a''_{yi,min}}{\Delta a'_{yi,min}}$		46.93%	46.02%	62.01%	62.05%	72.15%
		$\frac{\Delta a'_{yi,max} - \Delta a''_{yi,max}}{\Delta a'_{yi,max}}$		33.34%	44.26%	65.87%	66.69%	70.19%
	VSS ON/ NO VSS	$\frac{a'_{yi,min} - a_{yi,min}}{a'_{yi,min}}$		82.03%	77.53%	81.58%	74.88%	82.31%
		$\frac{a'_{yi,max} - a_{yi,max}}{a'_{yi,max}}$		79.89%	74.88%	81.41%	71.64%	80.29%
		$\frac{\Delta a'_{yi,min} - \Delta a_{yi,min}}{\Delta a'_{yi,min}}$		82.03%	79.28%	91.77%	91.49%	95.88%
		$\frac{\Delta a'_{yi,max} - \Delta a_{yi,max}}{\Delta a'_{yi,max}}$		79.89%	83.58%	89.20%	92.49%	94.50%

7. Conclusions

A practical design was proposed for a seismic protection system that is suitable for slender metallic frames and that is based on dissipative bracings that integrate MRD dampers.

The nonlinear relationship between the dissipative force and the piston speed of the MRD was replaced by a linear equation that involves the evaluation of an equivalent damping coefficient for the MRD. A rational method was adopted for the calculation of the equivalent damping coefficient by assimilating the MRD to a proportional damper that dissipated the same amount of energy during the seismic event.

The linearization of the dynamic behavior of the magnetorheological fluid dampers allowed for the implementation of a robust iterative process for the evaluation of the seismic response of the structure in correspondence with the different activation levels of the embedded MRDs. The response was first evaluated in the generalized coordinates domain, and then in terms of the acceleration and speed of the structural nodes with respect to the base of the structure. A current intensity of 0.5 A was found to be adequate to achieve the target 50% reduction in the acceleration-response peaks at each floor.

A true-scale model of the reference structure was then manufactured and installed on a shaking table, which provided the reference seismic excitation. The acceleration response was measured at each floor of the structure in three operative conditions: (1) A clean structure (no dissipative bracings installed); (2) With the vibration-suppression system off (dissipative bracings installed and MRD not activated); (3) With the vibration-suppression system on (dissipative bracings installed and MRD activated by a current of 0.5 A).

With respect to the "clean structure" case, the activation of the MRD led to an average reduction of nearly 70% for the peaks of the floor accelerations with respect to the base of the structure, as well as for the interstory accelerations. The higher level of abatement with respect to the design target can be justified by considering the intrinsic limitations of the numerical simulations that did not take into account the extra damping that was due to the friction between the structural elements at their interface joints. A slight increase in the modal damping of the clean structure with respect to the assumed value (1.5% for all modes) would have surely limited the discrepancy between the numerical expectations and the experimental outcomes. On the other hand, the proposed simulation strategy is mainly intended to provide a fast and practical tool for the robust design of the seismic mitigation system; in light of this consideration, any simplification is acceptable if it results in the safer design of the system.

The robustness and effectiveness of the conceived system are further proven by the recorded seismic response of the structure in the case of MRDs being installed and inactivated. The average reduction factor for the floor acceleration peaks and for the interstory acceleration peaks is nearly 48%, even if some floors have a reduction factor lower than 50%. This means that, even in the case of a blackout (highly probable during an earthquake) or the malfunction of the hardware that powers the MRDs, the system still maintains good abilities to dampen the structural response to the seismic excitation.

The achievement of significant vibration-suppression levels in the case of both activated and nonactivated devices represents a remarkable result, which shows not only the effectiveness of MRDs, but also the soundness of the design and simulation approaches that were adopted for the assessment of the seismic protection system's dynamics.

Funding: Part of the research described in this paper has been carried out within the framework of the Tellus Stability Project, which was funded by the Italian Ministry of Research, Instruction and University, in the framework of the FAR platform, and pursuant to Article 5 of Ministerial Decree No. 593/2000.

Institutional Review Board Statement: Not applicable.

Informed Consent Statement: Not applicable.

Data Availability Statement: All the relevant data used in this study have been declared in the paper's main text. Further data may be requested from the corresponding author.

Acknowledgments: The author would like to thank the staff of the ENEA-Casaccia research center for the extraordinary support for all the experimental activities that were aimed at the characterization of the seismic response of the reference structure.

Conflicts of Interest: The author declares no conflict of interest. The funders had no role in the design of the study; in the collection, analyses, or interpretation of data; in the writing of the manuscript, or in the decision to publish the results.

Nomenclature

a_y	seismic acceleration (along the *Y*-axis of the reference system in Figure 1)
a_{yj}	acceleration (along the *Y*-axis) of the j-th floor with respect to the ground
$a_{yj,max}$	maximum value of a_{yj}
$a_{yj,min}$	minimum value of a_{yj}
$\underline{\underline{D}}$	damping matrix of the structure in the case of MRD installed
$\underline{\Delta d}$	displacement vector of the nodes of the structure with respect to its base
$D_{eq,i}$	equivalent damping of the i-th bracing
$a(t)$	seismic acceleration at the base of the structure
$\underline{e}$	excitation direction
ID	identification (number)
$\underline{\underline{K}}$	stiffness matrix of the structure constrained at its base
$\underline{\underline{K_G}}$	generalized stiffness matrix
$\underline{\underline{M}}$	mass matrix of the structure
$\underline{\underline{M_G}}$	generalized mass matrix
MRD(s)	magnetorheological fluid damper(s)
MRF	magnetorheological fluids
PGA	peak ground acceleration
$\underline{q}$	generalized displacement vector
SDOF	single degree of freedom
T	duration of the seismic event
t	time variable
VSS	vibration-suppression system
v_{yj}	speed (along the *Y*-axis) of the j-th floor with respect to the ground
$v_{yj,max}$	maximum value of v_{yj}
$v_{yj,min}$	minimum value of v_{yj}
Δa_{yj}	interstory acceleration between the j-th and the (j – 1)th floors ($\Delta a_{yj} = a_{yj} - a_{y(j-1)}$)
$\Delta a_{yj,max}$	maximum value of Δa_{yj}
$\Delta a_{yj,min}$	minimum value of Δa_{yj}
ΔV_i	relative speed between the nodes connecting the i-th bracing taken along the axis of the bracing
Δv_{yj}	interstory speed between the j-th and the (j – 1)th floors ($\Delta v_{yj} = v_{yj} - v_{y(j-1)}$)
$\Delta v_{yj,max}$	maximum value of Δv_{yj}
$\Delta v_{yj,min}$	minimum value of Δv_{yj}
Greek symbols	
ε^2	quadratic error
$\underline{\underline{\Phi}}_V$	modal matrix
Γ	vector of the modal participation factors
$\bar{\lambda}$	geometrical scale factor of the reference structure
μ	error margin
$\underline{\underline{\sigma}}$	damping matrix of the reference structure (no dampers installed)
$\gamma\underline{\underline{\sigma_G}}$	generalized damping matrix
τ	precision threshold
ω_{Ni}	pulsation of the *i*-th mode
ζ	damping ratio

Appendix A. Formulation Adopted for the Evaluation of the Seismic Response and the Residual Inertia Criterion

Let us call S a generic structure that is fixed to the ground. Let I_B be the subset of the structural nodes that belong to the base of S, with I_S being the subset of the remaining nodes of S, and B being an extra node to which all the nodes of the base are rigidly linked ($B \notin I_S$, $B \notin I_B$). Referring to a Cartesian system ($R \equiv$ (O,XYZ)), which is fixed with respect to the ground, if a forced displacement (d_b) is imposed on B along the direction K, then, the $I_{n,S}$ nodes will all move by d_b in the same direction. In mathematical terms, if N is the number of active degrees of freedom of the $I_{n,S}$ nodes, the displacement $\underline{d}$ of the $I_S$nodes will be equal to $d_b \cdot \underline{e}$, where $\underline{e}$ is the (N $\times$ 1) vector that is characterized by unitary components in terms of the positions associated with the K direction, and by zero components in all the other positions To clarify this with an example, let us suppose that I_S is characterized by eight nodes, and that each of the nodes has two active degrees of freedom that are related to the translations along the X and Y axis of a given reference frame (N = 16). In this case, if the displacement (db) is imposed along the X direction, then the vector (e) would have the following expression:

$$(\underline{e})^T = \begin{bmatrix} 1 & 0 & 1 & 0 & 1 & 0 & 1 & 0 & 1 & 0 & 1 & 0 & 1 & 0 & 1 & 0 \end{bmatrix}$$

In the absence of external forces and dissipation, the dynamic response of S to an acceleration that is imposed at its base is governed by the following equation:

$$\begin{bmatrix} m_e & \underset{(1\times N)}{\underline{0}} \\ \underset{(N\times 1)}{\underline{0}} & \underset{(N\times N)}{\underline{\underline{M}}} \end{bmatrix} \begin{bmatrix} \ddot{d}_b \\ \ddot{\underline{d}} \end{bmatrix} + \begin{bmatrix} \underline{e}^T \underline{\underline{K}}\, \underline{e} & -\underline{e}^T \underline{\underline{K}} \\ -\underline{\underline{K}}\, \underline{e} & \underset{(N\times N)}{\underline{\underline{K}}} \end{bmatrix} \begin{bmatrix} d_b \\ \underset{(N\times 1)}{\underline{d}} \end{bmatrix} = \begin{bmatrix} 0 \\ \underset{(N\times 1)}{\underline{0}} \end{bmatrix}, \quad \text{(A1)}$$

where:

m_e is the inertia of the base of S along the displacement direction that is individuated by the vector, $\underline{e}$;
$\underline{\underline{M}}$ is the mass matrix of S related to the In,S nodes only;
$\underline{\underline{K}}$ is the stiffness matrix of S constrained at its base;
$\underline{d}$ is the displacement vector of the In,S nodes.

Let $\underline{\Delta d}$ be the displacement vector of the I_S nodes with respect to a reference system ($R' \equiv$ (O′, X′Y′Z′)), moving together with the base of S and having the axes X′, Y′, and Z′, which are oriented as X, Y, and Z, respectively. Since it results in $\underline{d} = d_b\underline{e} + \underline{\Delta d}$, the vector, $\begin{bmatrix} d_b \\ \underline{d} \end{bmatrix}$, can be expressed in the following way:

$$\begin{bmatrix} d_b \\ \underline{d} \end{bmatrix} = \begin{bmatrix} 1 & \underset{(1\times N)}{\underline{0}} \\ \underline{e} & \underset{(N\times N)}{\underline{\underline{I}}} \end{bmatrix} \begin{bmatrix} d_b \\ \underline{\Delta d} \end{bmatrix} \quad \text{(A2)}$$

where $\underline{\underline{I}}$ is the identity matrix of the size (N).
By substituting Equation (A2) into Equation (A1), we obtain:

$$\begin{bmatrix} m_e & \underset{(1\times N)}{\underline{0}} \\ \underset{(N\times 1)}{\underline{0}} & \underset{(N\times N)}{\underline{\underline{M}}} \end{bmatrix} \begin{bmatrix} 1 & \underset{(1\times N)}{\underline{0}} \\ \underline{e} & \underset{(N\times N)}{\underline{\underline{I}}} \end{bmatrix} \begin{bmatrix} \ddot{d}_b \\ \ddot{\underline{\Delta d}} \end{bmatrix} + \begin{bmatrix} \underline{e}^T \underline{\underline{K}}\, \underline{e} & -\underline{e}^T \underline{\underline{K}} \\ -\underline{\underline{K}}\, \underline{e} & \underline{\underline{K}} \end{bmatrix} \begin{bmatrix} 1 & \underset{(1\times N)}{\underline{0}} \\ \underline{e} & \underset{(N\times N)}{\underline{\underline{I}}} \end{bmatrix} \begin{bmatrix} d_b \\ \underline{\Delta d} \end{bmatrix} = \begin{bmatrix} 0 \\ \underset{(N\times 1)}{\underline{0}} \end{bmatrix}$$

or, in a more compact form:

$$\begin{bmatrix} m_e & \underset{(1\times N)}{\underline{0}} \\ \underline{\underline{M}}\,\underline{e} & \underset{(N\times N)}{\underline{\underline{M}}} \end{bmatrix}\begin{bmatrix} \ddot{d}_b \\ \Delta\ddot{\underline{d}} \end{bmatrix} + \begin{bmatrix} 0 & -\underline{e}^T\underline{\underline{K}} \\ \underset{(N\times 1)}{\underline{0}} & \underset{(N\times N)}{\underline{\underline{K}}} \end{bmatrix}\begin{bmatrix} d_b \\ \Delta\underline{d} \end{bmatrix} = \begin{bmatrix} 0 \\ \underset{(N\times 1)}{\underline{0}} \end{bmatrix}. \quad \text{(A3)}$$

Since $\ddot{d}_b$ is equal to the seismic accelerogram (a(t)), the second part of Equation (A3) can be rewritten as follows:

$$\underline{\underline{M}}\,\Delta\ddot{\underline{d}} + \underline{\underline{K}}\,\Delta\underline{d} = -\underline{\underline{M}}\,\underline{e}a(t) \quad \text{(A4)}$$

which allows for the evaluation of the seismic response of the structure (i.e., of the time histories of the displacements ($\Delta\underline{d}$(/speeds $\Delta\dot{\underline{d}}$/accelerations $\Delta\ddot{\underline{d}}$)) of the I_S nodes with respect to the base of S).

In the presence of viscous dissipation, all the considerations made so far may be repeated, which thus leads to the following expression of the equation that governs the seismic response of the structure:

$$\underline{\underline{M}}\,\Delta\ddot{\underline{d}} + \underline{\underline{\sigma}}\,\Delta\dot{\underline{d}} + \underline{\underline{K}}\,\Delta\underline{d} = -\underline{\underline{M}}\,\underline{e}a(t) \quad \text{(A5)}$$

where $\underline{\underline{\sigma}}$ is the viscous damping matrix of the system and, for the cases of practical interest, is proportional to the stiffness and mass matrixes.

Equation (A5) may be effectively solved in the generalized coordinates domain [34].

If $\underset{(N\times N)}{\underline{\underline{\Phi}}_V}$ is the matrix of the N modes of the structure constrained to its base, then the vector ($\Delta\underline{d}$) may be expressed as the product, $\underset{(N\times N)(N\times 1)}{\underline{\underline{\Phi}}_V\ \underline{q}(t)}$, where $\underline{q}(t)$ is the vector of the generalized coordinates of the system.

By placing this expression into Equation (A5), we obtain the following:

$$\underline{\underline{M}}\,\underline{\underline{\Phi}}_V\ddot{\underline{q}}(t) + \underline{\underline{\sigma}}\,\underline{\underline{\Phi}}_V(t)\dot{\underline{q}}(t) + \underline{\underline{K}}\,\underline{\underline{\Phi}}_V(t)\underline{q}(t) = -\underline{\underline{M}}\,\underline{e}a(t)$$

and, after pre-multiplying all the terms by the transpose of $\underline{\underline{\Phi}}_V$:

$$\underline{\underline{M_G}}\,\ddot{\underline{q}}(t) + \underline{\underline{\sigma_G}}\,\dot{\underline{q}}(t) + \underline{\underline{K_G}}\,\underline{q}(t) = -\ \Gamma\ \_a(t) \quad \text{(A6)}$$

where:

$\underline{\underline{M_G}} = \underline{\underline{\Phi}}_V^T\,\underline{\underline{M}}\,\underline{\underline{\Phi}}_V$ is the diagonal matrix of the generalized masses (mG,i (i = 1, . . . , N));

$\underline{\underline{K_G}}$ = is the diagonal matrix of the generalized masses (mG,i (i = 1, . . . , N));

$\underline{\underline{\sigma_G}} = \underline{\underline{\Phi}}_V^T\,\underline{\underline{\sigma}}\,\underline{\underline{\Phi}}_V$ is the matrix of the generalized damping (sG,i (i = 1, . . . , N)), which is also diagonal, according to Basile's hypothesis [34];

$\Gamma\ \_ = \underline{\underline{\Phi}}_V^T\,\underline{\underline{M}}\,\underline{e}$ is the vector of the modal participation factors (Gi (i = 1,...,N)).

Equation (A6) represents a set of decoupled differential linear equations that can be solved individually. The generic equation has the following expression:

$$m_{G,i}\ddot{q}_i(t) + \sigma_{G,i}\dot{q}_i(t) + K_{G,i}q_i(t) = -\Gamma_i a(t) \quad \text{(A7)}$$

and its solution is given by:

$$q_i(t) = -\frac{\Gamma_i}{m_{G,i}} \cdot \int_0^t h_i(t-\tau)a(\tau)d\tau, \quad \text{(A8)}$$

where $h_i(t)$ is the impulse response function of the i-th generalized coordinate:

$$h_i(t) = \frac{e^{-\zeta_i \omega_{Ni} t} \sin(\omega_{Di} t)}{\omega_{Di}}$$

where:

ω_{Ni} is the natural pulsation of the i-th mode;
$\zeta_i = \frac{\sigma_i}{2 m_{Gi} \omega_{Ni}}$ is the modal damping of the i-th mode;
ω_{Di} is the damped pulsation of the i-th mode, which is equal to $\omega_{Ni}\sqrt{1-\zeta_i^2}$.

Once the time histories of the generalized coordinates are evaluated, together with their first and second derivatives, the seismic response of the structure can be obtained by recalling that:

$$\underline{\Delta d} = \underline{\underline{\Phi}}_V \underline{q}(t),\ \underline{\Delta \dot{d}} = \underline{\underline{\Phi}}_V \underline{\dot{q}}(t),\ \underline{\Delta \ddot{d}} = \underline{\underline{\Phi}}_V \underline{\ddot{q}}(t). \quad \text{(A9)}$$

The evaluation of the time histories of *all* the generalized coordinates may, however, represent a demanding task from a computational standpoint, especially when N is significantly large. This is why, in general practice, only a subset of modes (and generalized coordinates) is selected to solve the dynamic equation of the structure with adequate approximation. Clearly, the greater the number of selected modes, the better the level of approximation will be. On the other hand, not all modes make the same contribution to the physical response of the structure, and, after a certain number of modes, the computational effort is not rewarded by a related increase in the results' precision. The residual inertia criterion represents a powerful tool to individuate the right number of modes to be used for the evaluation of the seismic response with a good level of approximation. The criterion is based on the following identity:

$$\underline{e}^T \underline{\underline{M}}\, \underline{e} = (\Gamma_1)^2 + (\Gamma_2)^2 + \ldots + (\Gamma_N)^2 = \sum_{i=1}^{N} (\Gamma_i)^2, \quad \text{(A10)}$$

according to which the total inertia of the structure along the direction of excitation ($\underline{e}^T \underline{Me}$) is equal to the sum of the squares of the modal participation factors. This result can be easily demonstrated by observing that the natural modes represent a base of the vector space ($\mathbb{R}^N$) and, therefore, the vector ($\underline{e}$) may be expressed as a sum of its modal components (e_1, $e_2, \ldots, e_N$):

$$\underline{e} = \underline{\Phi}_{V,1} e_1 + \underline{\Phi}_{V,2} e_2 + \ldots + \underline{\Phi}_{V,N} e_N \quad \text{(A11)}$$

The components may be determined by invoking the orthogonality of the modal base. If we pre-multiply each member of Equation (A11) by $\underline{\Phi}_{V,i}^T \underline{\underline{M}}$, we obtain:

$$\underline{\Phi}_{V,i}^T \underline{\underline{M}}\, \underline{e} = \underline{\Phi}_{V,i}^T \underline{\underline{M}}\, \underline{\Phi}_{V,1} e_1 + \underline{\Phi}_{V,i}^T \underline{\underline{M}}\, \underline{\Phi}_{V,2} e_2 + \ldots + \underline{\Phi}_{V,i}^T \underline{\underline{M}}\, \underline{\Phi}_{V,i} e_i + \ldots + \underline{\Phi}_{V,i}^T \underline{\underline{M}}\, \underline{\Phi}_{V,N} e_N \quad \text{(A12)}$$

Since it results in the following:

$$\underline{\Phi}_{V,i}^T \underline{\underline{M}}\, \underline{\Phi}_{V,j} = \begin{cases} m_{G,i} & \text{if } j = i \\ 0 & \text{if } j \neq i \end{cases}$$

we know, from Equation (A12), that $e_i = \frac{\underline{\Phi}_{V,i}^T \underline{\underline{M}}\, \underline{e}}{m_{G,i}}$; therefore, Equation (A11) may be rewritten as follows:

$$\underline{e} = \frac{\underline{\Phi}_{V,1}^T \underline{\underline{M}}\, \underline{e}}{m_{G,1}} e_1 + \frac{\underline{\Phi}_{V,2}^T \underline{\underline{M}}\, \underline{e}}{m_{G,2}} e_2 + \ldots + \frac{\underline{\Phi}_{V,N}^T \underline{\underline{M}}\, \underline{e}}{m_{G,N}} e_N. \quad \text{(A13)}$$

Equation (A10) may be derived from Equation (A13) by pre-multiplying each member by $\underline{e}^T \underline{\underline{M}}$ and then by recalling the orthogonality of the modal base.

If we select only $\widetilde{N}$ modes for the evaluation of the seismic response of the structure, according to Equation (A10), we consider only a part of the actual inertia of the structure along the direction of excitation:

$$\underline{e}^T \underline{\underline{M}}\, \underline{e} = I_e = \sum_{i=1}^{\widetilde{N}} (\Gamma_i)^2 + \sum_{i=\widetilde{N}+1}^{N} (\Gamma_i)^2 = \sum_{i=1}^{\widetilde{N}} (\Gamma_i)^2 + I_{e,residual} \tag{A14}$$

The level of approximation of the actual seismic response is correlated with the amount of structural inertia that participates in the response; it follows that, the lower the residual inertia, the higher the precision of the obtained results.

For slender metallic structures, such as the one considered in this paper, and irrespective of the direction of excitation ($\underline{e}$), the first 5–6 modes generally ensure that the residual inertia is lower than 3% of the actual one, which thus provides a more than satisfactory characterization of the seismic response.

Appendix B. LORD ® MRD-1005-3 Datasheet

RD-1005-3 Damper

Description

LORD RD-1005-3 damper is a compact, magneto-rheological (MR) fluid damper suitable for industrial suspension applications. Continuously variable damping is controlled by the increase in yield strength of the MR fluid in response to magnetic field strength.

Features and Benefits

Fast Response Time – responds in less than 15 milliseconds to changes in the magnetic field.

Easy to Use – provides simple electronics and straight forward controls.

Durable – provides excellent long term stability.

Storage

Damper should be stored at -40 to +100°C (-40 to +212°F).

RD-1005-3 damper is a monotube shock containing high-pressure nitrogen gas (300 psi). Handle with care and do not heat or puncture body.

Electrical Properties*

Input Current, Amp	
Continuous for 30 seconds	1 max
Intermittent	2 max
Input Voltage, Volt	12 DC
Resistance, ohms	
@ ambient temperature	5
@ 71°C (160°F)	7

*Data is typical and not to be used for specification purposes.

Typical Properties*

Compressed Length, mm (in)	155 (6.1)
Extended Length, mm (in)	208 (8.2)
Body Diameter, mm (in)	41.4 (1.63)
Shaft Diameter, mm (in)	10 (0.39)
Tensile Strength, N (lbf)	4448 (1000) max
Damper Forces, N (lbf) Peak to Peak	
5 cm/sec @ 1 A	>2224 (>500)
20 cm/sec @ 0 A	<667 (<150)
Operating Temperature, °C (°F)	71 (160) max

*Data is typical and not to be used for specification purposes.

Typical Force vs. Velocity

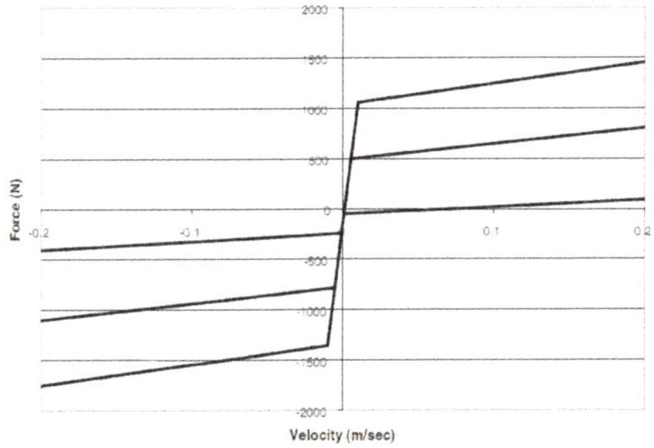

Offset at origin is due to gas precharge required for temperature compensation and to prevent cavitation.

Values stated in this technical data sheet represent typical values as not all tests are run on each lot of material produced. For formalized product specifications for specific product end uses, contact the Customer Support Center.

Information provided herein is based upon tests believed to be reliable. In as much as LORD Corporation has no control over the manner in which others may use this information, it does not guarantee the results to be obtained. In addition, LORD Corporation does not guarantee the performance of the product or the results obtained from the use of the product or this information where the product has been repackaged by any third party, including but not limited to any product end-user. Nor does the company make any express or implied warranty of merchantability or fitness for a particular purpose concerning the effects or results of such use.

LORD Wonder Box® Device Controller Kit

LORD Wonder Box® device controller kit is a companion product for the magneto-rheological (MR) fluid devices. The kit includes the device controller, 12 Volt DC power supply and two banana plugs.

The Wonder Box device controller kit provides closed loop current control to compensate for changing electrical loads up to the limits of the power supply. The Wonder Box controller may be operated as an interface device for PLC or computer control of MR fluid devices. The Wonder Box device controller kit is for indoor use only.

If manual operation is desired, a potentiometer is provided to control the current supplied to an attached device.

Operating Instructions

Connect the banana plugs to each wire of the MR fluid device. Insert the banana plugs into the Wonder Box controller output (B). Polarity is not important. Connect the power supply to the input connector (A). The red LED should light indicating proper operation of the unit.

Manual Control

Current to the device is adjusted by rotating the potentiometer (C) in the clockwise direction. An external voltage control input should not be connected to the BNC terminal (D).

External Voltage Control

With the potentiometer control set to zero (rotate fully counter clockwise), connect a 0-5 Volt DC control signal to the BNC terminal on the side of the Wonder Box controller (D). Output current to a device can now be controlled with this input signal. The input control signal can be switched up to 1 kHz.

The output current is linearly proportional to the input voltage. The output current will be 0.0 Amps when the control input is approximately 0.4-0.6 Volts at the BNC terminal (refer to Typical Performance graph).

Pulse Control

In Manual or External Voltage Control, current to the device may be switched on and off by depressing the switch on the side of the controller (E).

Technical Data

Overall Dimensions (LxWxH), mm (in)	63.5 x 27.9 x 88.9 (2.5 x 1.1 x 3.5)
Input Receptacle	2.1 mm Female
Pulse Width Modulation (PWM) Frequency, kHz	30
Output Current, Amp	2 max

Wonder Box Device Controller

LORD
AskUsHow™

References

1. Medeot, R. The European Standard on Anti-Seismic Devices. *IABSE Symp. Rep. N* **2010**, *97*, 39–46.
2. Castellano, M.G.; Infanti, S. Recent Applications of Italian anti-seismic devices. *WIT Trans. Built Environ.* **2009**, *104*, 333–341.
3. Nakamura, Y.; Okada, K. Review on seismic isolation and response control methods of buildings in Japan. *Geoenviron. Disasters* **2019**, *6*, 1–10. [CrossRef]
4. Wei, Z.; Xiao-ting, R. Semiactive Vibration Control Using a Magnetorheological Damper and a Magnetorheological Elastomer Based on the Bouc-Wen Model. *Shock. Vib.* **2014**, *2014*, 405421. [CrossRef]
5. Michele Dassisti, M.; Brunetti, G. Introduction to Magnetorheological Fluids. In *Encyclopedia of Smart Materials*; Olabi, A.-G., Ed.; Elsevier: Amsterdam, The Netherlands, 2022; pp. 187–202.
6. Zhang, Y.; Zhang, X.S.; Silva, R.P.; Ding, B.; Zhang, P.; Shao, G. Lithium-Sulfur Batteries Meet Electrospinning: Recent Advances and the Key Parameters for High Gravimetric and Volume Energy Density. *Adv. Sci.* **2022**, *9*, e2103879. [CrossRef] [PubMed]
7. Zhang, Y.; Zhang, P.; Zhang, S.; Wang, Z.; Li, N.; Silva, S.R.P.; Shao, G. A flexible metallic TiC nanofiber/vertical graphene 1D/2D heterostructured as active electrocatalyst for advanced Li–S batteries. *InfoMat* **2021**, *3*, 790. [CrossRef]
8. Kerboua, M.; Benguediab, M.; Megnounif, A.; Benrahou, A.H.; Kaoulala, F. Semi Active Control of Civil Structures, Analytical and Numerical Studies. *Phys. Procedia* **2014**, *55*, 301–306. [CrossRef]
9. Qiao, F.; Liu, J.; Dai, J.; Sun, L.; Guo, H. Modelling of MRD and Its Application for Seismic Reduction. In Proceedings of the 2010 International Conference on Modelling, Identification, and Control, Innsbruck, Austria, 17–19 July 2010; pp. 682–687.
10. Dyke, S.J.; Spencer, B.F.; Sain, M.K.; Carlson, J.D. Modeling and control of magnetorheological dampers for seismic response reduction. *Smart Mater. Struct.* **1996**, *5*, 565–575. [CrossRef]
11. Dyke, S.J.; Spencer, B.F.; Sain, M.K.; Carlson, J.D. Experimental Verification of Semi-Active Structural Control Strategies using Acceleration Feedback. In Proceedings of the 3rd International Conference on Motion and Vibration Control, Chiba, Japan, 1–6 September 1996; pp. 291–296.
12. Dyke, S.J.; Spencer, B.F. A Comparison of Semi-Active Control Strategies for the MR Damper. In Proceedings of the International Conference on Intelligent Information Systems, Bahama Island, Bahamas, 8–10 December 1997.
13. Jansen, L.M.; Dyke, S.J. Investigation of Nonlinear Control Strategies for the Implementation of Multiple Magnetorheological Dampers. In Proceedings of the ASCE Engineering Mechanics Conference, Baltimore, MD, USA, 13–16 June 1999.
14. Ying, Z.G.; Ni, Y.Q.; Ko, J.M. Non-Clipping Optimal Control of Randomly Excited Nonlinear Systems using Semi-Active ER/MR dampers. In Proceedings of the SPIE, Smart Systems and Materials-Smart Systems for Bridges Structures and Highways, the International Society for Optical Engineering, San Diego, CA, USA, 28 June 2002; pp. 209–218.
15. Zhang, J.; Roschke, P.N. Active control of a tall structure excited by wind. *J. Wind. Eng. Ind. Aerodyn.* **1999**, *83*, 209–223. [CrossRef]
16. Hiemenz, G.J.; Choi, Y.-T.; Wereley, N.M. Seismic Control of Civil Structures Utilizing Semi-Active MR Bracing Systems. In Proceedings of the SPIE, Smart Structures and Materials: Smart Systems for Bridges Structures and Highways, The International Society for Optical Engineering, Newport Beach, CA, USA, 20 April 2000; pp. 217–228.
17. Li, J.; Samali, B.; Ha, G. Fuzzy Sliding Mode Control of a Five Story Benchmark Model Equipped with Active Mass Driver (AMD). In Proceedings of the 6th International Conference on Motion and Vibration Control, Saitama, Japan, 20–23 August 2002; pp. 172–177.
18. Ni, Y.Q.; Liu, H.J.; Ko, J.M. Experimental Investigation on Seismic Response Control of Adjacent Buildings using Semi-Active MR Dampers. In Proceedings of the SPIE, Smart Systems and Materials-Smart Systems for Bridges Structures and Highways, the International Society for Optical Engineering, San Diego, CA, USA, 28 June 2002; pp. 334–344.
19. Xu, Z.D.; Shen, Y.P.; Guo, Y.Q. Semi-active control of structures incorporated with magnetorheological dampers using neural networks. *Smart Mater. Struct.* **2003**, *12*, 80–87. [CrossRef]
20. Kim, S.; Clark, W.W. Fuzzy logic semi-active vibration control. *Adapt. Struct. Mater. Syst.* **1999**, *16424*, 367–372.
21. Alli, H.; Yakut, O. Fuzzy sliding-mode control of structures. *Eng. Struct.* **2005**, *27*, 277–284. [CrossRef]
22. Website of the Consortium of the University Laboratories of Seismic and Structural Engineering. Available online: https://www.reluis.it/ (accessed on 1 March 2022).
23. European Commision, Eurocode 8: Design of Structures for Earthquake Resistance; BS EN 1998 SET; European Publication Office. 2006. Available online: https://data.europa.eu/doi/10.2788/91658 (accessed on 1 March 2022).
24. Femap® Website. Available online: https://www.plm.automation.siemens.com/ (accessed on 1 March 2022).
25. Lanczos, C. An iteration method for the solution of the eigenvalue problem of linear differential and integral operators. *J. Res. Natl. Bur. Stand.* **1950**, *45*, 255–282. [CrossRef]
26. Matlab® Website. Available online: https://www.mathworks.com/ (accessed on 3 February 2022).
27. Cremer, L.; Heckl, M. *Stucture-Borne Sound*; Springer: New York, NY, USA, 1988; Chapter 4.
28. Saatcioglu, M. High-Rise Buildings in Natural Disaster. In *Encyclopedia of Natural Hazards, Encyclopedia of Earth Sciences Series*; Bobrowsky, P.T., Ed.; Springer: Dordrecht, The Netherlands, 2013.
29. Lalanne, C. Mechanical Vibration and Shock Analysis. In *Mechanical Shock*, 2nd ed.; Wiley: Hoboken, NJ, USA, 2009; Volume 2.
30. Srinivasan, A.V.; Mc Farland, D.M. *Smart Structures, Analysis and Design*; Cambridge University Press: Cambridge, UK, 2010; Chapter 4.
31. LORD® MRD-1005-3 Datasheet. Available online: https://www.lord.com/ (accessed on 2 March 2022).
32. ENEA Resarch Center. Available online: https://www.casaccia.enea.it/ (accessed on 2 March 2022).

33. Siemens LMS Scadas/Test-Lab®. Available online: https://www.plm.automation.siemens.com/global/it/products/simcenter/scadas.html (accessed on 2 March 2022).
34. Girard, A.; Roy, N. *Dynamique des Structrues Industrielles*; Hermes Science Publications: Paris, France, 2003.

applied sciences

MDPI

Article

Design Optimization of a Hydrodynamic Brake with an Electrorheological Fluid

Zbigniew Kęsy [1], Ireneusz Musiałek [2] and Seung-Bok Choi [3,4,*]

1 Faculty of Mechanical Engineering, Kazimierz Pulaski University of Technology and Humanities in Radom, 26-600 Radom, Poland
2 Branch in Sandomierz, Jan Kochanowski University of Kielce, 25-369 Kielce, Poland
3 Department of Mechanical Engineering, The State University of New York, (SUNY Korea), 119 Songdo Moonhwa-Ro, Yeonsu-Gu, Incheon 21985, Republic of Korea
4 Department of Mechanical Engineering, Industrial University of Ho Chi Minh City (IUH), 12 Nguyen Van Bao Street, Go Vap District, Ho Chi Minh City 70000, Vietnam
* Correspondence: seungbok.choi@sunykorea.ac.kr

Abstract: This article describes the design optimization of a hydrodynamic brake with an electrorheological fluid. The design optimization is performed on the basis of mathematical model of the brake geometry and the brake's electrical circuit. The parameters of the mathematical models are selected based on experimental tests of the prototype brake. Six different objective functions are minimized during the design optimization. The functions are created taking into consideration the following factors: the braking torque, brake weight, electric power absorbed by the brake, and the torque rise time. The assumed design variables are: the number of blades and the radii (inner and outer) of the brake's working space. The optimization calculations are performed for two design variables intervals. The first interval is defined taking into consideration the accuracy of the mathematical model. The second, narrower interval is assumed for the tested prototypical brake. On the basis of the optimization calculation results, general guidelines are presented for the optimization of the hydrodynamic brakes with an ER fluid. In addition, the possibilities of optimizing the prototype brake are determined.

Keywords: electrorheological fluid; hydrodynamic ER brake; optimization method

Citation: Kęsy, Z.; Musiałek, I.; Choi, S.-B. Design Optimization of a Hydrodynamic Brake with an Electrorheological Fluid. *Appl. Sci.* **2023**, *13*, 1089. https://doi.org/10.3390/app13021089

Academic Editors: Antonio Concilio, Rosario Pecora, Ignazio Dimino, Salvatore Ameduri and Vikram G Kamble

Received: 5 December 2022
Revised: 5 January 2023
Accepted: 10 January 2023
Published: 13 January 2023

1. Introduction

Hydrodynamic brakes (HB) belong to the group of hydraulic components of machines, as do clutches and hydrodynamic torque converters. They are used in drive systems of machines, mainly due to their favorable weight-to-torque ratio on the input shaft and the long durability [1]. The basic way of controlling the HBs is by changing their working fluid filling degree. The lower the filling degree, the smaller the torque on the input shaft of the brake. However, the dependence of the torque on the filling degree is not linear. Moreover, this manner of control has a significant disadvantage, which is the unstable operation of HB at low filling degrees. The analysis of the operation of hydrodynamic components leads to the conclusion that controlling them is also possible by changing properties of the working fluid [2]. Thus, in this case, it is possible to use smart fluids, which react with a change in shear stress when exposed to changes in the electrical field (electrorheological fluids) or in the magnetic field (magnetorheological fluids) [3,4]. Currently, smart fluids are used in numerous devices and components of machines, e.g., in viscous clutches and brakes [5–7], shock absorbers [8–10], valves [11–13], sealings [14], cantilever sandwich beams [15,16], landing gear systems in planes [17,18], and industrial robots [19–21]. In addition, prototypes of hydrodynamic clutches with electrorheological fluids (ER) and magnetorheological fluids (MR) are constructed [22–25].

A frequent method of perfecting the design of machine components is the use of construction optimization. The construction optimization is performed on the basis of the component's mathematical model. The model parameters whose values are changed during the optimization process are called the design variables. The objective function is determined, and then its global extreme is sought on the basis of the requirements analysis assigned to the machine component (i.e., criteria) and to the mathematical model. To search for the global extreme of the objective function, multiple methods have been developed. Among the most frequently used methods are randomization methods, including the Monte Carlo Method and the Genetic Algorithm method [26–28]. It is advantageous to use structure optimization methods in the design process due to the complex design of hydraulic clutches with ER or MR fluids, which results from the necessity to generate an electric field inside. For mathematical modeling of clutches, brakes and torque converters, and one-, two-, and three-dimensional flows within the rotors are taken into consideration [29–31]. In the model based on the one-dimensional flow, it is assumed that the working fluid moves with the mean flow path with a uniform through-flow velocity [32]. The high accuracy of this model, comparable to the accuracy of two- and three-dimensional models, is obtained as a result of parameter selection based on experimental research. Due to their simplicity, these models are particularly suitable for use in design optimization [33].

Publications concerning the optimization of brakes and hydraulic clutches with smart fluids most frequently concern viscous clutches and brakes with MR fluids [34,35]. The article [36] describes the design optimization of a disc viscous clutch with ER fluids. The optimization is performed with the use of two optimization methods: the Monte Carlo method and the Genetic Algorithm method. The purpose of the design optimization is to seek such values of design variables, for which the viscous clutch with the ER fluid transmits a large torque, while the clutch's dimensions are small and its determined operating temperature is low. The optimization calculations are performed for two models including three and eight design variables. As a result of the design optimization, a disc viscous clutch with the ER fluid is constructed. In the publication [37], the object of the optimization process is a disc brake with a MR fluid. The brake was intended for use in a motorcycle. The objective function is the brake's weight, and the assumed constraints of the design variables are the braking torque and the temperature of the MR fluid during braking. The optimization is performed with the use of optimization procedures from the ANSYS program. In the article [38] the optimized object is a disc brake with one disc and different locations of the electromagnet coils and with three types of MR fluid. The constraints of the design variables concern the radius of the brake shaft and the maximal current flowing through the electromagnet coil. The optimization is conducted on the basis of the optimization method included in the ANSYS software. In the objective functions, factors taken into consideration are the brake's weight, the power supply, and the braking torque. The work [39] shows the use of an original Taguchi method in the optimization of a clutch with a MR fluid. The method allows identification of decision variables whose influence on the objective functions is the greatest.

Apart from the use for hydraulic clutches and brakes with ER fluids, design optimization has also been successfully used to optimize devices with smart fluids [11,40]. Despite several works on design optimization for ER brakes or MR brakes, a study on the optimal design of ER brakes considering practical feasibility, in which many design parameters should be considered, is considerably rare. Thus, in this work, to achieve this target, many principal design parameters, such as the radius and electrode gap of ER brake, and several braking performance and product criteria also, such as torque level, blade number, brake weight, electric power and braking time, are included for the optimization. To accomplish such a target, the HB optimization method is used, and subsequently, the possibilities of optimal selection of considered parameters of the optimized prototypical hydrodynamic brake with the ER fluid (PHB) are discussed in detail, focusing on the performance efficiency with several ratios: the blade number versus torque level versus brake weight. Consequently, the main technical contribution of this work is to calculate the optimized design and performance variables of the PHB that can be feasible in a practical environment. As far as the authors know, so far there is no report on this optimal design issue in the technology associated with ER fluid systems, such as hydrodynamic brakes.

2. Construction of the Hydrodynamic Brake

The HB has rotors with radial blades lying in planes passing through the rotor axis. These rotors have no inner rings. Such a design solution is frequently used in practice. The construction scheme of the HB is shown in Figure 1.

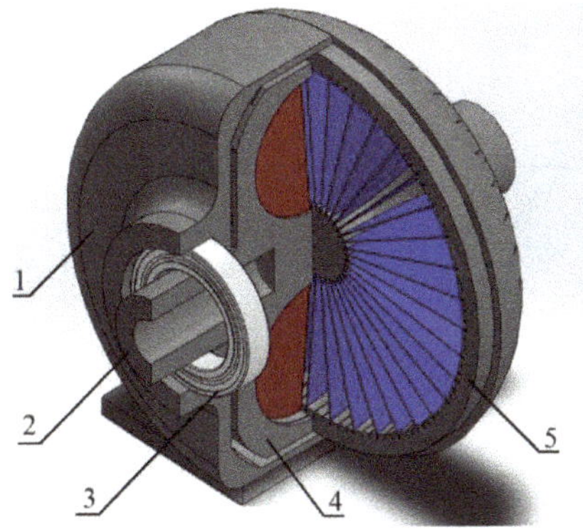

Figure 1. HB construction scheme: 1—casing, 2—input shaft, 3—bearing with sealing, 4—pump rotor, 5—turbine rotor connected to casing.

The HB is filled with the ER fluid. The pump rotor rotates with the angular velocity ω_1, and the turbine rotor is immobile ($\omega_2 = 0$). The meridional cross-sections of both HB rotors are identical. The casing and rotors are made of an electrical insulator material, while the blades are made of an electrical conductor material. The pump and turbine blades are alternately connected to the (+) and (−) poles of the high voltage power supply with the use of slip rings, brushes, and electric wires. The HB operates as follows: the rotating pump rotor sets the fluid in motion; the fluid then flows into the immobile turbine rotor and presses on the blades, which causes the reaction torque to occur. After supplying the high voltage U to the rotor blades, the electric field is created, and it causes an increase in the shear stresses τ within the ER fluid. This causes an increase in the fluid flow resistance and, simultaneously, a decrease in the torque M affecting the pump rotor. Thus, the HB operates differently than a viscous brake with an ER fluid, in which an increase of the high voltage U causes an increase in the torque. The designations of the basic dimensions of the HB are shown in Figure 2.

Table 1. PHB data.

R mm	r_w mm	d_1 mm	d_2 mm	g_b mm	n -	r_s mm	r_1 mm	r_2 mm	**ER Fluid LID 3354S**
50.5	15.5	1.5	7.3	1.0	38	37.4	28.6	44.4	

The following notations are used in Figure 2: R—the outer radius of the rotors, r_w—the inner radius of the rotors, n—the number of blades, g_b—the blade thickness, d_1—the smallest distance between blades, d_2—the largest distance between blades, r_s—radius, determining the origin of the mean streamline, r_1—the inner radius of the mean streamline, r_2—the outer radius of the mean streamline, ω_1—angular velocity of the pump rotor, ω_2—angular velocity of the turbine rotor. The design solution for PHB is shown in Figure 3.

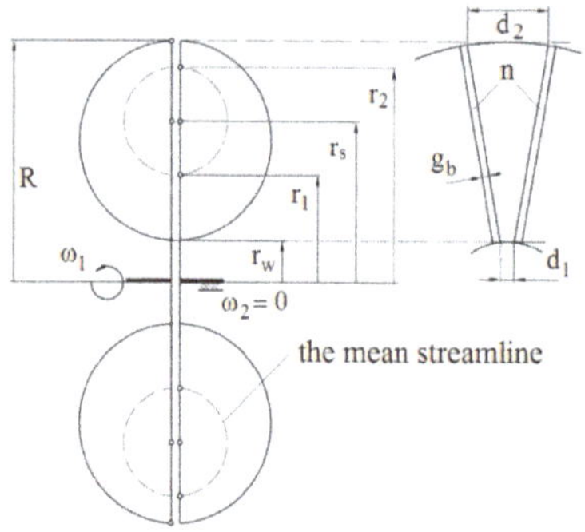

Figure 2. HB geometry for calculations (refer to Table 1).

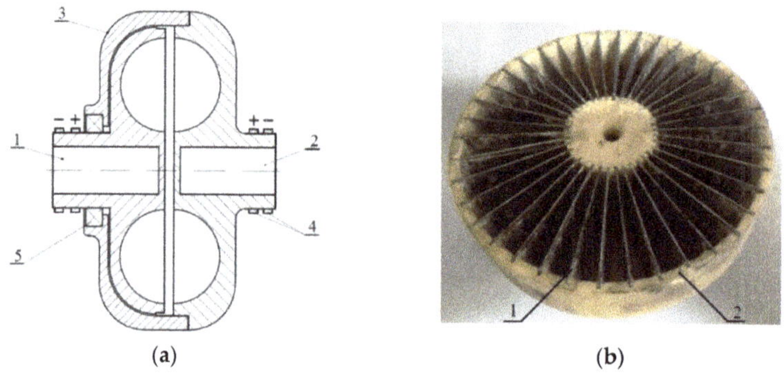

Figure 3. PHB with the ER fluid: (**a**)—design solution: 1—pump, 2—turbine, 3—casing, 4—slip rings supplying the high voltage, 5—sealing; (**b**)—view of the pump rotor after tests: 1—blade, 2—rotor casing.

The PHB rotors are made of 3D-printed nylon. They are equipped with flat steel sheet blades. The basic data concerning the PHB are shown in Table 1.

3. The Mathematical Model of the Hydrodynamic Brake

3.1. The Mathematical Model of the Hydrodynamic Brake Geometry

The parameters describing the geometry of the brake rotors are as follows: the outer radius of the rotors R, the inner radius of the rotors r_w, and the number of blades n. The radius R is larger than the radius r_w. The blade number n is even, so that in each of the channels formed by the blades, an electric field influences the ER fluid. Moreover, the blades cannot be set too close to each other due to the possibility of electric breakdown occurrence. Thus, the number of the blades should meet the following condition:

$$n < \frac{2\,\pi\,r_w}{g_b + h_{min}} \tag{1}$$

where g_b—blade thickness, h_{min}—the smallest gap width for which an electric breakdown occurs. The average blade distance d is calculated from the formula:

$$d = 0.5(d_1 + d_2) \tag{2}$$

wherein:

$$\begin{aligned} d_1 &= (2\pi r_w / n) - g_b \\ d_2 &= (2\pi R / n) - g_b \end{aligned} \tag{3}$$

where d_1—the smallest distance between blades, r_w—inner radius of the rotors, R—outer radius of the rotors, d_2—the largest distance between blades, n—number of blades.

In order to calculate the rotor volume V_t, the following formula is used:

$$V_t = 2\pi^2 \left(\frac{R - r_w}{2} \right)^2 \cdot \frac{R + r_w}{2} \tag{4}$$

where R—outer radius of the rotors, r_w—inner radius of the rotors.

The blades volume V_b is calculated on the basis of the following formula:

$$V_b = n \cdot \pi \left(\frac{R - r_w}{2} \right)^2 \cdot g_b \tag{5}$$

where n—number of blades, r_w—inner radius of the rotors, R—outer radius of the rotors, g_b—blade thickness.

The volume of the fluid V_{ER} contained in the HB is then:

$$V_{ER} = V_t - V_b \tag{6}$$

where V_t—the rotor volume, V_b—the blades volume. With the use of the Formulas (4)–(6), it is possible to formulate the weight G of the brake as follows:

$$G = V_{ER}\rho + V_b\rho_1 + \alpha V_t \rho_2 \tag{7}$$

where ρ—density of the working fluid, ρ_1—density of the blade material, ρ_2—density of the casing material, α—coefficient describing the casing density in relation to the rotor volume.

3.2. Mathematical Model of Hydraulic Interactions

In order to calculate the transferred torque M, the one-dimensional flow model of a hydrodynamic clutch is used [41,42]. The torque M for the HB is obtained from the equations of the one-dimensional flow model of a hydrodynamic clutch after assuming that $\omega_2 = 0$. It is assumed that the flow of the working fluid inside the hydrodynamic clutch is steady, continuous, and takes place along a single line of the mean stream. The radii r_1, r_2, r_s (Figure 3) determining the position of the mean streamline in the meridional cross-section are calculated on the basis of the following formulas:

$$\begin{aligned} r_s &= \sqrt{(R^2 + r_w{}^2)/2} \\ r_2 &= \sqrt{(r_s{}^2 + R^2)/2} \end{aligned} \tag{8}$$

where r_w—inner radius of the rotors, R—outer radius of the rotors. The unit torque dM in relations to the rotation axis of the rotor is only caused by the force dF caused by a change in speed c_u of the working fluid particles. The force dF and the unit torque dM are written as:

$$dF = \frac{d(mc_u)}{dt} \tag{9}$$

$$dM = dF\ r = \frac{d(m\ c_u\ r)}{dt} \tag{10}$$

where d/dt—differential operator, t—time, m—mass of the fluid particle, c_u—lifting speed, r—radius. The mass m of the fluid flowing over time dt can be calculated as:

$$m = \rho\ V = \rho\ Q\ dt \tag{11}$$

where V—fluid volume, t—time, Q—flow rate, c_m—meridional speed, A—rotor meridional cross-section. After taking into consideration the dependence (11) in relation to (10), the following is obtained:

$$dM = \rho\, Q\, dt\, \frac{d(c_u\, r)}{dt} = \rho\, Q\, d(c_u r) \tag{12}$$

where the product $c_u r$ is treated as a single variable. Integrating the dependence (12) from point 1 to point 2 positioned on the mean streamline leads to the following equation:

$$M = \rho\, Q \int_1^2 d(c_u\, r) = \rho\, Q [c_u\, r]_1^2 = \rho Q\, (c_{u2}\, r_2 - c_{u1}\, r_1) \tag{13}$$

where ρ—density of the working fluid, Q—flow rate, cu—lifting speed, r—radius.

When the blades of the hydrodynamic clutch are positioned in the planes passing through the rotor axes, then the rotor lifting speed is, respectively:

$$\begin{aligned} c_{u1} &= \omega_1 r_1 \\ c_{u2} &= \omega_2 r_2 \end{aligned} \tag{14}$$

where ω_1—angular velocity of the pump rotor, ω_2—angular velocity of the turbine rotor, r_1—the inner radius of the mean streamline, r_2—the outer radius of the mean streamline. For the determined operation conditions of the hydrodynamic clutch:

$$Q = c_m A \tag{15}$$

where c_m—meridional speed, A—rotor flow field. Considering the fact that for the HB, the angular velocity of the clutch turbine rotor ω_2 equals zero, on the basis of dependences (13)–(15), the following is obtained:

$$M = \rho\ c_m A\ \omega_1\, r_2^{\,2} \tag{16}$$

where ρ—working fluid density, c_m—meridional speed, A—rotor flow field, ω_1—angular velocity of the clutch pump rotor. After taking into consideration the blade thickness, the flow field of the HB is:

$$A = \pi\left(R^2 - r_s^{\,2}\right) - n\, g_b (R - r_s) \tag{17}$$

where n—number of blades, g_b—blade thickness. In order to calculate the meridional speed c_m, the head rise balance is considered, described as:

$$h = \Delta p / \rho\, g \tag{18}$$

where Δp—pressure change, ρ—working fluid density, g—gravitational acceleration. The balance of the head rise has the following form:

$$h_l - h_u - h_t = 0 \tag{19}$$

where h_l—head rise of the pump, h_u—impact loss, h_t—friction loss. For HBs with typical working fluids, the components of Equation (19) are written as follows:

$$h_l = \omega_1^{\,2} r_2^{\,2} / g \tag{20}$$

$$h_u = \omega_1^{\,2} (r_2^{\,2} + r_1^{\,2}) / 2g \tag{21}$$

$$h_t = \xi\, c_m^{\,2} / 2g \tag{22}$$

where ω_1—angular velocity of the pump rotor, g—gravitational acceleration c_m—meridional speed, r_1—the inner radius of the mean streamline, r_2—the outer radius of the mean streamline, ξ—friction losses factor.

During the HB modeling, in the presence of an electrical field, when considering the friction losses value h_t (described with Formula (22), it is necessary to take into account the decrease in pressure Δp_{ER} caused by the increase in the shear stresses τ. On the basis of the formulas used for calculation of the valves with ER fluids [11–13,43,44], it is assumed that:

$$\Delta p_{ER} = \frac{cl}{d}\tau_0 \tag{23}$$

where c—numerical coefficient, l—length of the channel formed by the blades, d—average distance between blades, τ_0—yield stress. After taking into consideration Formula (23) in Formula (18), and subsequently adding the result to Formula (22), the following is obtained:

$$h_t = \xi\frac{{c_m}^2}{2g} + \frac{c\,l\,\tau_0}{d\,\rho g} \tag{24}$$

where ξ—friction losses factor, c_m—meridional speed, g—gravitational acceleration, r_1—the inner radius of the mean streamline, r_2—the outer radius of the mean streamline, c—numerical coefficient, l—length of the channel formed by the blades, τ_0—yield stress, d—average distance between blades. The meridional speed c_m, which enables the calculation of the torque M on the basis of Formula (16), is obtained by substituting Formulas (20), (21), and (24) to the equation for the head rise balance (19) and subsequently solving the equation:

$$c_m = \left(\sqrt{{\omega_1}^2({r_2}^2 - {r_1}^2) - \varsigma\tau_0}\right)/\xi \tag{25}$$

wherein:

$$\varsigma = 2cl/(\rho d) \tag{26}$$

where ω_1—angular velocity of the pump rotor, r_1—the inner radius of the mean streamline, r_2—the outer radius of the mean streamline, τ_0—yield stress, ξ—friction loss coefficient, ς—channel coefficient, c—numerical coefficient, l—length of the channel formed by the blades, ρ—density of the working fluid, d—average distance between blades.

3.3. *Mathematical Model of the Electric Circuit of the Hydrodynamic Brake*

Electrorheological properties of the ER fluids are presented with the use of the following formulas:

$$\tau_0 = aE^2 \tag{27}$$

$$i_g = bE^2 \tag{28}$$

wherein:

$$E = U/d \tag{29}$$

where τ_0—yield stress, i_g—leakage current density, a, b—coefficients, E—the electric field intensity, U—voltage, d—average distance between blades. The dependence of τ_0 and i_g on the temperature Θ is not taken into account, because it is assumed that the HB operates at a constant temperature due to the use of a cooling system. The HB needs to be cooled, because all the energy supplied to the HB is converted to heat. The leakage current flowing between the electrodes is calculated on the basis of the formula:

$$i_c = S_e i_g \tag{30}$$

wherein:

$$S_e = n\pi(R - r_w)^2/4 \tag{31}$$

where S_e—surface of the electrodes, n—number of blades, r_w—inner radius of the rotors, R—outer radius of the rotors.

In the considered model, the increase of the high voltage U applied to the blades causes a decrease in the meridional speed c_m until it reaches zero. The meridional speed c_m

is described by Formula (25). The value of the voltage U_h for which $c_m = 0$ can be calculated on the basis of the equation:

$$\omega_1{}^2(r_2{}^2 - r_1{}^2) = \varsigma\tau_0 \tag{32}$$

where ω_1—angular velocity of the pump rotor, r_1—the inner radius of the mean streamline, r_2—the outer radius of the mean streamline, ς—channel coefficient, τ_0—yield stress. Having taken into consideration Formulas (27) and (29) in Equation (32), the following is obtained:

$$U_h = \sqrt{\omega_1{}^2(r_2{}^2 - r_1{}^2)\rho d^3/(2acl)} \tag{33}$$

where ω_1—angular velocity of the pump rotor, r_1—the inner radius of the mean streamline, r_2—the outer radius of the mean streamline, a, c—coefficients, l—length of the channel formed by the blades, ρ—working fluid density, d—average distance between blades. The maximal power P_e obtained from the power supply, while stopping the flow of the ER fluid in the brake, is:

$$P_{emax} = U_h i_{cmax} \tag{34}$$

where U_h—voltage value for which $c_m = 0$, i_{cmax}—leakage current calculated for $U = U_h$.

In order to assess the dynamics of the HB, the torque rise time M (after applying the high voltage to the electrodes U_h) is taken into consideration. The dynamics of the machine drive system containing the HB is not taken into account due to the fact that the total mass moment of inertia on the input shaft depends not only on the HB structure, but also on the structure of the drive system. The drive system structure, due to the general nature of the considerations, is not specified. The torque rise time for the HB are considered on the basis of the voltage courses in a simplified electrical circuit of the HB, as shown in Figure 4 [45].

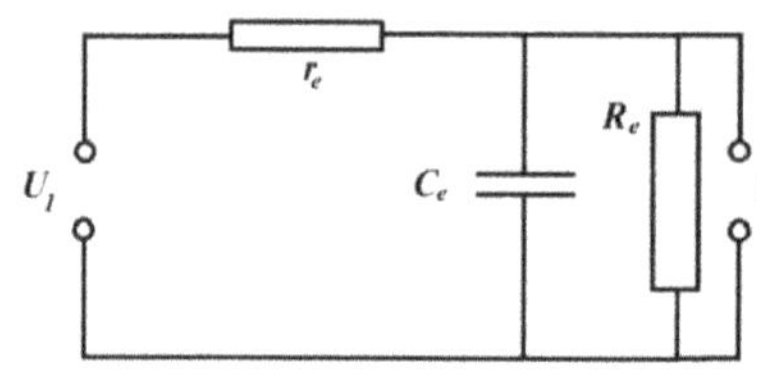

Figure 4. Electrical circuit of the HB: r_e—electrical resistance of the wires, R_e—electrical resistance of ER fluid, C_e—electric capacity of the HB, U_1—power supply voltage, U_2—voltage on the blades.

The blades of the HB, separated from each other by the ER fluid, form a capacitor whose capacity is C_e. The ER fluid provides the electrical resistance R_e during the leakage current flow i_c. After applying the supply voltage U_1, the electric current flows through the wires (the resistance of the wires is r_e) to the condenser with the capacity of C_e and charges the condenser. After the power supply U_1 is switched off, the electric current flows through the ER fluid whose resistance is R_e, and the condenser discharges. Changes of the voltage U_2 over time, with a step voltage increase U_1, is described by a first order system. It is assumed that the torque of the HB does not depend on time, but it is proportional to the voltage U_2. While the condenser is charging, the time constant of the first order system is $T_c = r_e \cdot C_e$, whereas while the condenser discharges, it is $T_d = R_e \cdot C_e$. Due to the fact that $r_e << R_e$, $T_d >> T_c$, the operation speed of the HB is assessed with the use of the time T described with the formula:

$$T = \beta\, R_e C_e \tag{35}$$

wherein:

$$R_e = U_h / i_{cmax} \tag{36}$$

$$C_e = \frac{\varepsilon_0 \varepsilon_r S_e}{d} \tag{37}$$

where β—number of time constants after which $U_1 \cong U_2$, U_h—value of the voltage for which $c_m = 0$, i_{cmax}—leakage current calculated for $U = U_h$, ε_0—electric permittivity of the vacuum, ε_r—relative permittivity of the ER fluid, S_e—surface of the electrodes, d—average distance between blades.

3.4. Coefficient Selection for the Mathematical Model of the Hydrodynamic Brake

Before using the developed HB model in optimization calculations, it is necessary to determine the values of the model's coefficients: α, ξ, c, a, b, β, ε_r. The values of the coefficients are determined on the basis of the design analysis and PHB experimental research. The coefficient α is calculated on the basis of the PHB weight, assuming that the modelled HB has an identical structure and proportional dimensions. Calculating the coefficient ξ demanded PHB bench tests without connecting the blades to the high voltage power supply. The test bench scheme is shown in Figure 5.

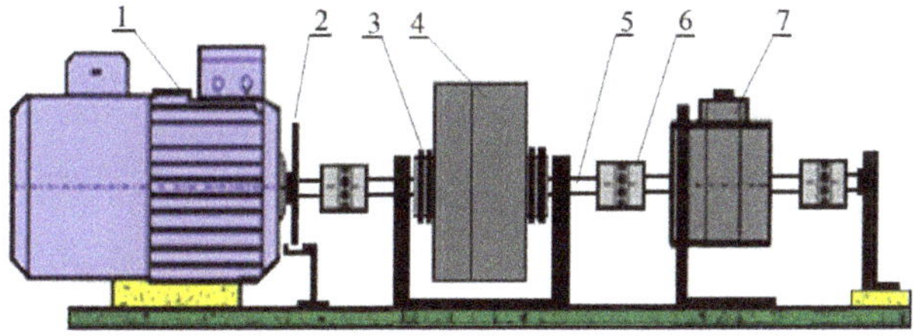

Figure 5. Test bench scheme: 1—electric motor, 2—speed sensor, 3—slip rings used to supply high voltage, 4—the tested brake with ER fluid, 5—shaft, 6—connecting clutch, 7—torque gauge.

The main elements of the test bench are: an electric drive motor and a torque gauge. The computer control system of the test bench rendered it possible to obtain the measurement values specified in the test program, as well as to collect and process measurement data. The values recorded during the test are: the angular velocity of the pump shaft ω_1, the braking torque M, and the temperature of the working fluid Θ. The temperature Θ is maintained at 35 °C by cooling the PHB with compressed air. Subsequently, the braking torque M is measured at a predetermined angular velocity of the PHB input shaft. The coefficient $\xi$is determined on the basis of Formulas (14) and (23). During the measurement, apart from the braking torque M, the rotational torques are of resistance to motion of the bearings and sealing. In further considerations, the torques of resistance are omitted due to their low values [42]. The value of the a coefficient is determined on the basis of the data from the manufacturer of the fluid LID3354S [46].

In order to determine the coefficient c for the immobile turbine shaft and for a specified angular velocity of the pump shaft ω_1, the braking torque M of the PHB is measured for a few selected voltages U. To generate the high voltage U, an electric power supply with a rated power of 18W is used. The maximal voltage U obtained during these measurements is U_{max} = 3 kV. The coefficient c is selected, so that for the voltage U_{max} = 3 kV, the torque calculated with the use of the dependence (22) is equal to the torque measured with the use of the test bench, i.e., M = 1.733 Nm, Figure 6.

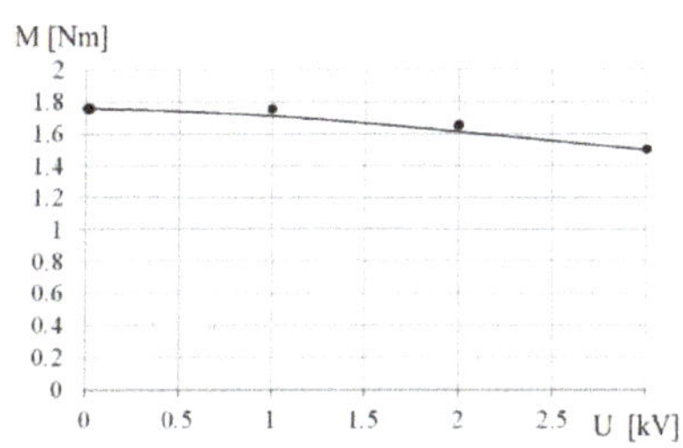

Figure 6. Method of determining the coefficient *c*: solid line—model, points—test.

The coefficient b is selected in order to obtain the measured total leakage current i_c = 6 mA for the voltage U = 3 kV. The value of the relative electric permittivity ε_r of the ER fluid is determined on the basis of [47,48]. The coefficient values of the HB mathematical model determined for ω_1= 200 rad/s and for the temperature Θ = 35 °C are presented in Table 2.

Table 2. Mathematical model coefficients for the researched PHB.

Coefficient	α -	ξ -	c -	a kPa mm²/kV²	b µA mm²/(cm² kV²)	β -	ε_r -
Value	2.30	6.64	3.93	0.41	35.31	3.00	5.00
Formula	(7)	(22)	(23)	(27)	(28)	(35)	(37)

Due to the chosen method of selecting the coefficients shown in Table 2, only the modelling error of the dependence of the torque M on the high voltage U can be assessed. The solid line shown in Figure 6, obtained on the basis of the mathematical model, according to the assumptions, passes through the points for U = 0 kV and U = 3 kV. The average relative error calculated for U = 1 kV and U = 2 kV is under 2%.

4. Optimization Method for the Hydrodynamic Brake Design

In order to obtain guidelines for optimization of the HB design (concerning the selection of design variables and objective functions), calculations are performed on the basis of the developed mathematical model. Two calculation groups are performed. One group includes the detailed PHB calculations, on the basis of which the design variables are selected. The other group contains the general HB calculations, on the basis of which the optimization criteria are selected. For the HB, the calculation data are randomly selected parameters from determined ranges.

4.1. Selection of Design Variables

The calculations for the PHB are performed for the data shown in Table 1. The torque M and the voltage U_h are calculated, while one of the parameters r_w, R, n is changed, and the remaining parameters are fixed. Examples of the calculation results for ω_1 = 200 rad/s, Θ = 35 °C are shown in Figures 7–9.

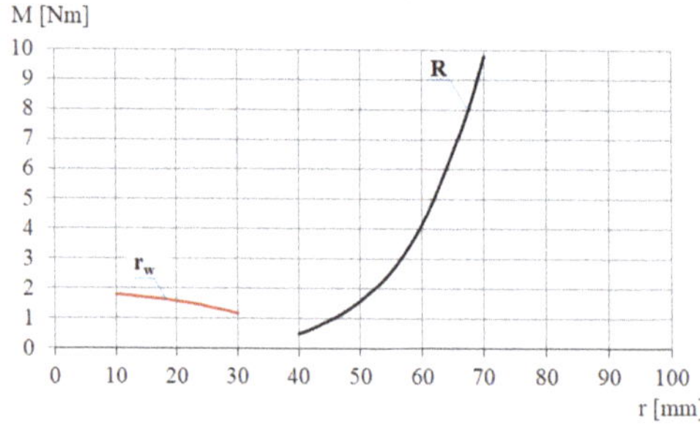

Figure 7. Dependence of the torque on the radius r_w and on the radius R.

Figure 7 presents the influence of the radius r_w and the radius Ron the braking torque M. Figure 8 shows the dependence of the braking torque M on the blade number n. Figure 9 shows the influence that the number of blades nin the PHB has on the value of the voltage U_h. As shown in Figure 7, increasing the inner radius r_w causes a decrease in the braking torque M, while increasing the outer radius R causes an increase in the torque M. This is related to the opposing influence of these radii on the flow velocity of the ER fluid within the rotors. Increasing the blade number n, Figure 8, causes a decrease in the torque M, which is connected to the decrease of the flow cross-sectional area A of the rotors. Figure 9 shows a decrease in voltage U_h as the number of blades increase, with a constant value of the torque M. The decrease is a result of decreasing average distance between blades d when the number of the blades increases. Decreasing the value of d causes an increase in the electric field strength E, which has a direct influence on the value of the yield stresses τ_0. A significant (and different) influence of the radii r_w and R, as well as the number of blades n, on the PHB characteristics justifies the purposefulness of selecting these parameters as design variables for optimization of the PHB design.

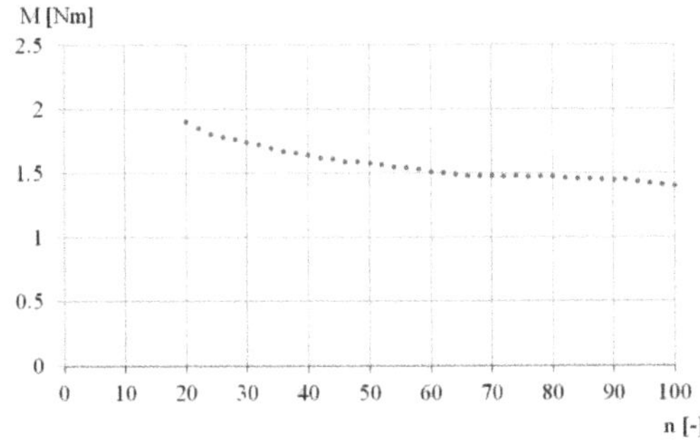

Figure 8. Dependence of the torque M on the blade number n.

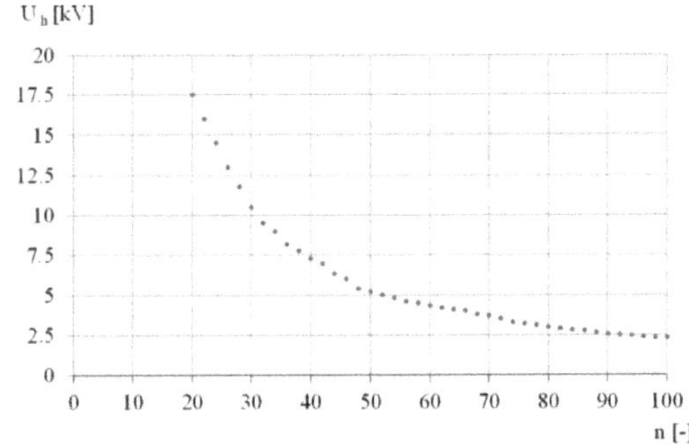

Figure 9. Dependence of the voltage U_h on the blade number n.

4.2. Selection of Optimization Criteria

The calculations included the dependence of weight G, electric power P_e, and time T on the torque M, as well as the dependence of the torque M, weight G, electric power P_e, and time T on the radii r_w, R, and the blade number n. The intervals of the HB parameters and additional conditions ensuring the absence of electric breakdowns are shown in Table 3.

Table 3. Intervals and conditions concerning the parameters.

Parameter Interval			Conditions	
r_w mm	R mm	n -	d_1 mm	U_h kV
$10 < r_w < 30$	$40 < R < 70$	$10 < n < 120$	$d_1 > h_{min} = 1$	$U_h < 20$ kV

The HB parameter intervals are determined arbitrarily, aiming at possibly small differences between the parameters of the calculated HB and the parameters of the PHB used to determine the coefficients of the mathematical model. This allows obtaining the necessary accuracy of the used mathematical model. The HB calculations are performed as follows:

- it is assumed that ω_1 = 200 rad/s;
- the radii r_w and R are randomly selected from the intervals given in Table 3;
- it is assumed that g_b = 0.02 R;
- calculations are performed according to the formulas of the mathematical model described in Section 2;
- the calculation results not meeting the requirements given in Table 3 are rejected.

The calculations are performed on the basis of a computer program written in Turbo Pascal 7. Examples of calculation results for 5000 draws and for ω_1= 200 rad/s and Θ = 35 °C are shown in Figures 10–18.

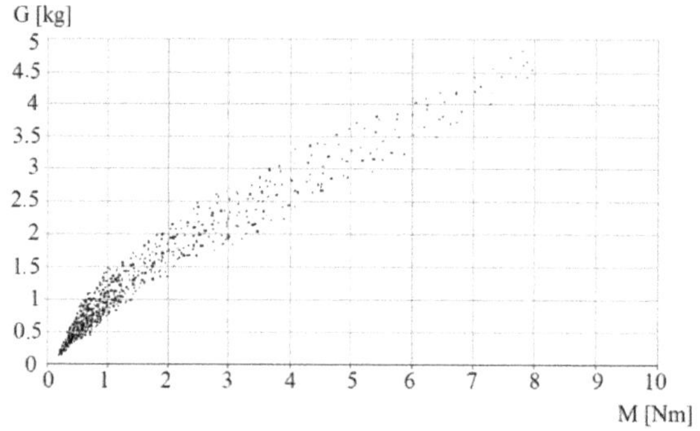

Figure 10. Dependence of the weight G on the torque M.

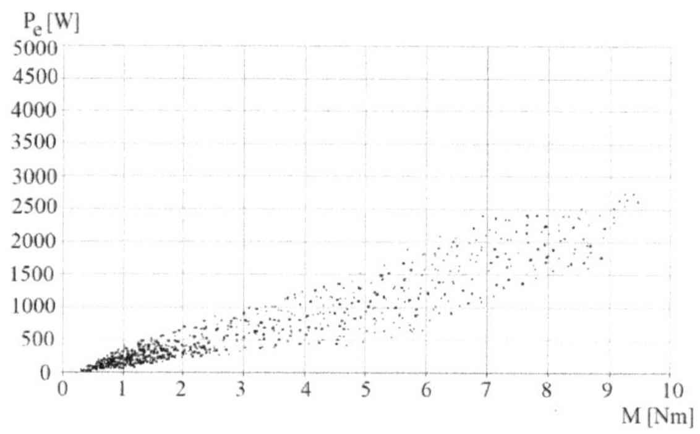

Figure 11. Dependence of the electric power P_e on the torque M.

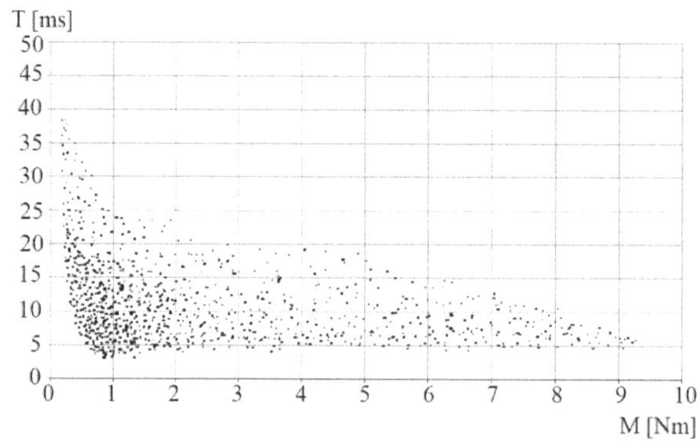

Figure 12. Dependence of the time T on the torque M.

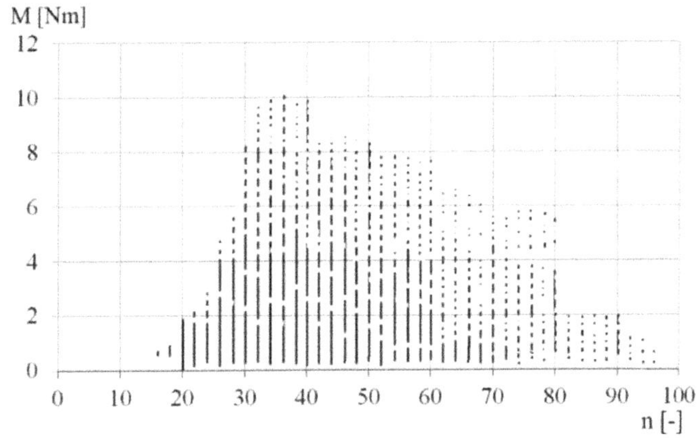

Figure 13. Dependence of the torque M on the number of blades n.

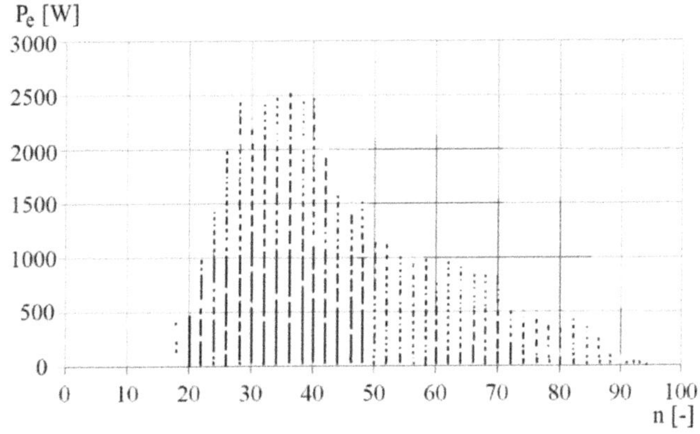

Figure 14. Dependence of the electric power P_e on the number of blades n.

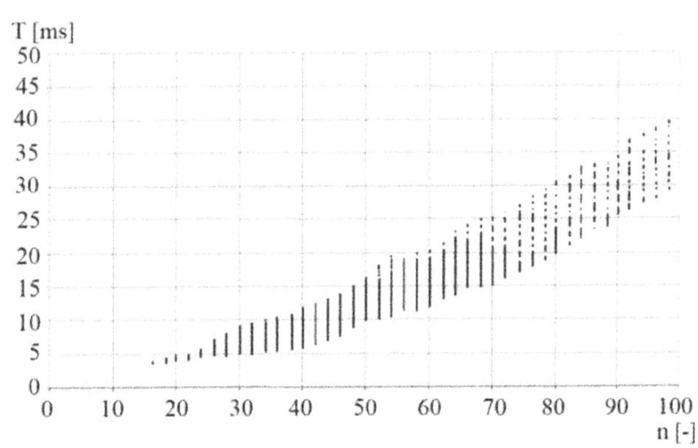

Figure 15. Dependence of the time T on the number of blades n.

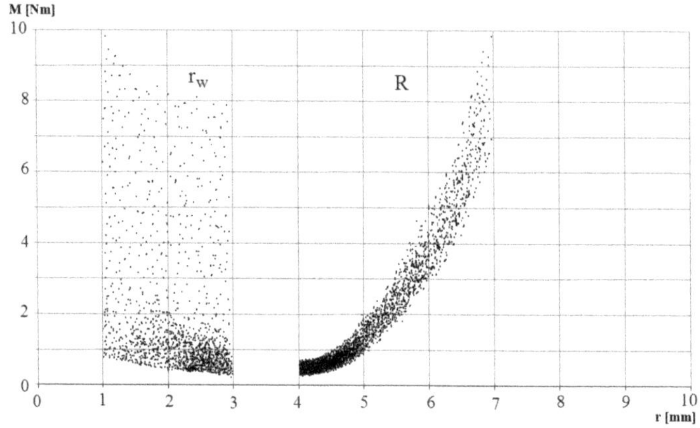

Figure 16. Dependence of the torque M on the radius r_w and on the radius R.

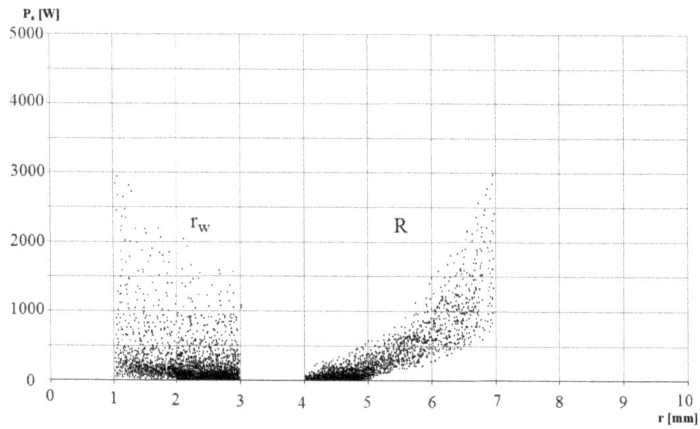

Figure 17. Dependence of the electric power P_e on the radius r_w and on the radius R.

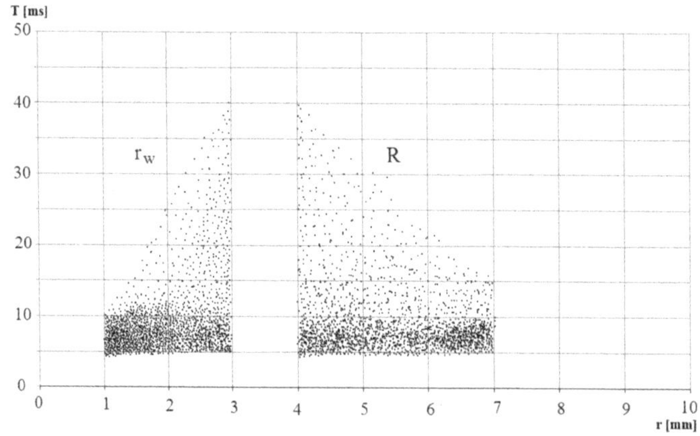

Figure 18. Dependence of the time T on the radius r_w and on the radius R.

Figures 10 and 11 show the proportional dependence of the weight G and electric power P_e on the torque M. Different values of the weight G and the power P_e occurring for the same values of the torque M are mostly a result of the difference in the number of blades n. For fixed values of the radius r_w and the radius R, the smaller the number of blades n, the smaller the weight G. In this case, the dependence of the power P_e on the number of blades is the opposite; the smaller the number of blades n, the higher the voltage U_h necessary to reach the yield stress τ_0 needed to stop the flow of the ER fluid (and, therefore, the power P_e is greater). Figure 12 shows that small values of time T occur for all values of the torque M. A wider range of the time values T for smaller torques M is a result of smaller dimensions of the HB and, therefore, smaller distances between blades and a larger capacity C_e. As shown in Figures 13 and 14, the dependence of the torque M and the electric power P_e on the number of blades are similar. The largest values of M and P_e occur within the range from n = 30 to n = 40. The increase of the average time values T with the increase in the number of blades n can be observed in Figure 15. It is a result of the smaller distances between blades and, therefore, larger capacities C_e on which the time T depends. As shown in Figures 16 and 17, the dependence of the torque M and the electric power P_e on the radius r_w and on the radius R are similar. As can be seen in Figure 18, by changing the value of the radius r_w and the radius R in the intervals presented in Table 3, one can obtain similar values of time T.

The dependences of weight G, electric power P_e, and time T on the torque M are not functions; thus, HBs with the same value of the torque M can have different weights G, electric powers P_e, and times T. Moreover, M, G, P_e, and T significantly (and in different ways) depend on the assumed design variables r_w, R, and n. Due to that, the torque M, weight G, electric power P_e, and time T_s are selected as optimization criteria for the design and are used to formulate the objective functions F. The selection of the criteria is consistent with the pursuit of mechanical engineering to increase efficiency, reduce weight, and save energy. The requirements concerning the assumed design optimization criteria are presented in Table 4.

Table 4. Requirements concerning the HB optimization criteria.

Name	Designation	Requirements	Main Calculation Formula Numbers
Torque	M	Maximal	(16), (17), (25), (26)
Weight	G	Minimal	(4)–(7)
Electric power	P_e	Minimal	(30), (31), (33), (34)
Time	T	Minimal	(35)–(37)

4.3. The Algorithm for Design Optimization of the Hydrodynamic Brake

On the basis of the criteria given in Table 4, the following objective functions are assumed:

$$\begin{aligned} F_G &= G - 0.5{\cdot}M \\ F_{G1} &= G/M \end{aligned} \tag{38}$$

$$\begin{aligned} F_P &= 0.1{\cdot}P_e - M \\ F_{P1} &= P_e/M \end{aligned} \tag{39}$$

$$\begin{aligned} F_T &= T - 0.3{\cdot}M \\ F_{T1} &= T/M \end{aligned} \tag{40}$$

The objective functions F_G, F_P, F_T,F_{G1}, F_{P1}, F_{T1} described by Formulas (38)–(40) cannot be written using simple formulas, in which there are the design variables, due to the complicated form of the applied mathematical model. The values of the criteria G, M, P_e, T were calculated on the basis of the formulas given in Table 4. The numerical coefficients appearing in the Formulas (38)–(40) are arbitrarily selected during initial calculations, so that minimal values of the objective functions F_{Gmin}, F_{Pmin}, F_{Tmin} are as close to zero as possible. The objective function, which is the difference between G and M, is also assumed in the articles [49,50]. The minimalization of the objective functions (38)–(40) leads to finding the design variables, for which the unit of the torque M will be assigned with, respectively, the lowest weight G, the lowest electrical power P_e, and the shortest time T. The design variables intervals r_w, R, n and constraints concerning d_1 and U_h are the ones given in Table 3. The optimization calculations for the HB are aimed at minimizing the selected objective function F described with one of the Formulas (38)–(40).The calculations are performed as follows:

- the decision variables r_w, R, and n are randomly selected from the assumed intervals with the use of a random number generator;
- the value of ω_1 is determined and g_b = 0.02R is calculated;
- calculations are performed according to the formulas of the mathematical model;
- the constraints of d_1 and U_h are checked;
- when the randomly selected values of the design variables meet the conditions of the constraints, the objective function F is calculated, its result is saved, and the calculations are repeated;
- the smaller value of the objective function from the previous F_p and present step of the calculations F_a is selected;
- the values of the design variables r_w, R, and n are saved, and on the basis of these values, an objective function of a smaller value is obtained.

The scheme of calculations carried out in one calculation step of a computer program is shown in Figure 19.

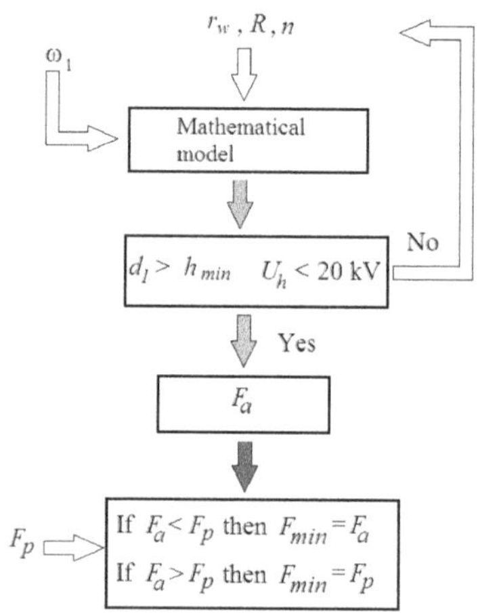

Figure 19. Algorithm of calculations carried out in one calculation step.

5. Design Optimization Results and Discussion

The results of the calculations shown in Figures 10–12 demonstrate that it is not possible to determine intervals of the torque M, for which there are minima of the considered objective functions. Due to that, initially, the design optimization of the HB is carried out for the entire range of decision variables. The results of the HB design optimization for 150,000 draws are shown in Table 5. The placement of the minima F_{min} of the objective function F on the background of the "Pareto chart" showing the dependence of the "lightness of construction" (described as 1/G) on the torque M is depicted in Figure 20. Table 6 contains the values of the ratios: G/M, P_e/M, T/M and G, P_e, T calculated on the basis of the data from Table 1. In order to assess the possibility of PHB optimization, calculations are made with the assumption that the braking torque M meets the requirement 1.700 Nm <M < 1.800 Nm, i.e., it is close to the torque of the experimentally tested PHB: M = 1.733 Nm. Assuming such a small range of torque M variations increases the possibility of finding the minima of the objective functions. A few percent change of the torque M during the HB selection for a machine's drive system is acceptable. Table 7 shows the results of these calculations for 150,000 draws. It is justified to assume the quotients of the criteria listed in Table 4 as the functions of the objectives, because the quotients G/M, P_e/M, T/M can be interpreted as efficiency related to 1 Nm. For instance, G/M = 0.497 kg/Nm means that 1 Nm of the braking torque M corresponds to 0.497 kg of the HB mass. Table 5 shows that the minimum of the objective function is achieved for torques M of approximately 7.5 Nm for the objective functions F_G, F_{G1}; 0.2 Nm for the objective functions F_{Pe}, F_{Pe1}; and 9.9 Nm for the objective functions F_T, F_{T1}. Therefore, the HBs having a large braking torque M have a low G/M ratio and a low T/M ratio, which is advantageous. On the contrary, the HBs having a low braking torque M have unfavorable values of G/M ratio and especially T/M ratio, but the value of the P_e/M ratio is advantageous. The minimal value of the G/M ratio does not correspond to the minimal value of the HB, due to the large differences between the values of the torques. However, the minimal values of the P_e/M and T/M ratios occur for the minimal values, respectively, P_e and T. Proximate values of the G/M, P_e/M, T/M ratios, defined for the objective functions as differences and quotients of criteria, prove the correctness of the performed calculations. Table 6 and Figure 20 show that the PHB is not an optimal brake in terms of the considered criteria. This is proven by the high values of the G/M, P_e/M, T/M ratios, as well as the placement of point 7 beyond the "Pareto front" (which is a set of points with the lowest values of 1/G). Table 7 shows that the minimal values of the G/M, P_e/M, and T/M ratios occur for the minimal values of, respectively, G, P_e, and T. The optimized PHBs with a lower weight also have lower mass moments of inertia. This results in the fact that these brakes have a short head rise time of the torque M. Table 7 also shows that the values of the braking torque M, differing from each other by less than 4%, are obtained for significantly different values of design variables. The relative differences between the highest and lowest values of individual design variables, related to the average values, are 80% for r_w, 12% for R, and 83% for n. Whereas, the relative differences between the highest and lowest values of the ratios G/M, P_e/M, and T/M are respectively 54%, 140%, and 146%.

Table 5. Optimization results for the whole intervals of decision variables.

F_{min}	M Nm	r_w mm	R mm	n -	G/M kg/Nm	P_e/M W/Nm	T/M ms/Nm	G kg	P_e W	T ms
F_{Gmin}	7.478	29.1	69.9	42	0.497	181.4	0.790	3.717	1356	5.9
F_{G1min}	7.473	29.9	69.4	38	0.446	231.6	0.681	3.333	1731	5.1
F_{Pmin}	0.218	28.8	40.9	98	1.136	38.5	201.821	0.248	8.42	44.4
F_{P1min}	0.222	28.9	41.1	100	1.150	38.2	206.113	0.255	8.45	45.8
F_{Tmin}	9.850	11.6	69.8	30	0.601	254.6	0.358	5.920	2508	3.5
F_{T1min}	9.886	10.6	69.9	30	0.611	313.3	0.489	6.040	3096	4.8

The initial values of PHB presented in Table 6 can be compared with the optimized values presented in Table 7. The comparison of the values of the G/M, P_e/M and T/M ratios (presented in Tables 6 and 7) show that, on the basis of the design optimization, it is possible to reduce the following factors: the weight of PHB 1.7 times, the maximum electric power 1.5 times, and the time T 2.5 times. However, these benefits cannot be gained simultaneously. The results shown in Table 7 confirm that it is advisable to improve the PHB design on the basis of design optimization methods. Due to the fact that the selection of decision variables is random, repeating the optimization calculation always provides different results. Moreover, it is not possible to unequivocally determine the optimal number of draws because there is always a possibility that the following draws will provide better results. However, it is necessary to take into consideration the fact that, during the design optimization, only the changes of radii r_w and R by tenths of a millimeter are practically significant.

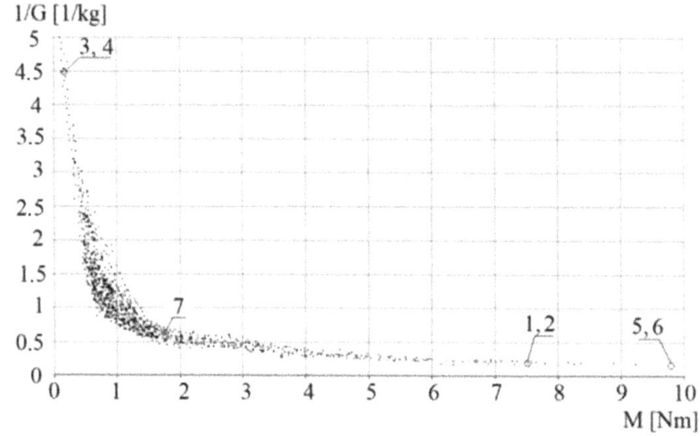

Figure 20. Optimization results on the background of the Pareto chart: 1—F_{Gmin}, 2—F_{G1min}, 3—F_{Pmin}, 4—F_{P1min}, 5—F_{Tmin}, 6—F_{T1min}, 7—PHB.

Table 6. Values of the ratios G/M, Pe/M, T/M and G, P_e, T for the optimized PHB.

M Nm	r_w mm	R mm	n -	G/M kg/Nm	P_e/M W/Nm	T/M ms/Nm	G kg	P_e W	T ms
1.733	15.5	50.5	38	1.013	97.9	5.72	1.756	170	9.9

Table 7. PHB optimization results for 1.700 Nm < M < 1.800 Nm.

F_{min}	M Nm	r_w mm	R mm	n -	G/M kg/Nm	P_e/M W/Nm	T/M ms/Nm	G kg	P_e W	T ms
F_{Gmin}	1.728	29.7	53.6	30	0.570	239.0	3.027	0.985	413	5.2
F_{G1min}	1.782	28.9	53.6	30	0.586	240.9	3.070	1.044	429	5.4
F_{Pmin}	1.749	29.6	55.8	88	0.837	64.8	14.603	1.464	113	25.5
F_{P1min}	1.753	29.6	55.9	86	0.823	66.5	13.986	1.443	116	24.5
F_{Tmin}	1.746	16.2	50.0	22	0.894	356.1	2.245	1.561	621	3.9
F_{T1min}	1.795	12.6	49.8	22	0.992	351.0	2.269	1.781	630	4.0

6. Conclusions

On the basis of the performed optimization calculations concerning the design of the HB with the ER fluid, the following conclusions have been achieved:

1. Performing the optimization of the HB with the ER fluid requires formulating a mathematical model, establishing constraints, and selecting design variables and the objective function. While planning the design optimization, it is advisable to aim at a simple mathematical model and a low number of design variables, especially when the optimization calculations are performed with the use of the Monte Carlo method (so that it is possible to perform a large number of calculations within an acceptable time). On the basis of the obtained results of the optimization of the HB with the ER fluid, it can be presumed that the assumptions are correct.
2. The form of the objective function has a significant influence on the optimization results. However, there are no strict rules as to how the objective function should be formulated. This is why the selection of the objective function is always arbitrary. It is advisable to formulate the objective function in a simple way, so that the results of optimization calculations are easy to interpret. Defining two different objective

functions on the basis of the same criteria increases the probability of finding the optimal solution. Due to that, it is justifiable to use the difference and quotient of only two criteria when optimizing the design of the HB with the ER fluid in order to determine the objective function.

3. The Monte Carlo method can be successfully used for design optimization of brakes with ER fluid for three design variables. Increasing the number of the design variables causes an exponential increase in the number of draws necessary to obtain the required accuracy of calculations. Not all optimization methods can be used for this purpose because the number of blades, which is a natural number, causes the calculations to be discrete.
4. In order to obtain credible and reliable results of the optimization calculations, it is advisable to select coefficients for the assumed mathematical model, as well as for the range of the mathematical model. It is especially important when the HB optimization calculations are performed with the use of a mean streamline model whose accuracy depends on the correctness of the selected coefficients.

It is finally remarked that the optimized method proposed based on the mathematical model can be used when the design of the transmission system is specified only. Therefore, it is a general optimization method that can be applied to optimally design several application systems utilizing ER fluid and MR fluid also.

Author Contributions: I.M. and Z.K. contributed conception and design of the study. Z.K. conducted optimization analysis and calculations. Z.K. and I.M. carried out tests. Z.K. and I.M. wrote the first draft of the manuscript. S.-B.C. contributed to idea analysis, manuscript revision, reading, and approving the submitted version. All authors have read and agreed to the published version of the manuscript.

Funding: The authors received no external funding.

Institutional Review Board Statement: Not applicable.

Informed Consent Statement: Not applicable.

Data Availability Statement: Not applicable.

Conflicts of Interest: The authors declare no potential conflict of interest with respect to the research, authorship, and publication of this article.

References

1. Brun, K.; Meyenberg, C.; Thorp, J. Hydrodynamic torque converters for oil & gas compression and pumping applications: Basic principles, performance characteristics and applications. In Proceedings of the Asia & Pump Symposium Marina Bay Sand, Singapore, 22–25 February 2016; pp. 1–15. [CrossRef]
2. Kęsy, Z.; Kęsy, A. Prospects for the control of a torque converter using magnetic fluid. In Proceedings of the IEEE International Colloquium on Innovative Actuators for Mechatronic Systems, London, UK, 18 October 1995; pp. 10-1–10-3.
3. Dong, Y.Z.; Seo, Y.; Choi, H.J. Recent development of electro-responsive smart electrorheological fluids. *Soft Matter* **2019**, *15*, 3473–3486. [CrossRef]
4. Qader, I.N.; Kök, M.; Dagdelen, F.; Aydogdu, Y.A. Review of smart materials: Researches and applications. *El-Cezeri Fen Ve Mühendislik Derg.* **2019**, *6*, 755–788. [CrossRef]
5. Bulough, W.; Johnson, A.R.; Tozer, R.; Makin, J. Electrorheological clutch, methodology, performance and problems in ERclutch based positioning mechanisms. *Intern. J. Mod. Phys. B13* **1999**, *14*, 2101–2108. [CrossRef]
6. Li, W.H.; Du, H. Design and experimental evaluation of a magnetorheological brake. *Int. J. Adv. Manuf. Technol.* **2003**, *21*, 508–515. [CrossRef]
7. Choi, S.B.; Cheong, C.C.; Kim, G.W. Feedback control of tension in moving tape using an ER brake actuator. *Mechatronics* **1997**, *7*, 53–66. [CrossRef]
8. Wereley, N.M.; Lindler, J.; Rosenfeld, N.; Choi, Y.T. Biviscous damping behavior in electrorheological shock absorbers. *Smart Mater. Struc.* **2004**, *13*, 743–752. [CrossRef]
9. Choi, S.B. Performance comparison of vehicle suspensions featuring two different electrorheological shock absorbers. *Proc. Inst. Mech. Eng. Part D J. Automob. Eng.* **2003**, *217*, 999–1010. [CrossRef]
10. Gołdasz, J.; Sapiński, B. *Insight into Magnetorheological Shock Absorbers*; Springer International Publishing: Cham, Switzerland, 2003. [CrossRef]

11. Rosenfeld, N.C.; Wereley, N.M. Volume-constrained optimization of magnetorheological and electrorheological valves and dampers. *Smart Mater. Struc.* **2004**, *13*, 1303–1313. [CrossRef]
12. Choi, S.B.; Choi, W.Y. Position control of a cylinder using a hydraulic bridge circuit with ER valves. *J. Dyn. Syst. Meas. Control* **2003**, *122*, 201–209. [CrossRef]
13. Kciuk, S.; Martynowicz, P. Special application magnetorheological valve numerical and experimental. *Solid State Phenom.* **2011**, *177*, 102–115. [CrossRef]
14. Szczęch, M. Theoretical analysis and experimental studies on torque friction in magnetic fluid seals. *Proc. Inst. Mech. Eng. Part J J. Eng. Tribol.* **2020**, *234*, 274–281. [CrossRef]
15. Lara-Prieto, V.; Parkin, R.; Jackson, M.; Silberschmidt, V.; Kęsy, Z. Vibration characteristics of MR cantilevers and sandwich beams: Experimental study. *Smart Mater. Struc.* **2003**, *19*, 015005. [CrossRef]
16. Berg, C.D.; Evans, L.F.; Kermode, P.R. Composite structure analysis of a hollow cantilever beam filled with electro–rheological fluid. *J. Intell. Mater. Struct.* **1996**, *7*, 494–502. [CrossRef]
17. Choi, Y.T.; Wereley, N.M. Vibration control of a landing gear system featuring electrorheological/magnetorheological fluids. *J. Aircr.* **2003**, *40*, 432–439. [CrossRef]
18. Mikułowski, G.; Holnicki-Szulc, J. Adaptive landing gear concept—Feedback control validation. *Smart Mater. Struc.* **2007**, *16*, 2146–2158. [CrossRef]
19. Białek, M.; Jędryczka, C.; Milecki, A. Investigation of thermoplastic polyurethane finger cushion with magnetorheological fluid for soft-rigid gripper. *Energies* **2021**, *14*, 6541. [CrossRef]
20. Fernández, M.A.; Chang, J.Y. Development of magnetorheological fluid clutch for robotic arm applications. In Proceedings of the 2016 IEEE 14th International Workshop on Advanced Motion Control (AMC), Auckland, New Zealand, 22–24 April 2016. [CrossRef]
21. Rdice, I.; Bergmann, J.H. Conceptual exploration of a gravity-assisted electrorheological fluid-based gripping methodology for assistive technology. *Bio-Des. Manuf.* **2019**, *2*, 145–152. [CrossRef]
22. Madeja, J.; Kęsy, Z.; Kęsy, A. Application of electrorheological fluid in a hydrodynamic clutch. *Smart Mater. Struc.* **2011**, *20*, 105005. [CrossRef]
23. Olszak, A.; Osowski, K.; Kęsy, A.; Kęsy, Z. Experimental researches of hydraulic clutches with smart fluids. *Int. Revive Mech. Eng.* **2016**, *10*, 364–372. [CrossRef]
24. Olszak, A.; Osowski, K.; Kęsy, Z.; Kęsy, A. Investigation of hydrodynamic clutch with MR fluid. *J. Intell. Mater. Struct.* **2018**, *30*, 155–168. [CrossRef]
25. Musiałek, I.; Migus, M.; Osowski, K.; Olszak, A.; Kęsy, Z.; Kęsy, A.; Kim, G.W.; Choi, S.B. Analysis of a combined clutch with an electrorheological fluid. *Smart Mater. Struc.* **2020**, *29*, 087006. [CrossRef]
26. Casella, G.; Robert, C.P. *Monte Carlo Statistical Methods*; LLC PubSpringer: New York, NY, USA, 2004.
27. Kęsy, A.; Kęsy, Z.; Kądziela, A. Estimation of parameters for a hydrodynamic transmission system mathematical model with the application of genetic algorithm. In *Evolutionary Methods in Mechanics*; Kluwer Academic Press: Amsterdam, The Netherlands, 2004.
28. Lempa, P.; Lisowski, E. Genetic algorithm in optimization of cycloid pump. *Technol. Trans. Mech.* **2013**, *110*, 229–236.
29. Andersson, S. Analysis of multi-element torque converter transmissions. *Int. J. Mech. Sci.* **1986**, *28*, 431–441. [CrossRef]
30. Asl, H.A.; Azad, N.L.; McPhee, J. Math-based torque converter modelling to evaluate damping characteristics and reverse flow mode operation. *Int. J. Veh. Syst. Model. Test.* **2014**, *9*, 36–55. [CrossRef]
31. Behreus, H.; Jaschke, P.; Steinhausen, J.; Waller, H. Modeling of technical systems: Application to hydrodynamic torque converters and couplings. *Math. Comput. Model. Dyn. Syst.* **2000**, *6*, 223–250. [CrossRef]
32. Jandasek, V.J. Design of single-stage, three element torque converter. *Des. Pract. Passeng. Cars Autom. Transm. SAE* **1970**, *5*, 201–226.
33. Kęsy, A. Mathematical model of hydrodynamic torque converter for vehicle power transmission system optimization. *Int. J. Veh. Des.* **2012**, *59*, 1–22. [CrossRef]
34. Nguyen, Q.H.; Lang, V.T.; Choi, S.B. Optimal design and selection of magneto-rheological brake types based on braking torque and mass. *Smart Mater Struct.* **2015**, *24*, 067001. [CrossRef]
35. Nguyen, Q.H.; Lang, V.T.; Nguyen, N.D.; Choi, S.B. Geometric optimal design of a magneto-rheological brake considering different shapes for the brake envelope. *Smart Mater. Struc.* **2013**, *23*, 015020. [CrossRef]
36. Żurowski, W.; Osowski, K.; Mędrek, G.; Musiałek, I.; Kęsy, A.; Kęsy, Z.; Choi, S.B. Design optimization of a viscous clutch with an electrorheological fluid. *Smart Mater. Struc.* **2022**, *31*, 095035. [CrossRef]
37. Sohn, J.W.; Jeon, J.; Nguyen, Q.H.; Choi, S.B. Optimal design of disc-type magnetorheological brake for midsized motorcycle: Experimental evaluation. *Smart Mater. Struct.* **2015**, *24*, 085009. [CrossRef]
38. Quoc, N.V.; Tuan, L.D.; Hiep, L.D.; Quoc, H.N.; Choi, S.B. Material characterization of MRF fluid on performance of MRF based brake. *Front. Mater.* **2019**, *6*, 125. [CrossRef]
39. Erol, O.; Gurocak, H. Interactive design optimization of magnetorheological-brake actuators using the Taguchi method. *Smart Mater. Struc.* **2011**, *20*, 105027. [CrossRef]
40. Gurubasavaraju, T.M.; Kumar, H.; Arun, M. Evaluation of optimal parameters o fMR fluids for damper application using particle swarm and response surface optimization. *J. Braz. Soc. Mech. Sci. Eng.* **2017**, *39*, 3683–3694. [CrossRef]

41. Kęsy, Z. Numerical analysis of torque carried by vehicle hydrodynamic clutch with electrorheological working fluid. *Int. J. Veh. Des.* **2005**, *38*, 210–221. [CrossRef]
42. Olszak, A.; Osowski, K.; Kęsy, Z.; Kęsy, A. Modelling and testing of a hydrodynamic clutch filled with electrorheological fluid in varying degree. *J. Intell. Mater. Struct.* **2019**, *30*, 649–660. [CrossRef]
43. Wolff-Jesse, C.; Fees, G. Examination of flow behavior of electrorheological fluids in the flow mode. *Proc. Inst. Mech. Eng. Part I J. Syst. Control Eng.* **1998**, *212*, 159–173. [CrossRef]
44. Nguyen, Q.A.; Jorgensen, S.J.; Ho, J.; Sentis, L. Characterization and testing of an electrorheological fluid valve for control of ERF actuators. *Actuators* **2015**, *4*, 135–155. [CrossRef]
45. Kęsy, Z.; Mędrek, G.; Olszak, A.; Osowski, K.; Kęsy, A. Electrorheological fluid based clutches and brakes. In *Reference Module in Materials Science and Materials Engineering*; Elsevier: Amsterdam, The Netherlands, 2019. [CrossRef]
46. SmartTechnologyLtd. UK. Information Brochure. 2004. Available online: www.smarttec.co.uk (accessed on 1 November 2022).
47. Kęsy, Z.; Mędrek, G.; Osowski, K.; Olszak, A.; Migus, M.; Musiałek, I.; Musiałek, K.; Kęsy, A. Characteristics of electrorheological fluids. In *Encyclopedia of Smart Materials*; Elsevier: Amsterdam, The Netherlands, 2020. [CrossRef]
48. Sturk, M.; Wu, X.M.; Wong, J.Y. Development and evaluation of a high voltage supply unit for electrorheological fluid dampers. *Veh. Syst. Dyn.* **1995**, *24*, 101–121. [CrossRef]
49. Karakoc, K.; Park, E.J.; Suleman, A. Design considerations for an automotive magnetorheological brake. *Mechatronics* **2008**, *18*, 434–447. [CrossRef]
50. Park, E.J.; Falcao DaLuz, L.; Suleman, A. Multidisciplinary design optimization of an automotive magnetorheological brake design. *Comput. Struct.* **2008**, *86*, 207–216. [CrossRef]

applied sciences

MDPI

Article

A Comparison Study on Magnetorheological Multi-Disc Clutches in Steady Continuous-Duty States from the Viewpoint of Electrical Energy Consumption and Spatial Temperature Distribution

Krzysztof Kluszczyński † and Zbigniew Pilch *,†

Faculty of Electrical and Computer Engineering, Cracow University of Technology, Warszawska Str. 24, 31-155 Cracow, Poland
* Correspondence: zbigniew.pilch@pk.edu.pl
† These authors contributed equally to this work.

Abstract: The paper is focused on magnetorheological (MR) clutches applied in industrial drive systems working in a steady continuous-duty state. The main goal of the carried out numerical and analytical analyses oriented towards electrical power consumption, copper losses (Joule heat) in an excitation coil, spatial temperature distributions and the highest temperature possible for an MR fluid is to compare MR clutches due to a different number of discs. The authors considered selected representative MR multi-disc clutches with one, two, three or four discs, developing clutching torque T_c equal to 20, 35 and 50 Nm. These clutches were constructed based on the in-house design that integrates analytical and field methods (further in the paper referred to as the integrated analytical-field design method) described in the literature. The thermal computer simulation results obtained with the help of the AGROS2D program, combined with findings achieved with the use of simplified physical reasonings, allow one to draw the conclusion that the most advantageous, recommended number of discs for a magnetorheological clutch from the viewpoint of various (both constructional and thermal) criteria is the number of discs: N = 2. This conclusion takes into account the results presented earlier in the literature: the choice is a compromise between decreasing the mass (volume) of the MR clutches and increasing both the electrical power consumption and the maximum temperature of MR fluids in a clutch working region as the number of discs, N, increases.

Keywords: MR fluid; MR multi-disc clutch; industrial continuous-duty drive systems; thermal clutch numerical calculations; electrical power consumption of MR clutches; MR clutch copper losses

Citation: Kluszczyński K.; Pilch Z. A Comparison Study on Magnetorheological Multi-Disc Clutches in Steady Continuous-Duty States from the Viewpoint of Electrical Energy Consumption and Spatial Temperature Distribution. *Appl. Sci.* **2022**, *12*, 7895. https://doi.org/10.3390/app12157895

Academic Editor: Gaetano Zizzo

Received: 1 July 2022
Accepted: 27 July 2022
Published: 6 August 2022

1. Introduction

Recently, there has been a systematic increase in interest in the use of magnetorheological (MR) fluids with viscosity controlled by magnetic or electric signals; these fluids can be found in a wide variety of devices (e.g., robots [1], testing stands [2], lifts [3] and two DoF brakes for investigating linear-rotary motors [4]) and seem to be particularly useful in automotive and motorcycle drive systems such as brakes and clutches [5,6]. In cars and motorcycles, there is a need for the frequent coupling and decoupling of internal combustion or electric drive motors with a vehicle's wheel-drive system, and during such transient states, the clutches operate with varying slips, which result in friction (MR fluid–solid bodies), slip heat generation and, in consequence, temperature increases in the MR fluids having a direct influence on the clutch life. In recent years, this problem has been widely discussed by many authors [6–9]. A power flow diagram for MR clutches in transient states (acting with slippage) is depicted in Figure 1a,b.

Magnetorheological (MR) clutches can also be used in industrial drive systems because of the ease of the coupling and decoupling process and the possibility of acting as an overload protection system [5]. Industrial drive systems work for long periods in a steady continuous-duty state with no slippage; hence, the generation of heat in the MR fluid associated with transient states does not play a significant role. In such cases electrical power consumption necessary for the production of a magnetic field and maintenance of a coupling state becomes the key issue for MR clutches. Electric power in an excitation coil connected to the flow of the excitation current is released in the form of Joule heat and becomes—next to the heat sources associated with friction in bearings—an additional heat source, increasing the temperature of the working clutch.

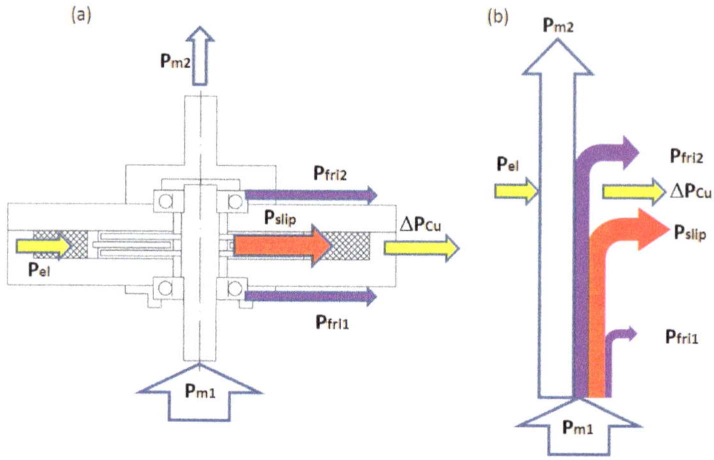

Figure 1. Cross-section of multi-disc MR clutch: (**a**) mechanical and electrical sources of heat and (**b**) power flow diagram in transient states or in steady state with slippage where P_{m1}—input mechanical power; P_{m2}—output mechanical power; P_{el}—input active electrical power; P_{slip}—slip power (related to slip heat generated in MR fluid in transient states); ΔP_{Cu}—copper losses (Joule heat); P_{fri1}—mechanical power losses in bearings of driving member; P_{fri2}—mechanical power losses in bearings of primary (driving) member.

For MR devices with a disc structure, commonly met in clutches, the possibility of a different number of discs is indicated, and each number of discs is associated with different properties.

In article [10], the authors considered in detail the problem of the optimal selection of the number of discs N in MR clutches (N is equal to the number of discs of a clutch primary (driving) member) in the industrial drive systems from the viewpoint of various criteria and limitations. In particular, the authors not only took into account the external dimensions of the clutches: their lengths and diameters, masses and volumes but also characteristic factors: the maximum torque per mass ratio T_c/m and the maximum torque per volume ratio T_c/V. The result of a detailed comparative analysis was the conclusion that the optimal (recommended) number of discs in a multi-disc magnetorheological clutch, taking into account the above-mentioned criteria, is equal to N = 2. In drawing this conclusion, the authors did not take into account the thermal operating conditions of the clutches. To their knowledge, in the available technical literature, the problem of power consumption by magnetorheological clutches in a steady continuous-duty states with no slippage, as well as copper losses in the excitation coil and spatial temperature distribution in such clutches (particularly, the maximum values of coils and MR fluid temperatures), has not been considered thus far. Hence, the main goal of this article is to consider the above-mentioned problems for MR clutches applied in the industrial continuous-duty drive systems.

The main view of an exemplary multi-disc MR clutch is depicted in Figure 2. This is the simplest case of a multi-disc MR clutch: the clutch having only two discs of a primary member (N = 2). The number of discs of a secondary member is equal to (N − 1).

Owing to its symmetry, it is sufficient to present only a quarter of the entire construction.

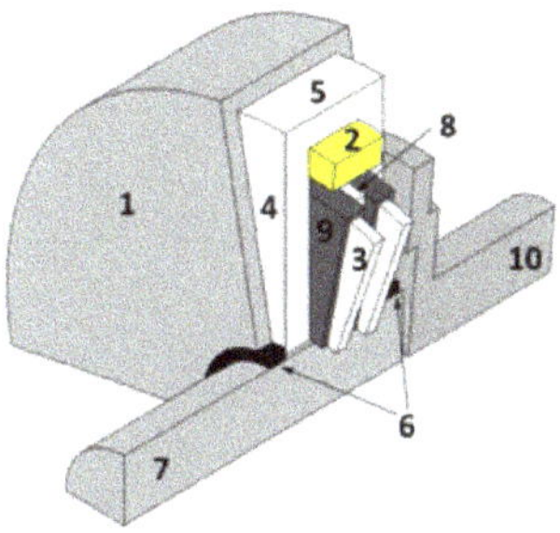

Figure 2. Exemplary two-disc MR clutch (N = 2)—general view and basic terminology.

Numbers 1, 2, 3, ..., 10 indicate the most important constructional parts: 1—non-magnetic housing; 2—coil; 3—discs of a primary member; 4—cover yoke; 5—cylinder yoke; 6—bearings; 7—non-magnetic shaft of a primary member; 8—discs of a secondary member; 9—MR fluid gaps; 10—non-magnetic shaft of a secondary member. Particular care must be taken to prevent the leakage of magnetorheological fluids in the coupling. It is necessary to use an appropriate sealing system, the selection of which is dictated by the operating conditions.

2. Materials and Methods

2.1. Electromagnetic Analysis

In [11,12] the experimentally verified integrated analytical-field design method of multi-disc magnetorheological (MR) clutches was described in detail. Using this method, 12 variants of MR clutches, differing in the number of discs N and the values of the rated clutching torque T_c were designed and their various features and properties essential from the viewpoint of manufacturing costs and industrial applications were compared in [10]. To render this comparison fully reliable and meaningful from a technical point of view, a number of assumptions were made, the most important of which were the following:

- The thickness of the MR fluid gap g is the same ($g = 1$ mm);
- Discs, cylinder yokes and cover yokes are made of the same magnetic steel;
- Shaft and mounting rings, separating the coil from the working region, are made of the same non-magnetic steel;
- Current density j_{Cu} is assumed the same (j_{Cu} = 4.5 A/mm^2) [13];
- Excitation current I is assumed the same ($I = 0.6$ A);
- The magnetic flux density in MR fluid-gaps B_0 is kept the same despite variations in the geometries of clutches ($B_0 = 0.7$ T);
- The maximum magnetic field density B_{mx} is kept the same ($B_{mx} = 1.2$ T) despite variations in the geometries of clutches (the most saturated point lies within the cover yoke at a length approximately equal to the external radius of primary member discs).

As mentioned earlier, the designed clutches were compared in terms of the following criteria: overall dimensions, geometric proportions (lengths and radii of clutches), total masses m and total volumes V and factors including clutching torque per mass ratio T_c/m and clutching torque per volume ratio T_c/V. The conclusion resulting from this comparison, taking into account the costs of active materials on the one hand and the complexity of the manufacturing process on the other hand, was the statement that the optimal, recommended number of discs is N = 2.

This paper, as a logical continuation of the previously presented considerations, compares exactly the same variants of MR clutches with regards to the electrical power consumption necessary to maintain the coupling state, as well as spatial temperature distribution caused by copper losses in the excitation coil with resistance R_c.

The electrical power consumption, P_1, equal to copper losses caused by the flow of excitation current, I (Joule effect), is determined by the following formula:

$$P_1 = \Delta P_{Cu} = I^2 R_c \tag{1}$$

$$R_c = \rho \cdot \frac{l}{S_{Cu0}} \tag{2}$$

$$S_{Cu0} = \frac{\pi \cdot d_{Cu0}^2}{4} \tag{3}$$

where ρ—copper resistivity; l—wire length needed to wind the coil; d_{Cu0}—diameter of the bare conductor; S_{Cu0}—cross-section of the bare conductor.

Due to the fact that the assumption of the same current density j_{Cu} in the excitation coils of various variants is made, the cross-section of the conductor S_{Cu0} is the same for all considered cases. The values of the resistances R_c for all point variants are, therefore, determined by the lengths of the wire l needed to wind the coils. This length is equal to the sum of the lengths of the individual turns, and these turns have different lengths depending on which coil layer they belong to (Figure 3). The geometrical cross-sections of all point variants (N = 1, 2, 3, 4 discs and T_c = 20, 35, 50 Nm (12 various variants)) have been presented graphically in [10] in a combined drawing, and this drawing is shown in this paper as Figure 4 with a variant numbering from **V1** to **V12**.

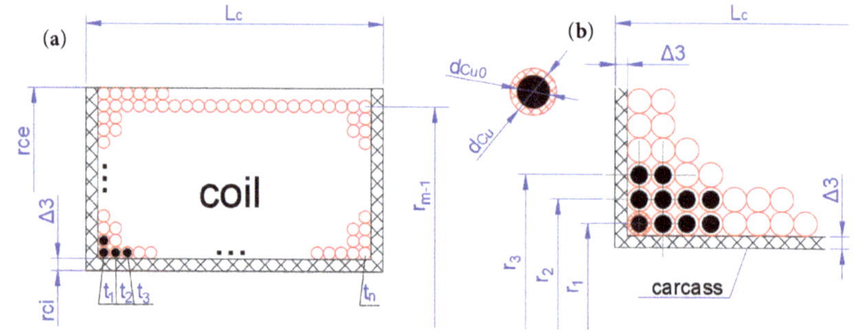

Figure 3. The cross-section of clutch coil: (**a**) dimensions of coil and (**b**) dimensions of conductors.

Let us now focus on the geometrical dimensions of the clutch elements and the winding data, the knowledge of which is necessary to determine the resistances of the individual coils. Letter markings of the clutch geometric dimensions and winding data (the number of coil turns z, the number of turns per one layer n and the number of layers m), necessary for determining the resistance of the coils, are shown in Figure 3 [10] on the example of an MR clutch with the number of discs: N = 2.

The design dimensions common for all the considered variants include the following: Δ_3, d_{Cu}, d_{Cu0}, S_{Cu} and S_{Cu0} (Figure 3). The design dimensions, which are different for individual variants and depend on the rated values of the clutching torque, T_c, and the number of discs, N, are summarized in Table 1 (these values were determined on the basis of the integrated analytical-field design method described in [10,12]).

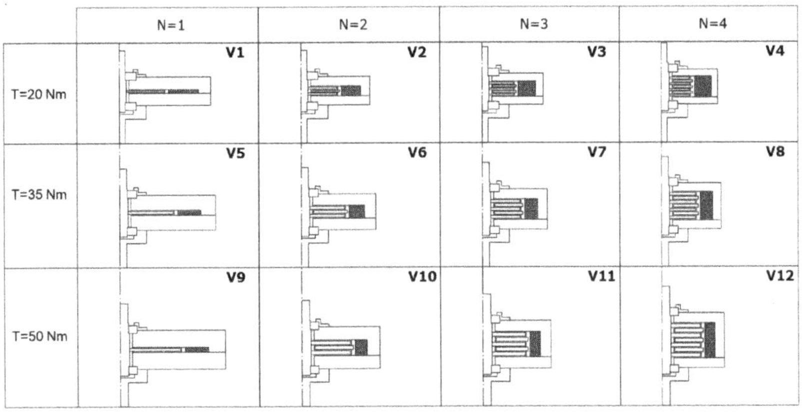

Figure 4. Graphical overview of clutch cross-sections for all 12 designed variants: $T_c = 20, 35, 50$ Nm, N = 1, 2, 3, 4.

Table 1. List of design geometrical dimensions and winding data different for 12 individual variants.

Variants	V1	V2	V3	V4	V5	V6
r_{ci} (mm)	51.9	42.1	37.2	34.2	61.5	49.7
r_{ce} (mm)	83.9	62.3	56.0	51.9	85.9	65.9
L_c (mm)	4	10	16	22	5	13
z (-)	366	694	1026	1365	368	695
m (-)	62	39	36	34	47	31
n (-)	6	18	29	41	8	23
Variants	**V7**	**V8**	**V9**	**V10**	**V11**	**V12**
r_{ci} (mm)	43.9	40.2	68.7	55.4	48.9	44.7
r_{ce} (mm)	58.1	53.4	93.1	71.6	63.1	57.9
L_c (mm)	21	29	5	13	21	29
z (-)	1028	1363	370	697	1030	1364
m (-)	27	25	47	31	27	25
n (-)	39	55	8	23	39	55

The formulas for the lengths of the wires needed to wind the coils differ depending on whether the last layer is full; i.e., whether the condition

$$z = m \cdot n \tag{4}$$

is fulfilled (Figure 3a) or whether the last layer is incomplete (Figure 3b), and there is a mathematical inequality.

$$n \cdot m > z. \tag{5}$$

If the condition (4) is met, we have the following:

$$l = \sum_{i=1}^{m} \left[2\pi \cdot \left(r_{ci} + \Delta_3 + (i-1) \cdot d_{cu} + \frac{d_{cu}}{2} \right) \right] \cdot n, \tag{6}$$

where r_{ci} denotes the internal radius of the coil, Δ_3 denotes the thickness of the carcass and d_{Cu} denotes the diameter of the insulated conductor.

If condition (5) is fulfilled, then another formula should be used:

$$l = 2\pi \cdot \sum_{i=1}^{m-1} \cdot \left(r_{ci} + \Delta_3 + (i-1) \cdot d_{cu} + \frac{d_{cu}}{2} \right) \cdot n + 2\pi \cdot \left(r_{ci} + \Delta_3 + (m-1) \cdot d_{cu} + \frac{d_{cu}}{2} \right) \cdot z_m \tag{7}$$

where the number of turns in the incomplete m-th layer is $z_m = z - (m-1)n$.

Formulas (1)–(3), (6) and (7) allow the resistance of the excitation windings coils to be determined in an accurate manner as the sum of the resistances of the turns of the individual layers.

When designing transformers and electrical machines, we very often use simplified formulas, based on the concept of the average turn length l_{tav}, calculated on the basis of the overall dimensions of the coil, i.e., inner radius of the coil r_{ci} and outer radius of the coil r_{ce}.

$$l_{tav} = 2\pi \cdot \frac{(r_{ci} + r_{ce})}{2}. \tag{8}$$

Using the above relation, the formula for the coil resistance excitation winding resistance can be greatly simplified. In Table 2, the coil resistances R_c and the electrical power consumption P_1 determined on the basis of exact Formulas (6) and (7) as well as their approximate values determined on the basis of simplified Formula (8) are listed.

Table 2. Coil resistances and electrical power consumption for variants **V1–V12** (exact values: R_c, P_{el} and approximate values determined on the basis of simplified formulas (R_{ca} and P_{ela})).

Variants	V1	V2	V3	V4	V5	V6
R_c (Ohm)	20.89	29.61	38.88	47.69	22.79	32.52
R_{ca} (Ohm)	20.83	30.35	39.98	49.16	22.73	33.65
P_{el} (W)	7.52	10.66	13.99	17.17	8.20	11.71
P_{ela} (W)	7.49	10.93	14.39	17.70	8.18	12.11
Variants	**V7**	**V8**	**V9**	**V10**	**V11**	**V12**
R_c (Ohm)	42.22	51.32	24.55	35.86	46.49	56.32
R_{ca} (Ohm)	43.87	53.34	25.09	37.08	48.29	58.54
P_{el} (W)	15.20	18.47	8.84	12.91	16.73	20.27
P_{ela} (W)	15.79	19.20	9.03	13.35	17.38	21.07

As observed, the approximate values calculated on the basis of the simplified formulas differ from the values by a maximum of 3.9%.

Exact values of coil resistance R_c and electrical power consumption P_{el} vs. the number of discs N are presented graphically in Figure 5.

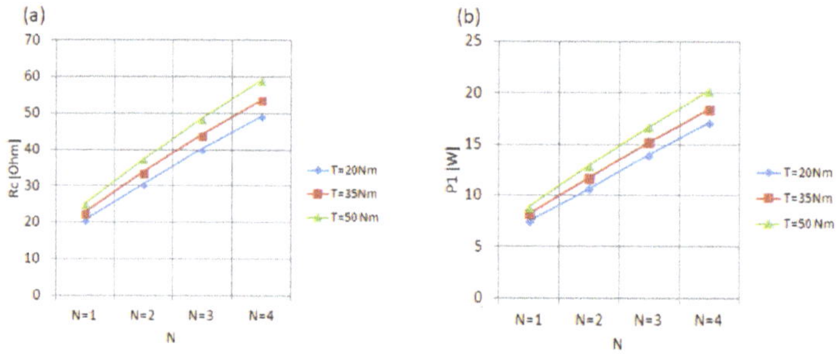

Figure 5. (**a**) Values of coil resistance R_c and (**b**) electrical power consumption P_{el} vs. number of discs N.

The curves presented in Figures 6 and 7 allow for a physical interpretation of the results obtained. The graphs $z = f(N)$ for $T_c = 20$, 35 and 50 Nm in Figure 6a show that the number of turns z needed to a develop magnetic field in the MR fluid gaps with the average value B_o (it should be remembered that the excitation current in all variants is the same and equals $I = 0.6$ A) increases rapidly with the number of discs N and the number of fluid gaps 2 N and is practically independent of the values of clutching torque T_c (notice that the three curves in Figure 6a coincide).

The graphs $l_{tav} = f(N)$ for $T_c = 20, 35$ and 50 Nm in Figure 6b show that as the number of discs increases, the average length of the turns of the individual coils drops, which is due to the fact that the internal diameter of the coil r_{ci} decreases (see Figure 7a, presenting $r_{ci} = f(N)$ for $T_c = 20, 35$ and 50 Nm) and, moreover, that the coil becomes increasingly slender (see Figure 7b, presenting the thickness of the coil $r_{ce} - r_{ci} = f(N)$ for $T_c = 20, 35$ and 50 Nm) .

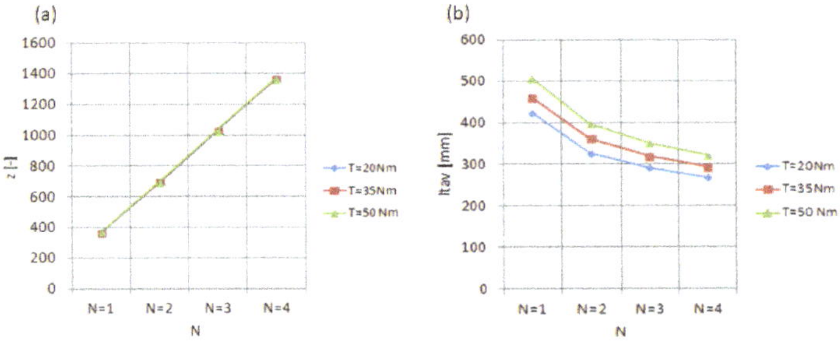

Figure 6. (**a**) Number of turns z and (**b**) average length of the turns l_{tav} for individual excitation coils.

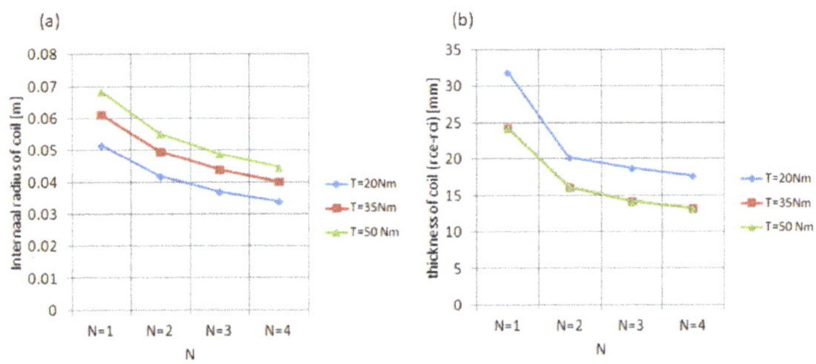

Figure 7. Drop in internal radius of excitation coil and in its thickness: (**a**) internal radius of excitation coil r_{ci} vs. number of discs N; (**b**) thickness of excitation coil ($r_{ce} - r_{ci}$) vs. number of discs N.

Obviously, the rapid increase in the number of turns together with the increasing number of discs is of decisive importance and the average length of the turns (decreasing as the number of discs increases) is not able to compensate for the effects of such a rapid increase in the number of turns, the consequence of which is an increase in the resistances of the excitation windings and the electrical power consumption.

Commenting finally on the results contained in Table 2 and shown graphically in Figure 5, it should be emphasized that the disadvantageous effect of a greater number of discs is an increase in the resistance values of the excitation coils, as well as an increase in clutch electrical energy consumption. The increase in copper losses is the reason why it is necessary to analyse the thermal state of clutches later in this article and to determine the influence of copper losses on the spatial temperature distributions in all the considered variants.

2.2. Thermal Analysis

The rapid increase in the electrical power consumed by the MR clutch (equal to copper losses) with the increased number of discs is the reason why the recommendation for the correct selection of the optimal variant of the device must take into account thermal calculations and the determination of the maximum operating temperature in a steady state. The maximum operating temperature of the clutch working region is important for the assessment of the clutch life, because high temperature may accelerate the aging of the magnetorheological fluid and even-after exceeding the permissible temperature can lead to the degradation of its properties [14].

Heat calculations are carried out according to the power flow diagram presented in Figure 8b. The only heat source taken into account is copper loss: the excitation winding is located in the immediate vicinity of the working region containing the MR fluid; hence, the loss in copper most strongly affects its fluid temperature. Thermal field simulations in a steady state were performed for all 12 variants **V1–V12** in the AGROS2D program [15] (computer program in which magnetic calculations were previously carried out). Table 3 summarized values of the coefficients related to the heating process for the materials used in the clutch construction and for the air. It is worth reminding that the discs of primary and secondary members, cylinder yokes and cover yokes are made of magnetic steel and the shaft, bearings and housing are made of non-magnetic steel.

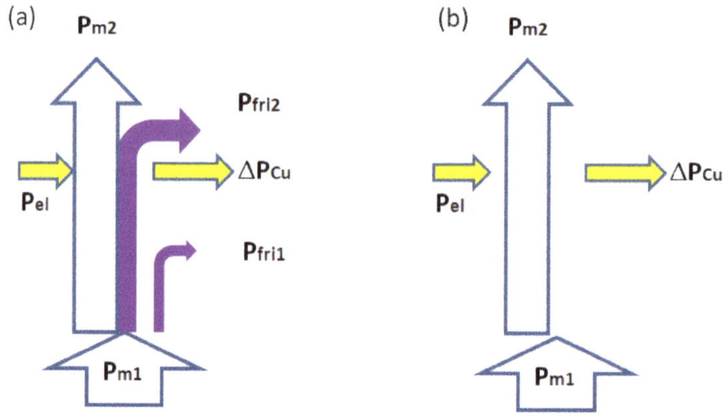

Figure 8. Power-flow diagram (**a**) in steady continuous-duty state with no slippage: (**b**) case study considered in the paper (mechanical losses P_{fri1} and P_{fri2} are neglected; heat is transferred mainly to surrounding air and because of that it slightly influences temperature in clutch working regions including MR fluid).

The results of the thermal calculations in the form of spatial temperature distributions for selected representative variants **V1, V4, V9** and **V12** are presented in Figure 9.

The next Figure 10 shows an exemplary temperature drop along the axis AB which runs through the point with the highest temperature in the clutch (center of the coil) and through the point with the highest temperature in working region ϑ_{max}. This is the highest temperature possible for the MR fluid.

Table 3. Thermal coefficients for constructional materials and air (where α—thermal expansion coefficient; λ—thermal conductivity; C_h—heat convection coefficient; c—heat capacity; ρ—mass density).

-	Material	α [1/K]	λ [W/m·K]	C_h [W/m²·K]	c [J/kg·K]	ρ [kg/m³]
1	air	3.43×10^{-3}	$\lambda_1 = 0.0257$	$C_{h1} = 10$	1005	1.205
2	non-magnetic steel	1.44×10^{-5}	$\lambda_2 = 16.26$	-	502.1	8027.2
3	copper	1.66×10^{-5}	$\lambda_3 = 372$	-	380	8800
4	sealing plastic	1.10×10^{-4}	$\lambda_4 = 0.25$	-	1260	1014
5	magnetic steel	1.34×10^{-5}	$\lambda_5 = 54$	-	465	7833
6	MR fluid	6.00×10^{-4}	$\lambda_6 = 0.8$	-	680	3640

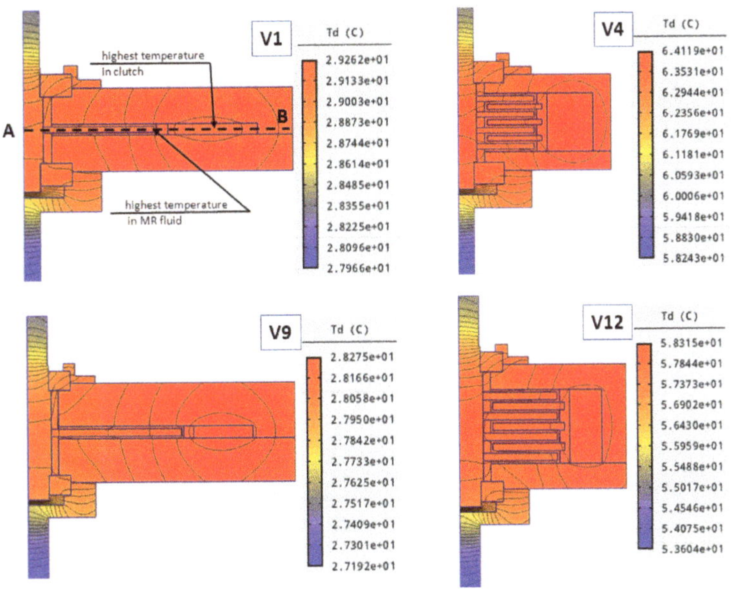

Figure 9. Spatial temperature distribution in steady state for selected representative variants: **V1**, **V4**, **V9** and **V12**.

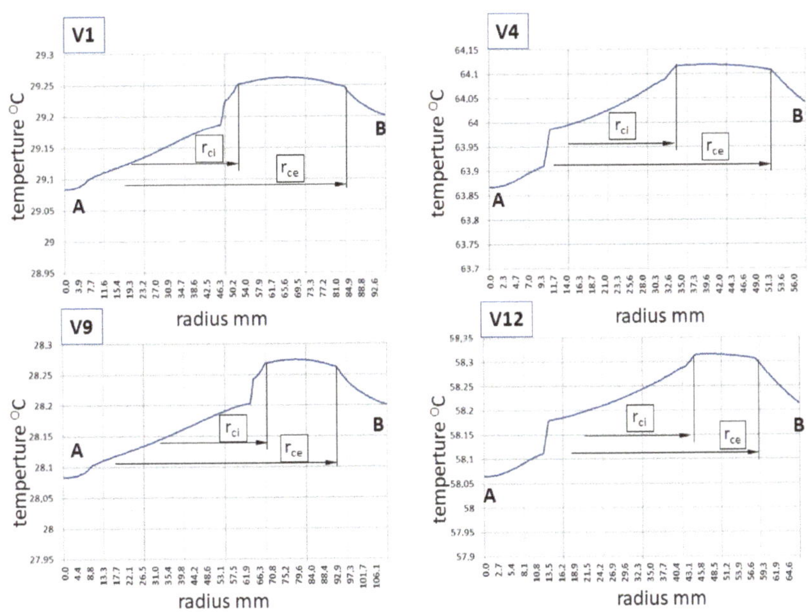

Figure 10. Temperature drop along axis AB in a steady state for selected representative variants: **V1**, **V4**, **V9** and **V12**.

The temperature drop for the considered 12 cases **V1–V12** along the axis AB in a steady state is about 0.12–0.15 K. Maximum temperatures ϑ_{max} in the working region for all the 12 considered cases are shown in Figure 11.

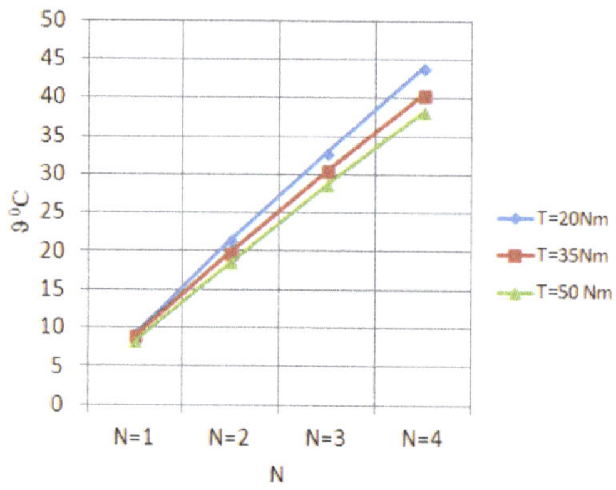

Figure 11. Maximum temperature ϑ_{max} in the working region (the highest possible temperature for MR fluid) vs. the number of discs N for $T_c = 20$, 35 and 50 Nm).

As predicted, we can conclude that the greater the number of discs, the higher the maximum temperature, which is directly evidenced by Figure 5b, showing a substantial increase in copper losses as the number of discs increases.

It needs an explanation as to why, with the same number of discs, the higher maximum temperature is associated with the lower torque clutches and not with the higher torque clutches (related to higher copper losses). The explanation can be based on simplified approximate reasoning from the description of the process of heating a homogeneous body treated as a source of heat that transfers the accumulated heat by convection through the surface of S_{air} to the surrounding air:

$$\Delta P_{Cu} dt = c \cdot m \cdot d(\Delta\vartheta) + C_h \cdot S_{air} \cdot \Delta\vartheta dt, \tag{9}$$

where λ—thermal conductivity; C_h—heat convection coefficient; c—heat capacity; m—mass; S_{air}—heat transfer area. The steady-state maximum temperature of such a homogeneous body as ϑ_{hmax} is described by the following formula:

$$\vartheta_{hmax} = \frac{\Delta P_{Cu}}{C_h \cdot S_{air}} + \vartheta_0, \tag{10}$$

where $\vartheta_0 = 20$ °C (293.15 K) is the temperature of the surrounding air.

In the case of the considered magnetorheological clutches, the area marked in Figure 12 can be considered as the surface of heat transfers to the environment:

$$S_{air} = 2\pi \cdot \left((r_{ce} + Y')^2 - r_{bc}^2 \right) + 2\pi \cdot (r_{ce} + Y') \cdot (L_C + 2Y') \tag{11}$$

where r_{ci}—internal radius of the coil; r_{ce}—external radius of the coil; r_{bc}—external radius of the bearings coupling; Y'—cylindrical and cover yoke thickness.

It is the outer area of the cylindrical yoke and 2 outer areas of the cover yokes. How they are changed with the increasing number of discs and increasing clutching torque is shown in Figure 13a,b.

As observed in Figure 13a, the area A_{conv} decreases for all variants with the increase in the number of discs, but with a fixed number of discs (Figure 13b), the area for a higher torque clutch is larger than that for a lower torque clutch.

This clearly explains the problem posed at the beginning of the discussion: why, with the same number of discs, the maximum temperature of the higher-powered clutches is lower.

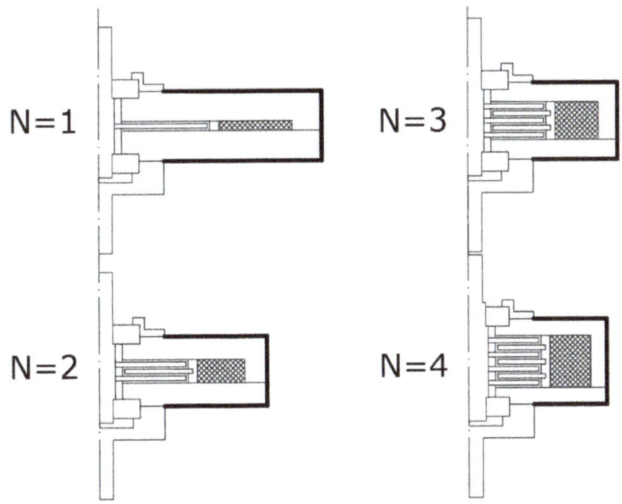

Figure 12. The surface of heat transfer to the environment (surrounding air).

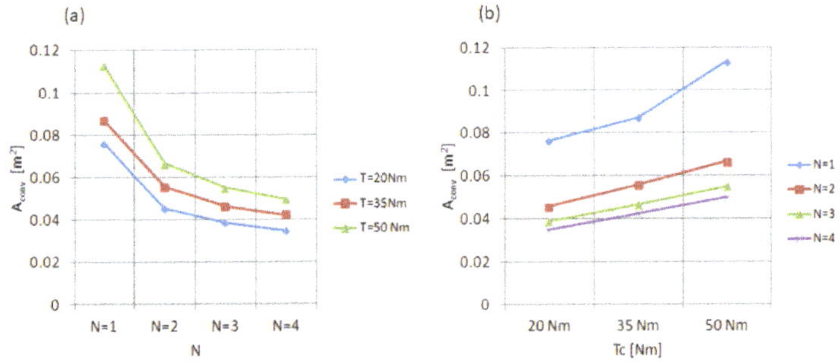

Figure 13. Heat convection areas: (**a**) areas A_{conv} vs. number of discs N for $T_c = 20$, 35 and 50 Nm; (**b**) areas A_{conv} vs. clutching torque T_c for the number of discs N = 1, 2, 3, 4.

This is evidently confirmed by Figure 14a, showing the course of the ratio $\Delta P_{Cu} / A_{conv}$ (see Equation (10)) as a function of the number of discs for clutches of different torques and which set of functions has the same form as the set of functions characterizing the maximum value of the MR fluid temperature shown again for direct comparison in Figure 14b.

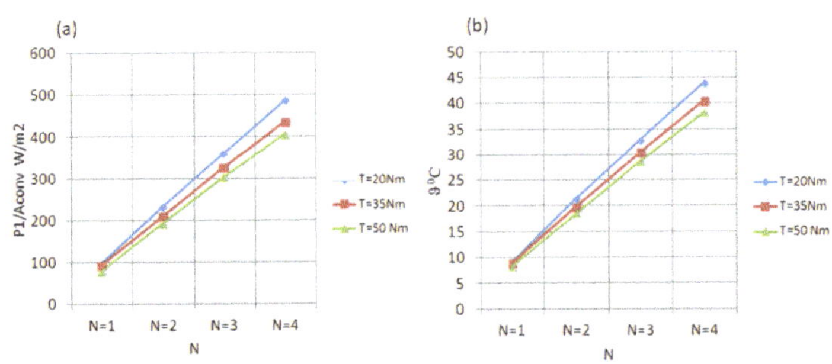

Figure 14. Comparison of function courses: (**a**) the course of ratio $\Delta P_{Cu} / A_{conv}$ vs. the number of discs for $T_c = 20$, 35 and 50 Nm; (**b**) maximum temperature ϑ_{hmax} vs. number of discs N for $T_c = 20$, 35 and 50 Nm.

Even better, the compliance of the curves describing the maximum temperature, T, with the curves presenting the ratio $\Delta P_{Cu} / A_{conv}$ can be seen in the 3D charts (Figure 15).

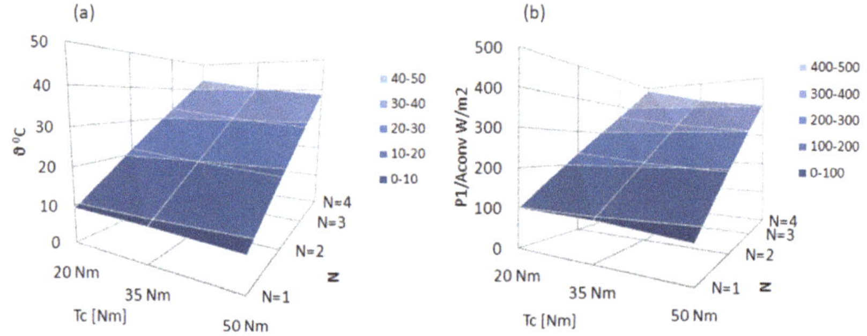

Figure 15. Three-dimensional charts: (**a**) maximum temperature vs. clutching torque and number of discs; (**b**) ratio $\Delta P_{Cu} / A_{conv}$ ratio vs. clutching torque and the number of discs.

The above-presented results of the thermal analysis allow one to confirm the conclusion given in [10,12] that the optimal recommended number of discs is equal to N = 2 and that it is a compromise option not only due to the values of characteristic factors (torque per mass ratio and torque per volume ratio) but also due to heat conditions. The operating temperature of the magnetorheological fluid is up to 130 °C [16]. Above this temperature, the properties of the liquid degrade and the magnetic properties are lost.

3. Results

The MR multi-disc clutches for industrial drive systems working in a steady continuous-duty state were compared from the viewpoint of electrical energy consumption and temperature spatial distribution. The electric power consumption P_{el}, excitation coil resistance R_c, power copper losses ΔP_{Cu}, spatial temperature distributions and the maximum temperature of the magnetorheological fluid ϑ_{hmax} were determined for twelve representative variants on the basis of field numerical calculations. It was explained in detail, in physical terms, why as the number of discs increases, the maximum temperature of the magnetorheological fluid increases too and also why, for the same number of discs in higher torque clutches, the temperature of the magnetorheological fluid is slightly lower than in smaller torque clutches. The general conclusion presented is that the most advantageous, recommended number of discs for a magnetorheological clutch from the viewpoint of electric, magnetic, mechanical and thermal criteria is the number of discs, N = 2. This conclusion takes into account the results of the analysis performed in an earlier article [10]: the choice is a compromise between the decreasing mass (volume) of the MR clutches and increasing both electrical power consumption and the maximum temperature of the MR fluid in the clutch working region as the number of discs N increases.

The authors focused on a steady continuous-duty state of industrial drive systems. The next step will be to use the presented approach based on computer field simulations combined with simplified physical reasoning to analyse transient thermal states and steady-states during slippage.

Author Contributions: Conceptualization, K.K. and Z.P.; methodology, K.K.; software, Z.P.; validation, Z.P. and K.K.; formal analysis, Z.P. and K.K.; writing—original draft preparation, Z.P.; writing—review and editing, Z.P.; visualization, Z.P.; supervision, K.K. All authors have read and agreed to the published version of the manuscript.

Funding: This research was conducted at the Faculty of Electrical and Computer Engineering, Cracow University of Technology. Funded by research funds for 2022.

Institutional Review Board Statement: Not applicable.

Informed Consent Statement: Not applicable.

Data Availability Statement: Not applicable.

Conflicts of Interest: The authors declare no conflict of interest. The funders had no role in the design of the study; in the collection, analyses, or interpretation of data; in the writing of the manuscript; or in the decision to publish the results.

Abbreviations

The following abbreviations are used in this manuscript:

MR clutches	Magnetorheological clutches

References

1. Pisetskiy, S.; Kermani, M.R. A Concept of a Miniaturized MR Clutch Utilizing MR Fluid in Squeeze Mode. In Proceedings of the IEEE/RSJ International Conference on Intelligent Robots and Systems (IROS), Las Vegas, NV, USA, 24 October 2020–24 January 2021; pp. 6347–6352. [CrossRef]
2. Xiong, H.; Luo, Y.; Ji, D.; Ren, H.; Weid, D.; Liu, W. Analysis and evaluation of temperature field and experiment for magnetorheological fluid testing devices. *Adv. Mech. Eng.* **2021**, *13*. [CrossRef]
3. Piech, Z., IV; Szeląg, W. Elevator Brake with Magneto-Rheological Fluid. US Patent, US 8,631,917 B2, 21 January 2014.
4. Kowol, P. Hamulce Magnetoreologiczne o Jednym i Dwóch Stopniach Swobody. (Magnetorheological Brakes with 1 or 2 DoF). Ph.D. Thesis, Silesian University of Technology, Gliwice, Poland, 2007. (In Polish)
5. East, W.; Turcotte, J.; Plante, J.; Julio, G. Experimental assessment of a linear actuator driven by magnetorheological clutches for automotive active suspensions. *J. Intell. Mater. Syst. Struct.* **2021**, *32*, 955–970. [CrossRef] [PubMed]
6. Wang, D.; Zi, B.; Zeng, Y.; Xie, F.; Hou, Y. An investigation of thermal characteristics of a liquid-cooled magnetorheological fluid-based clutch. *Smart Mater. Struct.* **2015**, *24*, 055020. [CrossRef]
7. Rahul, R.; Pugazhenthi, R.; Eugene, J.; Antony, A.; Viswanath, B. Mathematical Modelling of Temperature Rise in Clutch and Design, Analysis and Fabrication of Cooling System for Clutch. *Int. J. Sci. Eng. Res.* **2015**, *6*, 603–611.
8. Agyeman, P.K.; Tan, G.; Alex, F.J.; Peng, D.; Valiev, J.; Tang, J. Mathematical The study on thermal management of magnetorheological fluid retarder with thermoelectric cooling module. *Case Stud. Therm. Eng.* **2021**, *28*, 101686. [CrossRef]
9. Jędryczka, C.; Szeląg, W.; Myszkowski, A. Model of coupled electromagnetic, hydrodynamic, thermal and mechanical motion phenomena in axial symmetry magnetorheological fluid torque transducer. *Electr. Rev.* **2010**, *R86*, 195–200. (In Polish)
10. Kluszczyński, K.; Pilch, Z. The Choice of the Optimal Number of Discs in an MR Clutch from the Viewpoint of Different Criteria and Constraints. *Energies* **2021**, *14*, 6888. [CrossRef]
11. Kowol, P.; Pilch, Z. Analysis of the magnetorheological clutch working at full slip state. *Electr. Rev.* **2015**, *1*, 110–113. [CrossRef]
12. Kluszczyński, K.; Pilch, Z. Integrated analytical-field design method of multi-disc magnetorheological clutches for automotive applications. *Bull. Pol. Acad. Sci. Tech. Sci.* **2021**, *69*. [CrossRef]
13. Standard Specification for Standard Nominal Diameters and Cross-Sectional Areas of AWG Sizes of Solid Round Wires Used as Electrical Conductors. ASTM International. 2014. Available online: http://www.astm.org/Standards/B258.htm (accessed on 3 July 2022).
14. Pilch, Z. Analysis of Established Thermal Conditions for Magnetorheological Clutch for Different Loading Conditions. In *Analysis and Simulation of Electrical and Computer Systems*; Springer International Publishing: Berlin/Heidelberg, Germany, 2015; pp. 197–213. [CrossRef]
15. Available online: http://www.agros2d.org/ (accessed on 6 June 2022).
16. MRF-140CGMRFluid. Available online: https://lordfulfillment.com (accessed on 3 July 2022).

MDPI

A New Design Model of an MR Shock Absorber for Aircraft Landing Gear Systems Considering Major and Minor Pressure Losses: Experimental Validation

Byung-Hyuk Kang [1], Jai-Hyuk Hwang [2] and Seung-Bok Choi [3,*]

1 Department of Mechanical Engineering, Inha University, Incheon 22212, Korea; 1357op@gmail.com
2 School of Aerospace and Mechanical Engineering, Korea Aerospace University, Goyang 10540, Korea; jhhwang@kau.ac.kr
3 Department of Mechanical Engineering, The State University of New York, Korea (SUNY Korea), Incheon 21985, Korea
* Correspondence: seungbok.choi@sunykorea.ac.kr

Featured Application: Aircraft Shock Absorber with Controllable Damping Force.

Abstract: This work presents a novel design model of a magnetorheological (MR) fluid-based shock absorber (MR shock absorber in short) that can be applied to an aircraft landing gear system. When an external force acts on an MR shock absorber, pressure loss occurs at the flow path while resisting the fluid flow. During the flow motion, two pressure losses occur: the major loss, which is proportional to the flow rate, and the minor loss, which is proportional to the square of the flow rate. In general, when an MR shock absorber is designed for low stroke velocity systems such as an automotive suspension system, the consideration of the major loss only for the design model is well satisfied by experimental results. However, when an MR shock absorber is applied to dynamic systems that require high stroke velocity, such as aircraft landing gear systems, the minor loss effect becomes significant to the pressure drop. In this work, a new design model for an MR shock absorber, considering both the major and minor pressure losses, is proposed. After formulating a mathematical design model, a prototype of an MR shock absorber is manufactured based on the design parameters of a lightweight aircraft landing gear system. After establishing a drop test for the MR shock absorber, the results of the pressure drop versus stroke/stroke velocity are investigated at different impact energies. It is shown from comparative evaluation that the proposed design model agrees with the experiment much better than the model that considers only the major pressure loss.

Keywords: magnetorheological (MR) fluid; MR shock absorber; aircraft landing gear; valve path; major and minor pressure losses; impact energy

Citation: Kang, B.-H.; Hwang, J.-H.; Choi, S.-B. A New Design Model of an MR Shock Absorber for Aircraft Landing Gear Systems Considering Major and Minor Pressure Losses: Experimental Validation. *Appl. Sci.* **2021**, *11*, 7895. https://doi.org/10.3390/app11177895

Academic Editors: Antonio Concilio, Salvatore Ameduri, Ignazio Dimino, Vikram G. Kamble and Rosario Pecora

Received: 2 August 2021
Accepted: 24 August 2021
Published: 27 August 2021

1. Introduction

The landing gear system of an aircraft consists of the main landing gear and the nose gear. The nose gear is mainly located under the nose of the aircraft. The role of the nose gear is steering the aircraft during takeoff, landing, and taxiing and is not used for direct impact energy dissipation during landing. The main landing gear braces the aircraft's fuselage by dissipating most of the impact energy generated by the weight and sink speed of the aircraft during landing. For a stable landing, the shock absorbers of the main landing gear must be designed to dissipate sufficient shock energy. An oleo-pneumatic shock absorber (abbreviated as oleo-strut) is the most popular one used in modern aircraft landing gear systems. A structure called a metering pin is used inside the oleo-strut to vary the orifice area, and hence the damping force, in response to the piston displacement. Because the oleo-strut is the passive damper, its performance is limited. Therefore, the research on active and semi-active shock absorbers for the aircraft landing gear systems is being undertaken through several different approaches. A magnetorheological (MR) shock absorber is an attractive candidate to achieve optimal landing performance in almost all landing conditions. In general, a valve mechanism for the fluid flow is used for the MR shock absorber [1]. It is well known that MR fluid reacts to the magnetic field as a mixture of carbonyl-iron powder (CIP) and silicone oil. In the presence of the magnetic field, the in-fluid CIP particles are aligned in the direction of the magnetic field to form chain structures. The structure creates yield stress, resisting the flow of the MR fluid, which is a characteristic of the Bingham-plastic flow [2,3]. Thanks to the Bingham-plastic characteristics in response to the magnetic field, it is possible to achieve the desired damping force, resisting the fluid flow by controlling the magnetic field. Therefore, MR fluid is being actively used for several dynamic devices or systems that require a controllable damping force under different operating conditions. These systems include seismic dampers in civil structures, vehicle suspension systems, seat dampers, and lateral dampers for trains. In order to accurately predict the pressure drop (or damping force) of an MR shock absorber, several design models have been proposed so far [4–8].

One of the critical design parameters of MR shock absorbers is to determine the gap (or flow path) in which the flow resistance of the MR fluid is controlled by the magnetic core. In the flow path, two occur: the major pressure loss and the minor pressure loss. The major loss is caused by the fluid's dynamic viscosity generating a hydraulic resistance force in the gap. On the other hand, the minor loss is caused by inertia due to the density of the fluid generated in the inlet and outlet of the gap, which sharply bents the flow path and the developing region. The faster the operating speed of the MR shock absorber, the greater the effect of the minor loss because it is sensitive to the fluid velocity. In previous works on the design of MR shock absorbers, only the major pressure loss has been considered because the application systems are operated in the low operating speed range (or low stroke velocity), which is less than 0.5 m/s. Recently, several works on MR shock absorbers applicable to aircraft landing gear systems have been proposed. In some works [9,10], MR shock absorbers for aircraft landing gear systems were designed, analyzed, and experimentally validated under low-frequency conditions, and showed suitable damping force. An aircraft landing gear system with an MR shock absorber was designed, and the landing performance via the skyhook controller was evaluated in [11]. Robust adaptive control was proposed for the aircraft landing gear system equipped with an MR damper based on the adaptive hybrid control and sliding mode control algorithm to take account of the parameter uncertainties [12]. Moreover, the authors of this work formulated a full-scale aircraft model with 6-DOF dynamics integrated with an MR shock absorber and evaluated the landing performance through the modified skyhook controller and the inverse model of the mechanical energy [13,14]. However, only the major pressure loss was considered in these works.

Consequently, the technical contribution of this work is to propose a novel mathematical design model for an MR shock absorber, which can be applicable to high-speed operating systems, including aircraft landing gear systems, by considering both the major and minor pressures losses. As a first step to achieve this target, a structural configuration of an MR shock absorber for the aircraft landing gear system is presented, and its working principle is explained. The governing equations of motions are then derived, considering the pressure drop and damping force at compression and rebound conditions. Subsequently, a prototype of the MR shock absorber is designed and manufactured on the basis of the governing equations of motions and the design parameters of a commercial lightweight aircraft landing gear system. To validate the proposed design model, a drop test, which can generate a high stroke velocity of 2.5 m/s, is established, and the results on the pressure drop versus the stroke/stroke velocity are experimentally measured at various impact energy conditions representing different stroke velocities. The results achieved from the design models considering the major loss only, the major and minor loss, and the experiment are compared to validate an excellent agreement of the proposed design model with the experiment.

calculated,

(2)

(3)

matical Modeling for Aircraft MR Shock Absorber

'ng Force of MR Shock Absorber

1 shows the basic configuration of an aircraft landing gear system and a lumped he MR shock absorber. For MR shock absorbers, not only the pneumatic and rorce of the oleo-strut, but also the controllable MR force, are added. MR force can change the damping force depending on the input current generated by the controller, which is effective for landing performance. There are several different types of MR shock absorber that can be devised for landing gear systems. In this work, a cylindrical tube type, shown in Figure 2, is proposed.

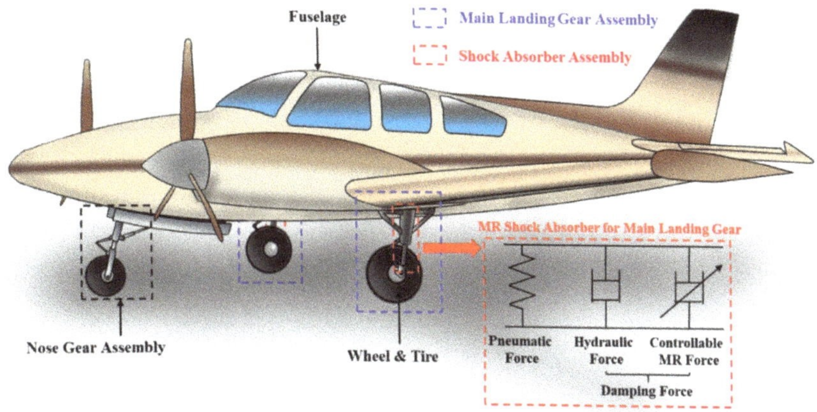

Figure 1. Configuration of aircraft landing gear and MR shock absorber.

The left of the diagram is the direction of the aircraft fuselage, and the right is the direction of the ground where the tire and wheel are located. The inside of the shock absorber is divided into an upper chamber, a lower chamber, and a gas chamber, with the MR valve and separator as boundaries. The upper and lower chambers are filled with MR fluid, and the gas chamber can be filled with air or nitrogen. Figure 2b shows a free-body diagram of the internal pressure and the external force acting on the MR shock absorber. From the figure, the main strut of the MR shock absorber is fixed to the aircraft fuselage, and the piston makes a relative motion depending on the stroke displacement, s, by the ground reaction. The external force in the direction of the piston axis is called a strut force, F_S. When the strut force is applied, the MR fluid flows through the MR valve based on the relative motion of the piston, and the damping force is generated to resist the strut force. In addition, the flow causes a change in the amount of fluid in the lower chamber, and the separator induces a pressure change in the gas chamber while making relative movements inside the piston to compensate for the changing fluid volume and the pneumatic pressure increases. The upper, lower, and gas chamber pressures are defined as P_1, P_2, and P_{gas}, respectively. The MR shock absorber is a form in which a damper and a spring are combined. In other words, the difference between P_1 and P_2 generates a damping force responsible for the damper's function, and P_{gas} generates the pneumatic pressure, which acts as the gas spring. The difference between P_1 and P_2 is the pressure drop, ΔP, which can be expressed as follows:

$$\Delta P = P_1 - P_2 \tag{1}$$

where P_1 and P_2 are the pressure of the upper and lower chambers, respectively. Using Equation (1) and the condition that P_{gas} is equal to P_2 in the quasi-equilibrium state, the strut force, F_s, the damping force, F_d, and the pneumatic force, F_{gas}, a respectively, as follows:

$$F_s = F_d + F_{gas}$$

$$F_d = \frac{\pi}{4} D_1{}^2 \cdot \Delta P$$

$$F_{gas} = \frac{\pi}{4} D_2{}^2 \cdot P_{gas}$$

where D_1 is the inner diameter of the main strut, D_2 is the outer diameter of the piston, and P_{gas} is the pneumatic pressure. Due to the volume ratio of the separator and the piston displacements, it is noted that the pneumatic force is expressed as the product of the pneumatic pressure and the cross-sectional area of the outer diameter of the piston, not the inner diameter. In this study, the description of the pneumatic pressure is omitted because the pressure drop for the major and minor losses will be focused on.

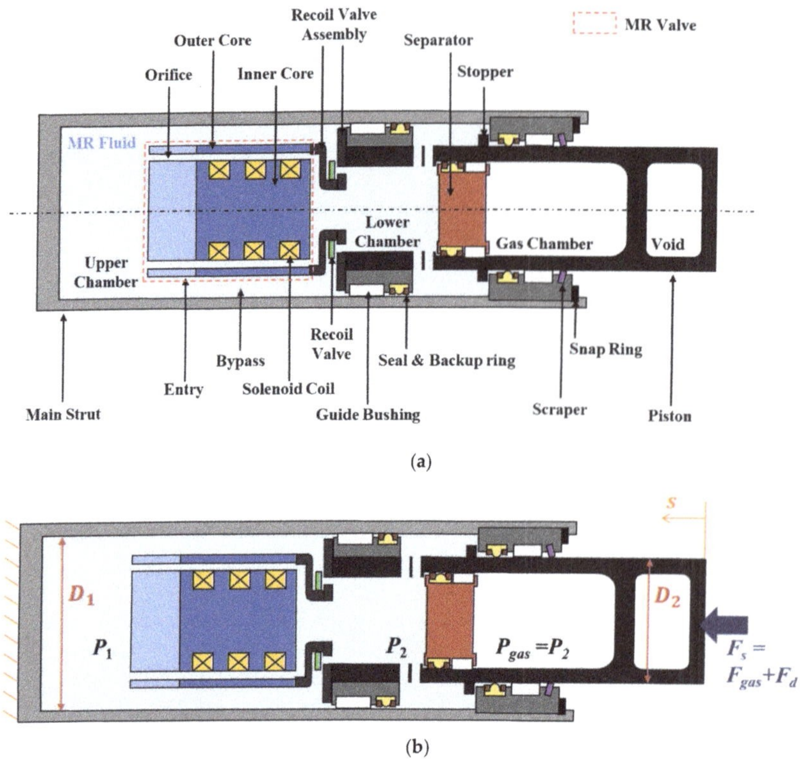

Figure 2. Schematic diagram of MR shock absorber: (**a**) configuration; (**b**) free body diagram.

2.2. MR Valve Design Considering Pressure Drop of Major and Minor Losses

Figure 3 shows the configuration of the MR valve, the magnetic field direction for the input current, and the pressure drop curves inside the orifice along the z-axis. MR valves are divided into entry and MR cores. The MR core is a type of electromagnet that causes the MR fluid to form a magnetic pole by the electric current input of the solenoid coil that is wound around the MR core. The region where the magnetic field path overlaps the MR fluid in the orifice is defined as the magnetic pole. Low carbon steel is mainly used for the MR core to form the path of a magnetic field well. In Figure 3, the state in which the current is applied to the coil is described as on-state, and the state with no current is described as off-state. The MR core designed in this work has three solenoid coils, and magnetic poles are placed in four locations. In Figure 2b, the pressure of P_1 and P_2 exist on both sides of the MR valve; the pressure in the region smaller than 0 is P_1 and the pressure in the region larger than L is P_2, based on the z-axis in the graph of Figure 3. The pressure curve for the z-axis in the figure represents the total pressure drop under the condition that the piston compresses, where ΔP_{major} and ΔP_{minor} are hydraulic pressure drops, considering the major and minor loss, respectively. ΔP_y indicates the pressure drop generated by the yield stress of the MR fluid. Such a pressure drop produces the flow rate of the MR fluid, and the flow rate, Q, is calculated as a function of the piston stroke velocity as follows:

$$Q = \frac{\pi}{4} {D_1}^2 \cdot \dot{s} \tag{5}$$

where $\dot{s}$ is the stroke velocity of the piston. The signs of the stroke velocity and flow rate are set to positive during piston compression and negative during piston rebound. For precise control of controllable force, the fluid must be of a laminar flow inside the MR valve, and the magnetic pole should be located in the fully developed region [4]. When operating at low piston velocity, it is common to design without considering the developing region in Figure 3. However, a general aircraft shock absorber has to dissipate the impact energy generated within 1 s at a sink speed (descent rate) of 3.05 m/s, so its operating speed must be rapid [15]. Therefore, when designing an MR shock absorber for the aircraft landing gear system, the auxiliary part must be installed to locate the core in the fully developed region. The name of this auxiliary part is 'Entry', provided in Figures 2a and 3. The Reynolds number in the annular orifice must be calculated in order to determine the entry length for the annular orifice. The Reynolds number, Re, for the annular pipe, taking into account the hydraulic diameter and wetted perimeter, is given by:

$$Re = \frac{2\rho \cdot Q}{\pi \mu \cdot D_o} \tag{6}$$

where D_o is the orifice mean circumference (diameter) and ρ and μ are density and dynamic viscosity of the MR fluid, respectively; the Reynolds number for the annular pipe must be less than 2000 for laminar flow [16]. In the case of laminar flow, the entry length, L_e, can be calculated using the following empirical formula [17]:

$$L_e = 0.0322 \max(Re) \cdot t_o \tag{7}$$

where t_o is orifice gap size. The length of the entry for the MR valve can be determined with the formula. The recoil valve in Figure 2a performs the same function as the check valve and implements asymmetry of the damping force during compression and rebound. Figure 4 shows the flow paths and the orifice and bypass velocity profiles under compressive and rebounding conditions. The recoil valve is closed by flow, only through the orifice during compression (Figure 4a). The recoil valve is opened by parallel flow through the orifice and bypass during rebound (Figure 4b). Figure 4c shows the velocity profile as the fluid flows through the orifice. Because the inner and outer cores do not move relative to each other, the velocity profile is in the form of an annular Poiseuille flow (APF). On the other hand, Figure 4d shows the velocity profile when the fluid flows through the bypass. The relative movement occurs between the inner wall of the main strut and the outer wall of the MR valve in response to the piston stroke, so that the velocity profile is in the form of the annular Poiseuille–Couette flow (APCF). The damping force calculation by APF and APCF is to be determined in Sections 2.3 and 2.4.

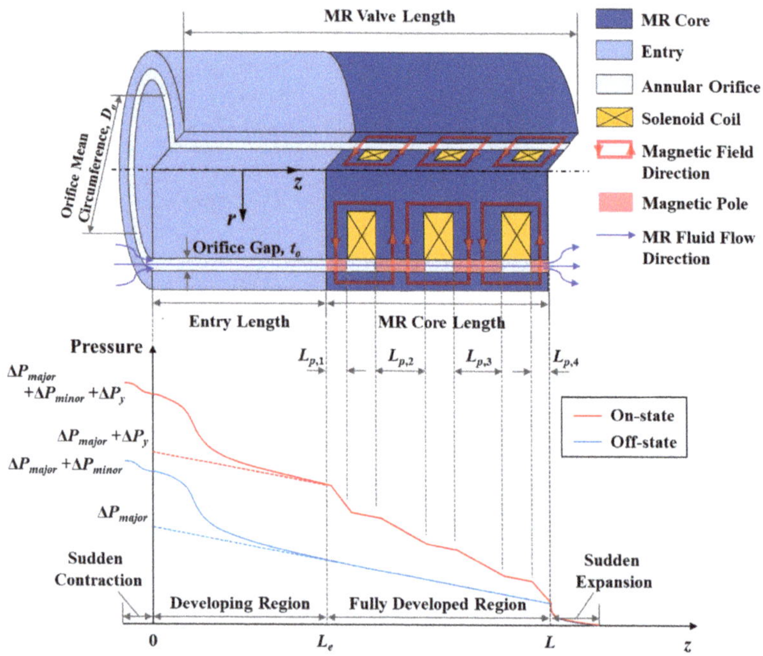

Figure 3. MR valve configuration and pressure drop in the annular orifice under compression condition.

2.3. Pressure Drop and Damping Force under Compression Conditions

When the piston is compressed, the fluid flows through the flow path, as shown in Figure 4a. Designed to close the recoil valve during piston compression, no fluid flows on the outer wall of the valve, including the bypass. That is, the pressure drop is generated by the velocity profile for off-state shown in Figure 4c because the fluid flows only through the orifice inside the valve. Derived from the Navier–Stokes equation, the pressure drops for the major and minor loss in the case of compression, ΔP_{major} and ΔP_{minor}, are given by:

$$\Delta P_{major} = C_1 \cdot Q_o \tag{8}$$

$$\Delta P_{minor} = C_2 \cdot {Q_o}^2 \tag{9}$$

where C_1 and C_2 are the major and minor pressure loss coefficients, respectively, and Q_o represents the flow rate into the orifice. Q_o is equal to Q under compressed conditions because the recoil valve is closed so that the fluid flows only through the orifice. The viscosity causes the major loss, and the inertia of MR fluid causes the minor loss due to the density of the fluid. Therefore, the minor loss cannot be ignored in a system with large fluid inertia such as an aircraft shock absorber, which has a high stroke velocity. C_1 and C_2 are calculated as follows:

$$C_1 = \frac{1}{8} f \cdot Re \frac{\mu \cdot L}{\pi D_o \cdot {t_o}^3} \tag{10}$$

$$C_2 = \frac{1}{2} \Sigma k_{o,c} \frac{\rho}{(\pi D_o \cdot t_o)^2} \tag{11}$$

where f is the friction factor for laminar flow; L is the orifice and bypass length of the MR valve; and $\Sigma k_{o,c}$ is the total loss coefficient, the sum of loss coefficients determined by the geometric shape of the flow path for orifice flow under compression conditions. The product of friction factor and Reynolds number, $f{\cdot}Re$, in Equation (10) can be calculated by the equation of geometry ratio for the annular orifice [18]:

$$f{\cdot}Re = \frac{128{t_o}^2}{{D_o}^2 + {t_o}^2 - 2D_o{\cdot}t_o/(\ln(D_o + t_o) - \ln(D_o - t_o))} \tag{12}$$

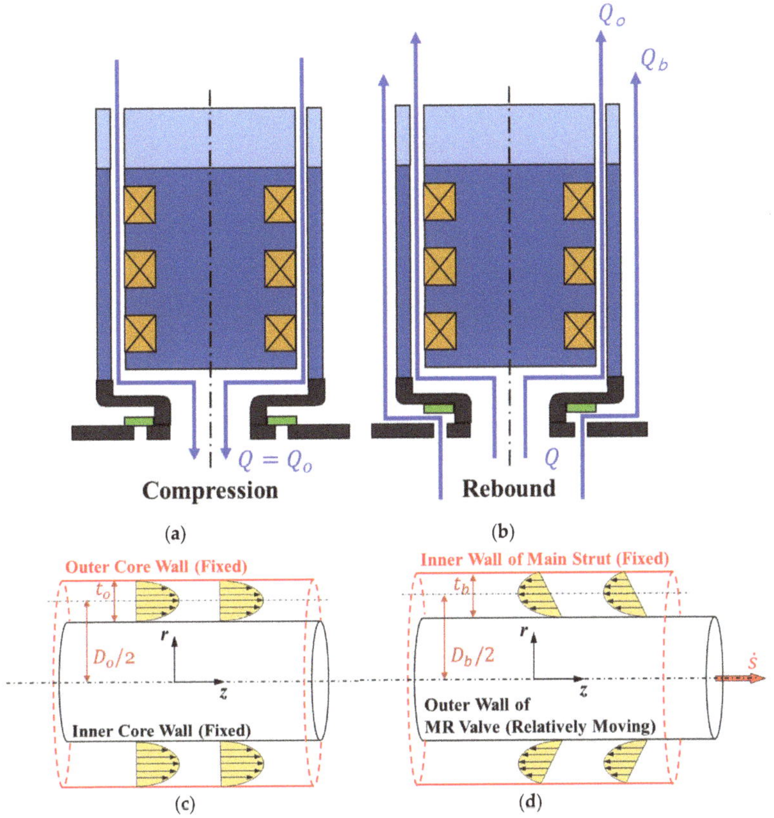

Figure 4. Fluid flow in compression and rebound conditions: (**a**) flow path in compression, valve closed; (**b**) flow path in rebound, valve open; (**c**) fluid velocity profile of the orifice: annular Poiseuille flow (APF); (**d**) fluid velocity profile of the bypass: annular Poiseuille–Couette flow (APCF).

As shown in Figure 3, the total loss coefficient from the MR valve is calculated by the sum of the loss coefficient of the sudden contraction, the developing region, and the sudden expansion. MR fluid can be modeled using the Bingham-plastic-viscous model. Therefore, in the off-state case, the pressure drop is originated by the dynamic viscosity and the initial yield stress of MR fluid. In the case of on-state, the magnetic poles generate an additional pressure drop due to the magnetic field, which is defined as the pressure drop due to yield stress. Assuming that the initial yield stress that acts without the magnetic field is infinitesimal, and the core is designed so that all magnetic fields in the magnetic poles act at similar levels depending on the input current, then the pressure drop for the yield stress, ΔP_y, can be expressed by the following simple equation:

$$\Delta P_y = \left(c(\overline{H}) \frac{\tau_y(\overline{H})}{t_o} \sum_j^k L_{p,j} + 3.07 \frac{\tau_0}{t_o} \left(L - \sum_j^k L_{p,j} \right) \right) \mathrm{sgn}(\dot{s}) \tag{13}$$

where τ_0 is the initial yield stress at zero magnetic field, characteristic of the Bingham-plastic-viscous model; $\tau_y(\overline{H})$ is the yield stress, a function of magnetic intensity; $\overline{H}$ represents the mean value of magnetic intensity at all magnetic poles; k is the total number of magnetic poles; and $L_{p,j}$ is the length of the j-th magnetic pole; 'sgn' denotes signum function. $c(\overline{H})$ was proposed by G. Yang et al., and the approximated function related to the orifice geometry, the flow rate, and the yield stress are as follows [4]:

$$c(\overline{H}) = 2.07 + \frac{12\mu \cdot |Q_o|}{12\mu \cdot |Q_o| + 0.4\pi D_o \cdot t_o{}^2 \cdot \tau_y(\overline{H})} \quad (14)$$

Because the stroke velocity is positively defined during piston compression, by Equations (8)–(14), the total pressure drop occurring at the orifice in the compression case, ΔP_c, can be summarized as follows:

$$\begin{aligned} \Delta P_c &= \Delta P_{major} + \Delta P_{minor} + \Delta P_y \\ &= C_1 \cdot Q + C_2 \cdot Q^2 + \left(c(\overline{H}) \frac{\tau_y(\overline{H})}{t_o} - 3.07 \frac{\tau_0}{t_o} \right) \sum_j^k L_{p,j} + 3.07 \frac{\tau_0}{t_o} L \end{aligned} \quad (15)$$

Furthermore, the damping force during piston compression, $F_{d,c}$, can be calculated as follows:

$$F_{d,c} = F_{hyd} + F_{MR} \quad (16)$$

where F_{hyd} and F_{MR} are, respectively, hydraulic force and controllable MR force during piston compression, and they can be expressed as follows:

$$F_{hyd} = \frac{\pi}{4} D_1{}^2 \cdot (\Delta P_{major} + \Delta P_{minor}) \quad (17)$$

$$F_{MR} = \frac{\pi}{4} D_1{}^2 \cdot \Delta P_y \quad (18)$$

2.4. Pressure Drop and Damping Force under Rebound Conditions

In the first piston compression during landing, most of the impact energy is dissipated in the shock absorber. Therefore, the hydraulic force during rebound can ignore the minor loss term because the stroke speed is slower than that during piston compression. However, in this work, the effect of minor loss, even in rebound cases, is analyzed. A lower damping force is required for safe landing, which is calculated by comparing the compression case to achieve a high-speed stroke during rebound. The bypass flow path is constructed to implement the lower damping force so that the fluid can flow in parallel, as shown in Figure 4b. When the MR shock absorber rebounds, the parallel flow occurs. Because the recoil valve is opened and fluid flows simultaneously through the orifice and bypass, the flow velocity profiles at the orifice in Figure 4c and bypass in Figure 4d generate the pressure drop. Based on Equations (8)–(14), during the piston rebound, the total pressure drop, $\Delta P_{o,r}$, acting from the orifice by hydraulics and yields is as follows:

$$\Delta P_{o,r} = C_1 \cdot Q_o - C_3 \cdot Q_o{}^2 - 3.07 \left(\frac{\tau_y(\overline{H}) - \tau_0}{t_o} \sum_j^k L_{p,j} + \frac{\tau_0}{t_o} L \right) \quad (19)$$

where C_3 is the minor pressure loss coefficient for orifice flow, expressed similarly to Equation (11), as follows:

$$C_3 = \frac{1}{2} \Sigma k_{o,r} \frac{\rho}{(\pi D_o \cdot t_o)^2} \quad (20)$$

where $\Sigma k_{o,r}$ is the total loss coefficient for the orifice flow under rebound conditions. Because the magnetic field does not act on bypass, the total pressure drop, $\Delta P_{b,r}$ from the bypass can be expressed by considering only the pressure loss of hydraulic and initial yield stress:

$$\Delta P_{b,r} = C_4 \cdot (Q - Q_o) - C_5 \cdot (Q - Q_o)^2 - 3.07 \frac{\tau_0}{t_b} L \tag{21}$$

where t_b is the bypass gap size; C_4 and C_5 are, respectively, the major and minor pressure loss coefficients for APCF in the bypass. Because MR fluid flows parallel through the orifice and bypass, the coefficients of the major and minor pressure loss of the bypass must be considered. Additionally, APCF at the bypass must be considered because APF and annular Couette flow are formed simultaneously by the relative motion of the stroke from the inner wall of the main strut and MR valve. Figure 4d illustrates the APCF well. The coefficients of major and minor pressure loss for the APCF, C_4, and C_5, are as follows:

$$C_4 = \frac{12\mu \cdot L}{\pi D_b \cdot t_b{}^3} - \frac{24\mu \cdot L}{(\pi D_1 \cdot t_b)^2} \tag{22}$$

$$C_5 = \frac{1}{2} \Sigma k_{b,r} \frac{\rho}{(\pi D_b \cdot t_b)^2} \tag{23}$$

where D_b is the bypass mean circumference and $\Sigma k_{b,r}$ is the total loss coefficient for bypass flow under rebound conditions. The pressure drop generated in the orifice and bypass during rebound is the same; Equation (19) is equal to Equation (21). Through some mathematical expansion, the flow rate of the orifice, $Q_{o,r}$, and the total pressure drop, ΔP_r for piston rebound are calculated as:

$$Q_{o,r} = \begin{cases} a_1 + \sqrt{a_1{}^2 + a_2} \operatorname{sgn}(C_5 - C_3) & , a_2 < 0 \\ 0 & , a_2 \geq 0 \end{cases} \tag{24}$$

$$\Delta P_r = C_4 (Q - Q_{o,r}) - C_5 (Q - Q_{o,r})^2 - 3.07 \frac{\tau_0}{t_b} L \tag{25}$$

In the above, the variables a_1 and a_2 are used to simplify the expression and are defined as follows:

$$a_1 := \frac{2C_5 \cdot Q - C_1 - C_4}{2(C_5 - C_3)} \tag{26}$$

$$a_2 := \frac{C_4 \cdot Q - C_5 \cdot Q^2}{C_5 - C_3} + \frac{3.07}{C_5 - C_3} \left(\frac{\tau_y(\overline{H}) - \tau_0}{t_o} \sum_j^k L_{p,j} + \frac{t_b - t_o}{t_b \cdot t_o} \tau_0 \cdot L \right) \tag{27}$$

If the pressure drop caused by the flow rate is smaller than the pressure drop caused by the yield stress, depending on the magnetic field, a block-up phenomenon occurs in the orifice [4]. If this occurs, MR fluid does not flow through the orifice and can only flow through the bypass. Equation (24) well represents the corresponding phenomenon. If only the major loss is considered in calculating the pressure drop in the rebound case, the quadratic term of the flow rate disappears. Based on Equations (19) and (21), the flow rate of the orifice, $Q_{o,maj}$, and the total pressure drop, $\Delta P_{r,maj}$, considering only the major loss for the piston rebound, can be calculated:

$$Q_{o,maj} = \begin{cases} \frac{C_4}{C_1 + C_4} Q + \frac{3.07}{C_1 + C_4} \left(\frac{\tau_y(\overline{H}) - \tau_0}{t_o} \sum_j^k L_{p,j} + \frac{t_b - t_o}{t_b \cdot t_o} \tau_0 \cdot L \right) & , Q_{o,maj} < 0 \\ 0 & , Q_{o,maj} \geq 0 \end{cases} \tag{28}$$

$$\Delta P_{r,maj} = C_4 \cdot (Q - Q_{o,maj}) - 3.07 \frac{\tau_0}{t_b} L \tag{29}$$

The damping force, $F_{d,r}$, taking into account hydraulics and yield stress in the rebound condition, can be calculated as follows:

$$F_{d,r} = \frac{\pi}{4} {D_1}^2 \cdot \Delta P_r \quad (30)$$

2.5. *Total Pressure Drop and Damping Force*

The total pressure drop, ΔP_{total}, and the damping force, F_d, generated in the MR shock absorber, considering both the major and minor losses, are as follows:

$$\Delta P_{total} = \begin{cases} \Delta P_c & Q \geq 0 \\ \Delta P_r & Q < 0 \end{cases} \quad (31)$$

$$F_d = \frac{\pi}{4} {D_1}^2 \cdot \Delta P_{total} = \begin{cases} F_{d,c} & \dot{s} \geq 0 \\ F_{d,r} & \dot{s} < 0 \end{cases} \quad (32)$$

Moreover, the pressure drop, ΔP_{maj}, and damping force, $F_{d,maj}$, generated in the MR shock absorber, considering only the major loss, are as follows:

$$\Delta P_{maj} = \begin{cases} \Delta P_{major} + \Delta P_y & Q \geq 0 \\ \Delta P_{r,maj} & Q < 0 \end{cases} \quad (33)$$

$$F_{d,maj} = \frac{\pi}{4} {D_1}^2 \cdot \Delta P_{maj} \quad (34)$$

3. Design Parameters

3.1. *Characteristic Evaluation of MR Fluid*

The damping force of the aircraft shock absorber utilizing MR fluid is sensitive, not only to the density of the fluid, but also to the yield stress for the magnetic field and the dynamic viscosity. Therefore, the dynamic viscosity and the yield stress for the MR fluid are measured through an MR viscometer. The density is calculated by measuring the mass and volume and is 3.510 g/cm^3 for the MRF-140CG fluid from LORD Corporation. As a result of measuring via MR viscometer in a temperature environment of 23 degree Celsius, the dynamic viscosity and the initial shear stress of MRF-140CG is estimated to be 0.290 $Pa \cdot s$ and 166.3 Pa, respectively. The yield stress, depending on the magnetic intensity measured by MR viscometer, is fitted to the following polynomial expression:

$$\tau_y(\overline{H}) = -0.005\overline{H}^3 + 0.900\overline{H}^2 + 298.5\overline{H} + 166.3 \quad (35)$$

where the unit of the shear stress and magnetic intensity are Pa and kA/m, respectively.

3.2. *Magnetic Analysis for MR Core*

Magnetic analysis is performed using the ANSYS MAXWELL program. It refers to the magnetic flux density-magnetic intensity curve (BH curve) required for the magnetic analysis of the LORD Corporation's datasheet for MRF-140CG [19]. The corresponding BH curve is used for each material, and the materials of the MR valve and the recoil valve are as follows: inner and outer cores are AISI1008 steel; the entry is aluminum7075-T651; the recoil valve parts are sus304 and aluminum7075-T651; and the coil uses polyester enameled copper wire (PEW) of AWG27 standard. The magnetic analysis conditions are set as follows: the solution type is magnetic transient; the basic mesh number of the orifice is 15; eddy effects and core loss are activated; and nonlinear residual is set as 10^{-6}. The magnetic intensity for MRF-140CG, obtained by the magnetic analysis, is curve-fitted, and the averaged magnetic intensity is determined as a polynomial equation, as follows:

$$\overline{H} = -6.057I^3 - 0.776I^2 + 57.29I \quad (36)$$

where I is the input current in Ampere units, and the unit of magnetic intensity is kA/m.

3.3. Minor Pressure Loss Analysis with CFD

It is almost impossible to theoretically calculate the minor pressure coefficients and loss coefficients in Equations (11), (20), and (23). Therefore, to identify the hydraulic resistance, such as the loss coefficient at the design process, direct experiments can be performed, or it is possible to refer to guidebooks experimentally calculated by many researchers for the duct and pipe flow specifications, as they have specific arbitrary cross-sections [20]. However, the loss coefficients for many of the annular MR valves and other parts designed in this study cannot be found in guidebooks. It is also impossible to determine losses experimentally in the design process. Consequently, in this work, the designed model's loss coefficients are estimated using the Fluent program of the ANSYS computational fluid dynamics (CFD) tool. The CFD analysis conditions are set as follows: the solver type is steady-pressure based; the viscous model is a realizable k-epsilon model with model constraints for C2-epsilon, TKE Prandtl number, and TDR Prandtl number set at 1.9, 1, and 1.2, respectively; basic mesh number of the orifice and the bypass are 15; the orifice and the bypass are set as the laminar zone; the solution methods are simple pressure-velocity coupling, least square cell based gradient, standard pressure, second order upwind momentum, first order upwind turbulent kinetic energy, and first order upwind dissipation rate; the residual of absolute criteria is set as 10^{-3}; the piston stroke speed when the piston is compressed is 2.50 m/s; and the piston stroke speed when the piston is rebounded is 1.33 m/s.

Using Equation (7), the maximum Reynolds number and entry length are calculated at 1113 and 87.8 mm when the sink speed is 3.05 m/s and the input current is not applied. At the same time, it is confirmed that the entry length is about 85 mm as a result of the CFD analysis, as shown in Figure 5, performed under the same conditions. Based on Equations (11) and (23), the pressure drop calculated by CFD includes both major and minor losses, so the total loss coefficients at the orifice and bypass, Σk_o and Σk_b, can be estimated as follows:

$$\Sigma k_o = 2\frac{(\pi D_o \cdot t_o)^2}{\rho}\frac{(\Delta P_{CFD} - C_1 \cdot Q_o)}{{Q_o}^2} \tag{37}$$

$$\Sigma k_b = 2\frac{(\pi D_b \cdot t_b)^2}{\rho}\frac{(\Delta P_{CFD} - C_4(Q - Q_o))}{(Q - Q_o)^2} \tag{38}$$

where ΔP_{CFD} is the CFD result of the pressure drop. Based on the CFD result in Figure 5a, under compression conditions, the entry length and total loss coefficient are set to 85 mm and 2.779, respectively. Through Equations (6) and (7), the Reynolds number and entry lengths of the orifice are calculated under the harshest rebound conditions. The Reynolds number and the entry length of the orifice are 195.31 and 15.41 mm, respectively. The entry length is 11.85% of the total MR valve length. Calculated based on equations in [21], the Reynolds number and the entry length of the bypass are 136.93 and 11.2 mm, respectively. The length of the entry is 8.62% of the total MR valve length. Additionally, under rebound conditions, there is no tendency for additional minor pressure drops in the entry region from Figure 5b. In conclusion, it is assumed that the entry length is short compared to the length of the total MR valve at the time of the rebound. Therefore, there is no need to consider the entry regions for the orifice and the bypass under rebound conditions. The total loss coefficients of the orifice and the bypass are calculated regarding Figure 5b and Equations (37) and (38). The flow rate analysis results of the orifice and the bypass by the CFD analysis to calculate the total loss coefficient are as follows: the flow rates of the orifice and the bypass are 1.886 cm^3/s and 1.508 cm^3/s, respectively, and the total loss coefficients of the orifice and the bypass are 3.233 and 14.25, respectively.

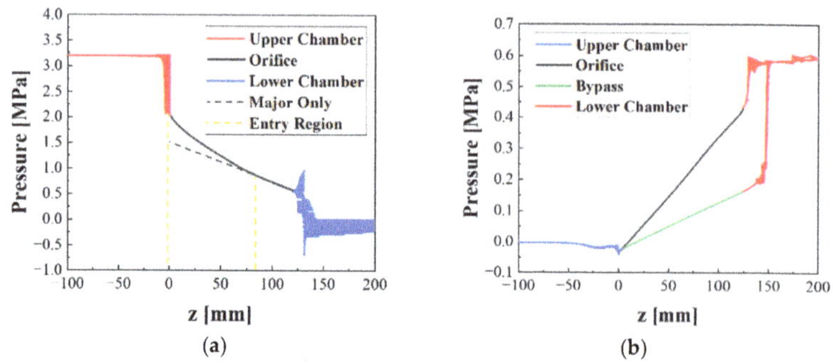

Figure 5. Pressure result calculated with CFD tool, ANSYS Fluent: (**a**) pressure change along the z-axis, compression condition, flow through the orifice, and comparison with major pressure loss; (**b**) pressure change along the z-axis, rebound condition, flow through orifice, and bypass simultaneously.

3.4. Design Parameters for MR Shock Absorber

In this work, the design parameters of an MR shock absorber are chosen on the basis of the lightweight landing gear system of a Beechcraft Baron B55 by considering the weight of 230 kg for each main landing gear. It is noted here that the original weight of the Beechcraft Baron B55 is 680 kg for each main landing gear. Hence, the design parameters of the prototype are appropriately adjusted, and the pressure drops are confirmed using the proposed design model. Table 1 provides the detailed design specifications of the MR shock absorber designed and manufactured in this work. It is noted that the optimization is performed on the design variables for the MR core to set the parameters using the equations in [22]. Based on the magnetic field analysis results and design parameters, the controllable forces are calculated as 0.9025 kN, 1.661 kN, and 2.095 at the input current of 0.5 A, 1.0 A, and 1.5 A, respectively, based on a stroke velocity of 2 m/s.

Table 1. Specifications of MR Shock absorber.

Parameter	Symbol	Value	Unit
Orifice diameter	D_o	43.75	mm
Orifice gap	t_o	2.450	mm
Bypass diameter	D_b	54.60	mm
Bypass gap	t_b	2.400	mm
Length of the orifice and bypass	L	130.0	mm
Entry length	L_e	85.00	mm
Length of the pole 1	$L_{p,1}$	5.500	mm
Length of the pole 2	$L_{p,2}$	11.00	mm
Length of the pole 3	$L_{p,3}$	11.00	mm
Length of the pole 4	$L_{p,4}$	5.500	mm
Wire diameter of the solenoid coil	d_{coil}	0.361	mm
Inner diameter of the solenoid coil	D_{coil}	20.00	mm
Height of the solenoid coil	h_{coil}	4.000	mm
Number of turns of the solenoid coil	N_{turn}	300	-
Inner diameter of the main strut	D_1	57.00	mm
Outer diameter of the piston	D_2	50.70	mm
Viscosity of MR fluid	μ	0.290	Pa·s
Density of MR fluid	ρ	3.510	g/cm^3
Initial yield stress of MR fluid	τ_0	166.3	Pa
Product of friction factor and Reynolds number for the orifice	$f \cdot Re$	95.98	-
Total loss coefficient for the orifice, compression	$\Sigma k_{o,c}$	2.779	-
Total loss coefficient for the orifice, rebound	$\Sigma k_{o,r}$	3.233	-
Total loss coefficient for the bypass, rebound	$\Sigma k_{b,r}$	14.25	-

4. Experimental Validation

4.1. Experimental Apparatus

The configuration of the drop device, the type of sensors, the DAQ device, and the model names used in the experiment are shown in Figure 6. The sprung and un-sprung masses of the constructed drop device are 230 kg and 15 kg, respectively. Experiments are categorized as follows: off-state pressure measurement with impact energy and pressure measurement corresponding to the input current when impact energy is fixed. The impact energy (*IE*) can be calculated through the sink speed at the moment of the touchdown of the un-sprung mass and the distance to the static equilibrium state:

$$IE = (m_s + m_u)(g \cdot z_{1,eq} + 0.5\dot{z}_1) - m_u \cdot g \cdot s_{eq} \tag{39}$$

where m_s and m_u are sprung and un-sprung masses, respectively; g is the gravity acceleration and is set to 9.806; $z_{1,eq}$ and s_{eq} are sprung mass and piston stroke displacements from the moment of touchdown to the static equilibrium state; $\dot{z}_1$ is the sink speed of the landing gear system at the moment of touchdown. The sink speed is estimated through the displacement and the acceleration of the sprung mass, measured with a laser sensor and an acceleration sensor attached to the sprung mass. In other words, the impact energy is converted into the form of the energy dissipated by the pressure loss in the MR shock absorber after impact, the energy dissipated by the friction between the MR shock absorber and the guide shaft of the drop device, and the elastic potential energy stored in the gas chamber and the tire as a spring energy form. This work compared the dissipated energy due to the pressure loss in the MR shock absorber estimated experimentally and the pressure drop model. ΔP_{total} and ΔP_{maj} are calculated for the experimentally estimated piston stroke velocities to validate the mathematically modeled pressure drop model. The piston stroke velocity is estimated through stroke displacement, and acceleration is measured via the wire sensor and accelerometers mounted at the sprung and un-sprung masses. ΔP_{total} and ΔP_{maj} can be calculated by Equations (31) and (33). In this paper, the pressure drop model, calculated by estimated piston stroke velocity, is called the $\Delta P{-}\dot{s}$ model. The pressure drop model is constructed in MATLAB, a numerical analysis program. Because the main content of this paper is a comparison of pressure drop in conformity with the presence or absence of the minor loss, the simulation on the landing efficiency using the landing gear model is omitted.

4.2. Comparison of Pressure Drop in Conformity with Impact Energy under Off-State

In this section, the drop test is performed by mounting the aircraft landing gear with the MR shock absorber onto the drop device to analyze the internal pressure response in conformity with impact energy under off-state conditions. The impact energy is applied and tested at 456.7 J, 937.5 J, and 1538 J. Figure 7 presents the inside pressure drop and estimated damping force of the MR shock absorber for, respectively, stroke displacement and velocity when the landing gear system is freely dropped. ΔP_{total} and ΔP_{maj} on the legend are the pressure drop models calculated using Equations (31) and (33) via the stroke velocity estimated through their respective experiments. Experimental data on the legend refers to the pressure drop measured by the actual measured upper and lower chamber pressures. The initial stroke displacement and velocity start from the origin of the coordinates, as shown in Figure 7. As the compression process progresses, the stroke displacement and velocity, pressure drop, and estimated damping force increase simultaneously. In Figure 7a,c,e, the response of the ΔP_{maj} model increases in error compared with the ΔP_{total} model as the impact energy increases. To compare the differences, the results of the dissipated energy for the pressure drop models are presented in Table 2. The difference between the impact energy and the dissipated energy comes from the elastic potential energy conserved through the pneumatic chamber and the tire. In Table 2, as the impact energy increases, the relative error of the dissipated energy is within 15% for the ΔP_{total} model. However, in the ΔP_{maj} model, the relative error gradually increases up to 54%. In

Figure 7b,d,f, as the stroke velocity increases, the difference between the ΔP_{total} model and the experiment does not increase, whereas the error from the experimental result increases compared to the ΔP_{maj} model; the difference in the slope due to the quadratic term of the flow rate increases as the stroke velocity increases. Table 2 also shows the error results numerically. In addition, Figure 7e,f is suitable for comparing the error of the pressure drop models with a maximum rebound speed of about 0.6 m/s after the first compression; the root mean square (RMS) of errors and relative errors for the experiment and pressure drop models are compared using data with a stroke velocity range of −0.1 m/s or less. For the ΔP_{total} model, the RMS values for the error and relative error with the experimental data are 35.69 kPa and 37.95%, respectively; for the ΔP_{maj} model, they are 16.22 kPa and 28.92%, respectively. The characteristics of the minor loss are not well displayed during rebound, which resulted in small RMS values of the ΔP_{maj} model. In other words, the pressure drop model can be simplified as Equations (28) and (29) by considering only the major loss at the time of rebound. In Figure 7e,f, the maximum stroke velocity during the second compression is about 0.5 m/s, which is suitable for comparing the error for the pressure drop model with low-speed compression. The RMS values of the error and relative error for the experiment and pressure drop models are compared using data within the stroke velocity range from 0.1 to 0.5 m/s. For the ΔP_{total} model, the RMS values for the error and relative error with the experimental data are 10.94 kPa and 4.507%, respectively; for the ΔP_{maj} model, they are 45.07 kPa and 15.02%, respectively. Furthermore, the dynamic viscosity and initial yield stress are estimated to be 0.2882 Pa·s and 230.3 Pa as a result of performing nonlinear least square parameter estimation based on Equation (15) using the experimental data, where the estimated values are similar to those measured with the MR viscometer. Because the plastic-viscosity model used in the mathematical modeling is constructed based on the measurement data of the MR viscometer, the error level is generally uniform in the low-stroke velocity interval of 0.5 m/s or less. In other words, the design formula for the general MR damper considering only the major loss may be used within an error range of 15% or less in the application utilized in the experiment under the stroke velocity of 0.5 m/s. The maximum pressure drop and the damping force at the maximum compression velocity are compared between the experimental results and the pressure drop models. The RMS of the maximum pressure drop error and relative error are calculated and compared. For the ΔP_{total} model at the stroke velocity of 0.5 m/s, the RMS values for the error and relative error with the experimental data are 15.49 kPa and 4.104%, respectively, and for the ΔP_{maj} model they are 54.51 kPa and 14.44%, respectively. Values are 35.13 kPa, 4.997%, 173.1 kPa, and 24.62% in sequence in Figure 7b; 50.79 kPa, 2.412%, 1011 kPa, and 48.01% in sequence in Figure 7d; and 426.6 kPa, 12.41%, 2.032 kPa, and 59.12% in sequence in Figure 7f. Therefore, it is necessary to consider the minor loss as the maximum stroke velocity increases in order to reduce the error.

Table 2. Dissipated energy in MR shock absorber for impact energy; RMS values of error for model-based and measured pressure data over stroke velocity range of 0.1 m/s or more.

Impact Energy (J)	Dissipated Energy					Pressure Drop			
	Experiment (J)	$\Delta P-\dot{s}$ Model (J)		Relative Error (%)		RMS Error (kPa)		RMS Relative Error (%)	
		ΔP_{total}	ΔP_{maj}	ΔP_{total}	ΔP_{maj}	ΔP_{total}	ΔP_{maj}	ΔP_{total}	ΔP_{maj}
456.7	276.6	263.5	200.6	4.748	27.46	16.72	109.2	5.650	19.37
937.5	737.2	672.5	406.1	8.781	44.92	70.28	419.1	13.54	28.07
1538	1339	1149	621.2	14.18	53.61	189.5	735.8	11.66	31.72

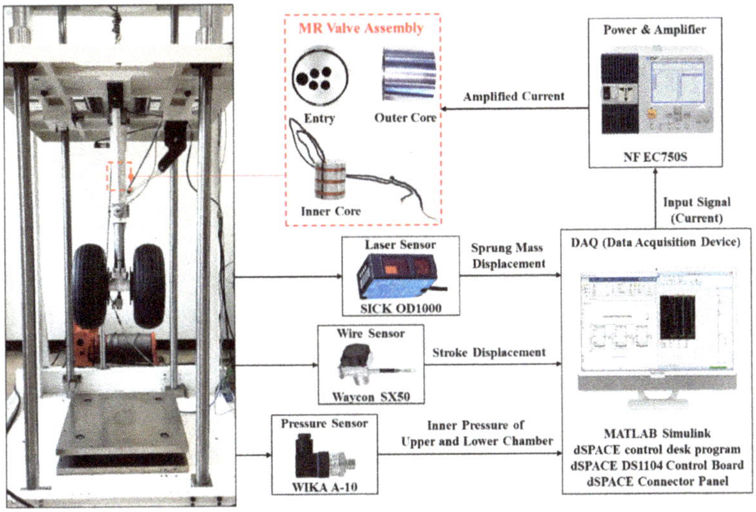

Figure 6. Vertical drop test device for performance evaluation of MR shock absorber.

4.3. Comparison of Pressure Drop and Landing Efficiency in Conformity with Input Current and Fixed Drop Height

In this section, the response of the MR shock absorber depending on the input current is analyzed experimentally with the impact energy fixed. The input current is applied 10 s before the drop to exclude a time delay with respect to the input current. The impact energy during touchdown is the same at 1538 J. Figure 8 presents the internal pressure drop of the MR shock absorber in conformity with the input current for the stroke displacement and velocity. As in the case of Figure 7, ΔP_{total} and ΔP_{maj} on the legend of Figure 8 are the pressure drop calculated for the estimated stroke velocity. 'Experiment' on the legend is the pressure drop measured through the pressure sensors during the drop test. When the impact energy is constant, the pressure drop in all stroke displacement ranges of both the ΔP_{total} and ΔP_{maj} models increases as the input current increases. As shown in Figure 8a,c,e, the maximum stroke displacement decreases as the input current increases due to the energy dissipated by the additional pressure drop generated by the yield stress. The dissipated energy is calculated in Table 3 for comparison with the input current. As the input current increases, the dissipated energy for the measured data and the ΔP_{total} model does not change significantly. Because the impact energy does not change, the potential and kinetic energies do not change, and the dissipated energy also does not. However, the dissipated energy increases as the input current increases in the case of the ΔP_{maj} model. The reason for this is that the area between the ΔP_{total} and ΔP_{maj} curve for the stroke decreases as the dissipated energy for the yield stress increases as the input current increases, as shown in Figure 8a,c,e. Regardless of all the input currents in Figure 8b,d,f, the ΔP_{total} model does not show a significant difference from the experimental result as the stroke velocity increases. On the other hand, the error between the ΔP_{maj} model-based pressure drop and the experimental result increase depending on the stroke velocity increase. In Table 3, the RMS values of the error and relative error are calculated over the stroke velocity range of 0.1 m/s or more. The input current and the pressure drop models are independent; the RMS errors and relative errors in Table 3 are due to the error of the magnetic analysis. As a result of the experiment, the controllable force is estimated at 0.6137 kN, 1.382 kN, and 2.128 kN for 0.5 A, 1.0 A, and 1.5 A, respectively, based on the stroke velocity of 2 m/s. The maximum compression velocity estimated by experiments from 0 A to 1.5 A in 0.5 A units is 2.390 m/s, 2.352 m/s, 2.359 m/s, and 2.338 m/s. The error in the estimated velocity occurred due to the friction force of the drop device and MR shock absorber, but it tends to decrease as the input current increases. The RMS values of the maximum pressure drop error and relative error are compared between the pressure drop model and the experimentally measured data. For the input current, the RMS values of the error and relative error between the experimental data and the ΔP_{total} model and the RMS values of the error and relative error between the experimental data and the ΔP_{maj} model are as follows, in sequence, respectively: 0.3399 MPa, 9.484%, 1.889 MPa, and 52.73% for 0.5 A; 0.3052 MPa, 7.922%, 1.863 MPa, and 48.37% for 1.0 A; and 0.3666 MPa, 9.047%, 1.905 MPa, and 47.03% for 1.5 A.

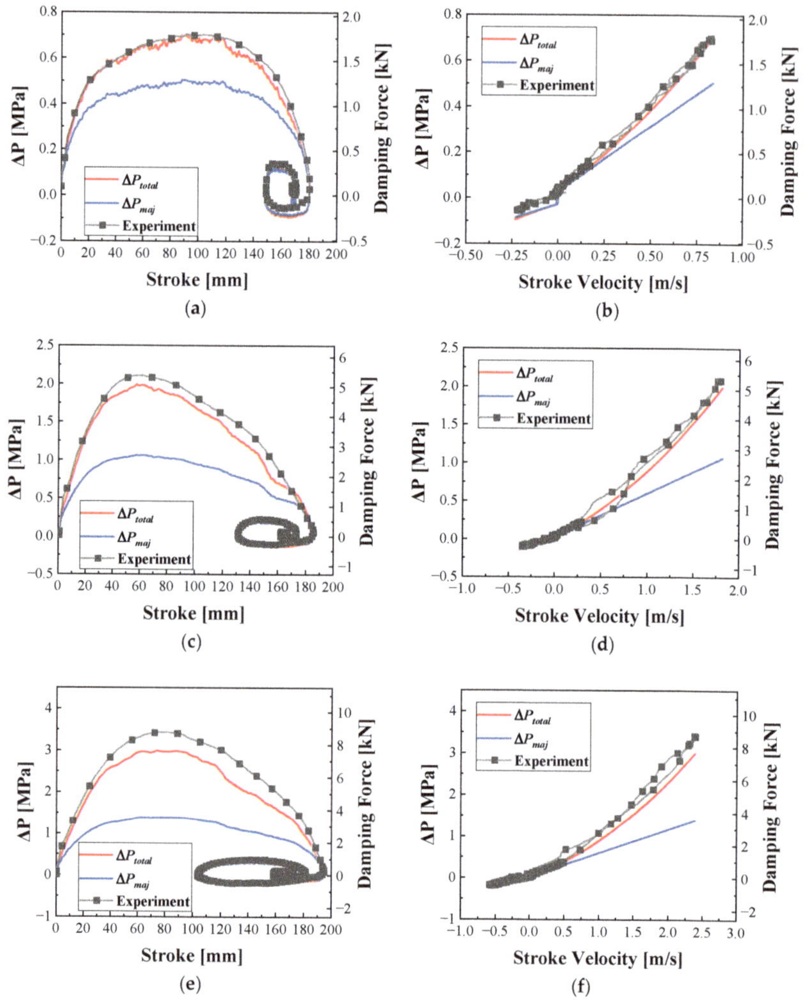

Figure 7. Pressure drop and estimated damping force considering ΔP_{total} and ΔP_{maj} models and experimental data: (**a**) for stroke and impact energy of 456.7 J; (**b**) for stroke velocity and impact energy of 456.7 J; (**c**) for stroke and impact energy of 937.5 J; (**d**) for stroke velocity and impact energy of 937.5 J; (**e**) for stroke and impact energy of 1538 J; (**f**) for stroke velocity and impact energy of 1538 J.

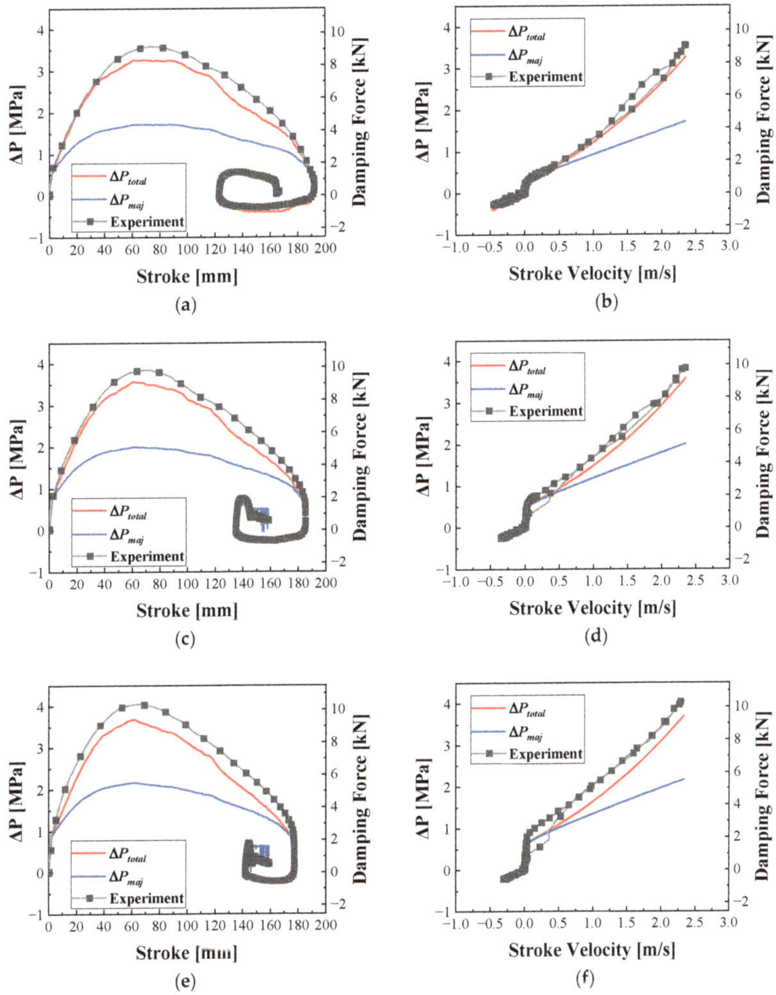

Figure 8. Pressure drop and estimated damping force considering ΔP_{total} and ΔP_{maj} models and experiment data: (**a**) for stroke and input current of 0.5 A; (**b**) for stroke velocity and input current of 0.5 A; (**c**) for stroke and input current of 1.0 A; (**d**) for stroke velocity and input current of 1.0 A; (**e**) for stroke and input current of 1.5 A; (**f**) for stroke velocity and input current of 1.5 A.

Table 3. Dissipated energy in MR shock absorber for input current; RMS values of error for model-based and measured pressure data over a stroke velocity range of 0.1 m/s or more.

Input Current (A)	Dissipated Energy					Pressure Drop			
	Experiment (J)	$\Delta P\text{–}\dot{s}$ Model (J)		Relative Error (%)		RMS Error (kPa)		RMS Relative Error (%)	
		ΔP_{total}	ΔP_{maj}	ΔP_{total}	ΔP_{maj}	ΔP_{total}	ΔP_{maj}	ΔP_{total}	ΔP_{maj}
0	1339	1149	621.2	14.18	53.61	189.5	735.8	11.66	31.72
0.5	1377	1253	757.2	8.984	45.01	172.4	749.5	10.80	28.70
1.0	1385	1253	815.7	9.564	41.10	245.6	920.5	16.36	33.66
1.5	1388	1208	822.3	12.99	40.75	381.5	1087	19.08	37.02

5. Conclusions

In this study, a novel design model for an MR shock absorber for an aircraft landing gear system was proposed, and its effectiveness was experimentally validated. Unlike the conventional model, which features only the major pressure loss, the proposed model was mathematically formulated by considering both the major and minor losses to take account of high-stroke velocity. After deriving the governing equation of motions associated with the pressure drop and damping force, an appropriate size of MR shock absorber was manufactured on the basis of the design parameters of a commercial landing gear system. Because the MR shock absorber for the landing gear is rapidly compressed during a drop, the proposed design model is validated through a drop test, generating different impact energies (stroke velocities). The pressure drop model, with respect to the impact energy and input current, was then compared with the measured result from the drop test. It has been identified from the comparative analysis that the pressure drop model considering of both the major and minor losses show an RMS error of 4.5% compared to the experimentally measured pressure drop. On the other hand, in the case of the pressure drop model considering the major loss only, an RMS error of 15% is identified. This result directly indicates that, when an MR shock absorber is designed for high stroke velocity, both the major and minor losses should be considered to accurately predict the pressure drop and hence the damping force. Finally, in the near future, the proposed model will be used to evaluate the landing efficiency of the aircraft landing gear system with an MR shock absorber in the absence and presence of the feedback controller.

Author Contributions: Conceptualization, J.-H.H., S.-B.C. and B.-H.K.; methodology, J.-H.H. and B.-H.K.; data analysis, B.-H.K.; investigation and validation, S.-B.C. and B.-H.K.; writing—original draft preparation, B.-H.K.; writing—review and editing, S.-B.C. All authors have read and agreed to the published version of the manuscript.

Funding: This work was supported by the Technology Innovation Program (Intelligent landing gear with variable damping force for 1500 lb class) (10073291) funded by the Ministry of Trade, Industry, and Energy (MOTIE, Korea).

Institutional Review Board Statement: Not applicable.

Informed Consent Statement: Not applicable.

Data Availability Statement: Not applicable.

Conflicts of Interest: The authors declare no conflict of interest.

References

1. Nguyen, Q.; Choi, S.; Lee, Y.; Han, M. An analytical method for optimal design of MR valve structures. *Smart Mater. Struct.* **2009**, *18*, 095032. [CrossRef]
2. Wereley, N.; Cho, J.; Choi, Y.; Choi, S. Magnetorheological dampers in shear mode. *Smart Mater. Struct.* **2007**, *17*, 015022. [CrossRef]
3. Carlson, J.; Sproston, J. Controllable fluids in 2000-status of ER and MR fluid technology. In Proceedings of the 7th International Conference on New Actuators, Bremen, Germany, 19–21 June 2000; pp. 126–130.
4. Yang, G.; Spencer, B., Jr.; Carlson, J.; Sain, M. Large-scale MR fluid dampers: Modeling and dynamic performance considerations. *Eng. Struct.* **2002**, *24*, 309–323. [CrossRef]
5. Kim, B.-G.; Yoon, D.-S.; Kim, G.-W.; Choi, S.-B.; Tan, A.S.; Sattel, T. Design of a novel magnetorheological damper adaptable to low and high stroke velocity of vehicle suspension system. *Appl. Sci.* **2020**, *10*, 5586. [CrossRef]
6. Oh, J.-S.; Jeon, K.; Kim, G.-W.; Choi, S.-B. Dynamic analysis of semi-active MR suspension system considering response time and damping force curve. *J. Intell. Mater. Syst. Struct.* **2021**, 1045389X20983920. [CrossRef]
7. Phu, D.X.; Mien, V.; Choi, S.-B. A New Switching Adaptive Fuzzy Controller with an Application to Vibration Control of a Vehicle Seat Suspension Subjected to Disturbances. *Appl. Sci.* **2021**, *11*, 2244. [CrossRef]
8. Wang, X.; Huang, D.; Qin, N.; Chen, C.; Zhang, K. Modeling and Second-Order Sliding Mode Control for Lateral Vibration of High-Speed Train With MR Dampers. *IEEE Trans. Intell. Transp. Syst.* **2021**, 1–10. [CrossRef]
9. Batterbee, D.; Sims, N.; Stanway, R.; Wolejsza, Z. Magnetorheological landing gear: 1. A design methodology. *Smart Mater. Struct.* **2007**, *16*, 2429. [CrossRef]

10. Batterbee, D.C.; Sims, N.D.; Stanway, R.; Rennison, M. Magnetorheological landing gear: 2. Validation using experimental data. *Smart Mater. Struct.* **2007**, *16*, 2441. [CrossRef]
11. Han, C.; Kang, B.-H.; Choi, S.-B.; Tak, J.M.; Hwang, J.-H. Control of landing efficiency of an aircraft landing gear system with magnetorheological dampers. *J. Aircr.* **2019**, *56*, 1980–1986. [CrossRef]
12. Luong, Q.V.; Jang, D.-S.; Hwang, J.-H. Robust adaptive control for an aircraft landing gear equipped with a magnetorheological damper. *Appl. Sci.* **2020**, *10*, 1459. [CrossRef]
13. Kang, B.-H.; Yoon, J.-Y.; Kim, G.-W.; Choi, S.-B. Landing efficiency control of a six degrees of freedom aircraft model with magneto-rheological dampers: Part 2—control simulation. *J. Intell. Mater. Syst. Struct.* **2020**, *32*, 1045389X20942593. [CrossRef]
14. Yoon, J.-Y.; Kang, B.-H.; Kim, J.-H.; Choi, S.-B. New control logic based on mechanical energy conservation for aircraft landing gear system with magnetorheological dampers. *Smart Mater. Struct.* **2020**, *29*, 084003. [CrossRef]
15. Regulations, Federal Aviation. In *Part 23 Airworthiness Standards: Normal, Utility, Acrobatic and Commuter Category Airplanes*; US Government Printing Office: Washington, DC, USA, 1989; Revised January.
16. Jones, O., Jr.; Leung, J. An Improvement in the Calculation of Turbulent Friction in Smooth Concentric Annuli. *ASME J. Fluids Eng.* **1981**, *103*, 615–623. [CrossRef]
17. Wojtkowiak, J.; Popiel, C.O. Inherently Linear Annular-Duct-Type Laminar Flowmeter. *ASME J. Fluids Eng.* **2006**, *128*, 196–198. [CrossRef]
18. Lundgren, T.; Sparrow, E.; Starr, J. Pressure Drop Due to the Entrance Region in Ducts of Arbitrary Cross Section. *ASME J. Basic Eng.* **1964**, *86*, 620–626. [CrossRef]
19. Lord Technical Data. In *MRF-140CG Magneto-Rheological Fluid*; Lord Corporation: Cary, NC, USA, 2012.
20. Idelchik, I.E. *Handbook of Hydraulic Resistance*; Hemisphere Publishing Corp.: Washington, DC, USA, 1986.
21. Bodoia, J.R.; Osterle, J. Finite difference analysis of plane Poiseuille and Couette flow developments. *Appl. Sci. Res.* **1961**, *10*, 265–276. [CrossRef]
22. Han, C.; Kim, B.-G.; Kang, B.-H.; Choi, S.-B. Effects of magnetic core parameters on landing stability and efficiency of magnetorheological damper-based landing gear system. *J. Intell. Mater. Syst. Struct.* **2020**, *31*, 198–208. [CrossRef]

applied sciences

MDPI

Article

Fuzzy Sliding Mode Control of Vehicle Magnetorheological Semi-Active Air Suspension

Gang Li *, Zhiyong Ruan, Ruiheng Gu and Guoliang Hu *

Key Laboratory of Vehicle Intelligent Equipment and Control of Nanchang City, East China Jiaotong University, Nanchang 330013, China; isshoni@126.com (Z.R.); ruihenggu@163.com (R.G.)
* Correspondence: ligang0794@163.com (G.L.); glhu@ecjtu.edu.cn (G.H.)

Abstract: In order to reduce vehicle vibration during driving conditions, a fuzzy sliding mode control strategy (FSMC) for semi-active air suspension based on the magnetorheological (MR) damper is proposed. The MR damper used in the semi-active air suspension system was tested and analyzed. Based on the experimental data, the genetic algorithm was used to identify the parameters of the improved hyperbolic tangent model, which was derived for the MR damper. At the same time, an adaptive neuro fuzzy inference system (ANFIS) was used to build the reverse model of the MR damper. The model of a quarter vehicle semi-active air suspension system equipped with a MR damper was established. Aiming at the uncertainty of the air suspension system, fuzzy control was used to adjust the boundary layer of the sliding mode control, which can effectively suppress the influence of chattering on the control accuracy and ensure system stability. Taking random road excitation and impact road excitation as the input signal, the simulation analysis of passive air suspension, semi-active air suspension based on SMC and FSMC was carried out, respectively. The results show that the semi-active air suspension based on FSMC has better vibration attenuating performance and ride comfort.

Keywords: MR damper; air suspension system; fuzzy sliding mode control; improved hyperbolic tangent model; parameter identification

Citation: Li, G.; Ruan, Z.; Gu, R.; Hu, G. Fuzzy Sliding Mode Control of Vehicle Magnetorheological Semi-Active Air Suspension. *Appl. Sci.* **2021**, *11*, 10925. https://doi.org/10.3390/app112210925

Academic Editors: Rosario Pecora, Antonio Concilio, Salvatore Ameduri, Ignazio Dimino, Vikram G. Kamble and César M. A. Vasques

Received: 12 September 2021
Accepted: 15 November 2021
Published: 18 November 2021

1. Introduction

The suspension system is the transmission device connecting the wheels and the car body, which mainly undertakes the body mass and ensures good contact between the tire and the ground, as well as reducing the uneven excitation from the road surface [1]. Compared with the traditional suspension system, the air spring suspension system has the advantages of being light weight, being low noise and having favorable vibration isolation performance. The air spring has the characteristics of non-linearity and self-adaptation, which makes the deviation frequency of the vehicle body mass in a stable state under the condition of load changes and obtains favorable ride comfort and handling stability [2–5].

As a new type of intelligent semi-active control device, the MR damper has become a research hotspot and can be well applied to semi-active air suspension, due to its good performance, i.e., simple structure, controllable damping force, low power consumption as well as rapid response time [6–9]. In industrial application, the accurate establishment of the mathematical model of MR damper is an important step for the design and analysis of the semi-active controller. However, the influences of movement condition of the piston, applied magnetic field, displacement amplitude and frequency on the dynamic performance of MR damper are inevitable, and the MR damper has the strong nonlinear characteristics, such as yield, hysteresis and saturation. It is difficult to establish the accurate dynamic model of the MR damper. So far, different mechanical models of MR damper have been proposed, which are mainly divided into two categories: non-parametric models and parametric models [10]. Parametric modeling of the MR damper is a process of replacing the mechanical behavior of the MR damper by the combination of a damping element, elastic element, mechanical element, etc. This type of model is relatively simple in structure, and there are many existing research studies, such as the Bingham model, Sigmoid model and hyperbolic tangent model, which are applied in semi-active control [11]. Non-parametric models, such as the differential equation model, neural network model and polynomial model, mainly simulate the mechanical properties of MR dampers by means of intelligent control [12]. However, these non-parametric models do not consider the rheological properties of MR fluid and have poor adaptability. Kowk et al. proposed the hyperbolic tangent model based on the study of the hyperbolic tangent function, which is mainly composed of three parallel structures of the hysteretic force element, damping element and spring element [13]. The hyperbolic tangent model can accurately describe the hysteretic characteristics of the MR damper by using the characteristics of the hyperbolic tangent function, but there are many parameters to be identified. Guo et al. constructed an improved hyperbolic tangent model, which applies another hyperbolic tangent function and is composed of viscous damping elements and elastic elements in series and parallel [14].

Appropriate control algorithms can make the semi-active air suspension almost achieve the same vibration attenuation effect as the active air suspension. In recent years, the sliding mode control (SMC) with strong anti-interference ability and good robustness attracted more and more attention in related fields. Choi et al. designed a sliding mode controller for the vehicle seat and considered the human body mass as the uncertainty factor of the system [15]. Through simulation analysis, it is concluded that the sliding mode controller has favorable robustness to the uncertainty of the system, and the ride comfort is also improved. Pang et al. proposed an improved adaptive sliding mode–based fault-tolerant control design, the results showing that the dynamics performance of half vehicle active suspensions with parametric uncertainties and actuator faults in the context of external road disturbances are improved [16]. Wang et al. presented a practical terminal sliding mode control framework based on an adaptive disturbance observer for the active suspension systems, which requires no exact feedback linearization about the suspension dynamics [17]. For the tracking control problem of vehicle suspension, Fei et al. derived a robust design method of adaptive sliding mode control [18]. Under this approach, the influence of parameter uncertainties and external disturbances on the system performance can be reduced, and the system robustness is also improved. Du et al. proposed a terminal sliding mode control approach for an active suspension system, which has an ability to reach the sliding surface in a finite time to achieve high control accuracy [19]. Combining the principle of fractional order control with sliding mode control, Nguyen et al. proposed a corresponding fractional order sliding mode control strategy for vehicle suspension and verified its effectiveness through simulation and experiment [20]. Zheng et al. combined the advantages of fuzzy control and sliding mode control and confirmed the effectiveness of the fuzzy sliding mode controller [21]. In addition, many other control strategies are applied to the control of suspension systems, which include switching control, adaptive robust control, linear–quadratic–Gaussian (LQG) control, and intelligent control methods [22–27].

Sliding mode variable structure control has good robustness, strong anti-interference ability, and can be well adapted to air suspension systems. However, when the actual controlled system switches the system structure at a higher frequency, the chattering phenomenon often occurs, which affects the accuracy of the sliding mode variable structure control and the system stability [28]. Fuzzy control can compensate for the uncertainty of the system and adjust the boundary layer of the sliding mode control by making fuzzy control rules to realize the imprecise control. In order to suppress the influence of chattering on the control precision and ensure the system stability, a FSMC controller was developed on the combination of fuzzy control and sliding mode variable structure control.

In this paper, the vibration test system was built to test and analyze the performance of the MR damper used in the air suspension system. Using the obtained test data, the genetic algorithm was applied to identify the parameters of the improved hyperbolic tangent model, and the inverse model of the MR damper was built, using an adaptive neuro-fuzzy inference system simultaneously. Subsequently, the air spring and MR damper were combined in vehicle suspension. The air spring has good nonlinear elastic characteristics and has different elastic characteristics under a heavy load and light load, which can improve ride comfort and handling stability. Then, a quarter car MR semi-active air suspension system model was established. In order to reduce the chattering problem in the sliding mode control and ensure system stability, a fuzzy sliding mode control strategy of semi-active air suspension was proposed. The fuzzy control can compensate for the uncertainty of the system and reduce the chattering problem in sliding mode control and ensure the stability and safety of the vehicle suspension. Finally, the numerical simulation was carried out to verify the effectiveness of the proposed control strategy under the conditions of excitations of the random road profile and impact road profile.

2. Damping Performance Test and Mechanical Modeling of MR Damper

2.1. Damping Performance Test of MR Damper

In order to test the damping characteristics of MR damper, a vibration test system was established as shown in Figure 1, which is mainly composed of a damping suspension test bench, DC power supply, electro-hydraulic servo controller, computer and MR damper. The DC power supply was used to supply the required DC current to the coil in the MR damper. The hydraulic station provided hydraulic power for the vibration test bench. The computer was used to collect experimental data, and the vibration test bench was controlled by the electro-hydraulic servo controller.

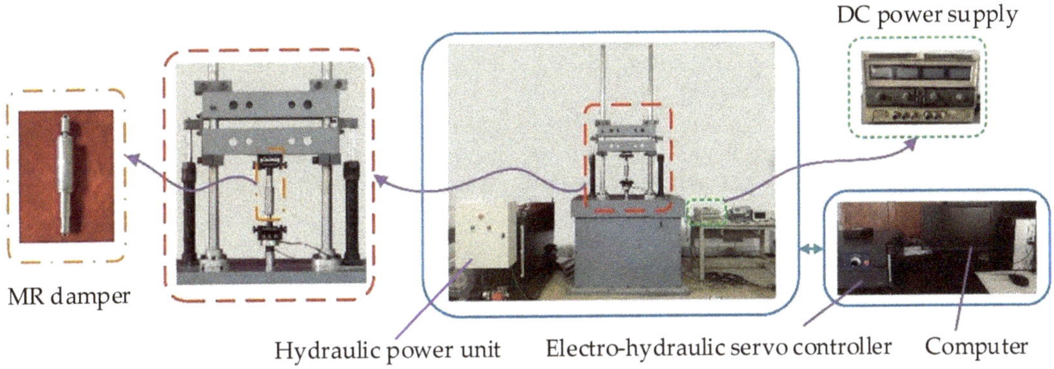

Figure 1. Vibration test system of MR damper.

In this experiment, a sinusoidal excitation signal with a frequency of 1 Hz and amplitudes of 5 mm, 7.5 mm, 10 mm was adopted. The currents were 0 A, 0.25 A, 0.5 A, 0.75 A and 1 A, provided by a DC regulated power supply. As shown in Figure 2, the variation of the output damping force with displacement and velocity of the MR damper under different excitation currents is obtained. Figure 2a shows that the output damping force of the MR damper increases with the increase in the excitation current. This is mainly because the magnetic flux density in the damping channel increases with the increase in the excitation current. The output damping force varies little with the displacement, and it is distributed symmetrically with respect to the relative equilibrium position. As shown in Figure 2b, the damping force of the MR damper increases with the increase in the excitation current and vibration speed, but the influence of the speed is much smaller than that with the current. The proposed MR damper has a wide range of damping force adjustments, and thus, it can be used in a vehicle suspension system.

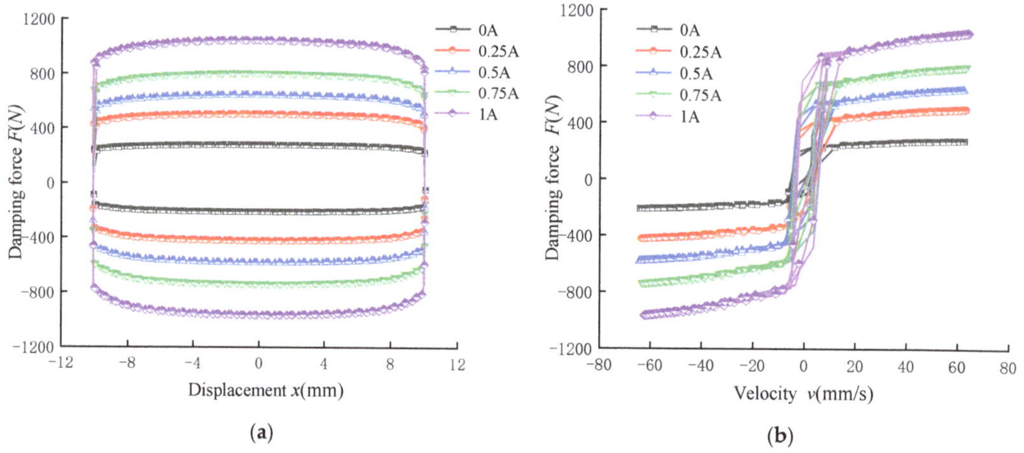

Figure 2. Mechanical characteristics of MR damper. (**a**) Force displacement, and (**b**) force velocity.

2.2. Improved Hyperbolic Tangent Model of MR Damper

The improved hyperbolic tangent model can comprehensively describe the damping characteristics of the MR damper. Compared with the hyperbolic tangent model, the physical quantities of this model have a clearer physical significance, and there are fewer unknown parameters, which makes it easy for subsequent parameter identification. The improved hyperbolic tangent model is shown in Figure 3, and its mathematical expression is as follows:

$$F = a_1 \tanh[a_2(\dot{x} + kx)] + a_3(\dot{x} + kx) + f_0 \tag{1}$$

where a_1 and k are both proportional factors, a_1 is related to the current and k affects the width of the hysteresis loop; a_2 and a_3 are damping coefficients that affect the curve variation trend of the pre-yield zone and post-yield zone, respectively; and f_0 denotes the offset damping force.

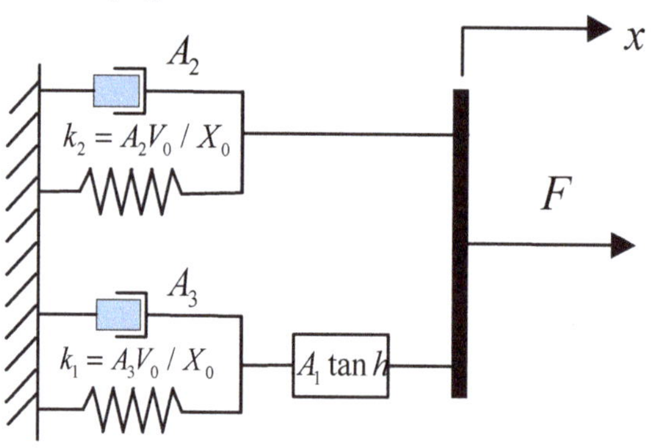

Figure 3. Improved hyperbolic tangent model.

2.3. Improved Parameter Identification of Hyperbolic Tangent Model

Figure 4 shows the parameter identification process using the genetic algorithm. Taking the mean square deviation of the difference between the actual damping force and the simulated damping force as the objective function, the smallest mean square deviation is the optimal solution, and the optimal model of the MR damper is obtained through genetic operation selection, crossover and mutation.

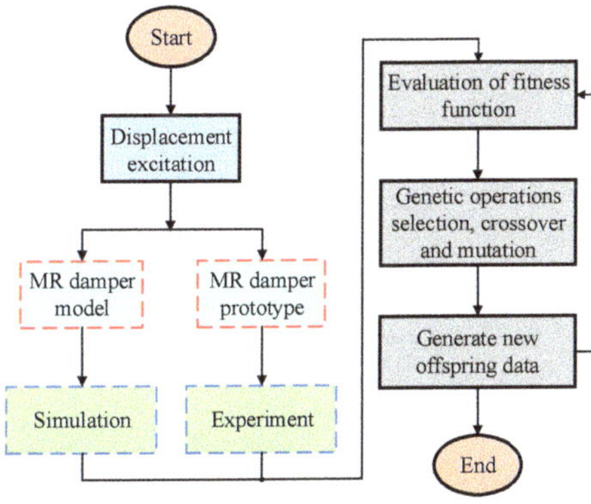

Figure 4. Parameter identification process of MR damper.

The fitness function of the improved hyperbolic tangent model is denoted as follows:

$$fitnessfunc = \frac{1}{m}\sum_{i=1}^{m}(F_{simi} - F_{\exp i})^2 \quad (2)$$

where F_{simi} and F_{expi} are the simulation damping forces and experimental damping forces of the ith point, respectively, and m is the number of experimental points.

The identification parameters were processed by using the real coding method. The scattered crossover method was used to determine the crossover probability by repeated experiments, and the crossover probability was set to 0.8. To improve the local search ability of the genetic algorithm, the Gaussian function was used for mutation with a scale of 0.5 as well as a shrink of 0.7, and the number of iterations was 100.

The improved hyperbolic tangent model has five parameters to be identified, which is expressed as follows:

$$\Theta = [a_1, a_2, k, a_3, f_0] \quad (3)$$

The test results of the mechanical characteristics of the MR damper under sinusoidal excitation with an amplitude of 10 mm and frequency of 1 Hz were selected as the identification data. Parameter identification was carried out for the data of 5 groups under different control currents so as to obtain parameter identification results under different current, as shown in Table 1. By comparing and analyzing the identification results of different current parameters, the variation rules of each parameter were found, and the fitting values of each parameter in the improved hyperbolic tangent model were also obtained.

Table 1. Parameter identification results of MR damper model.

Applied Current/(A)	Parameter Values				
	a_1	a_2	k	a_3	f_0
0	165.94	3890.24	0.3006	849.617	35.68
0.25	366.04	3859.98	0.3001	1548.13	39.69
0.5	483.51	3895.93	0.3004	2045.21	35.06
0.75	615.02	3859.26	0.3000	2573.25	35.07
1.0	805.25	3866.50	0.3001	3301.38	35.09

Observing Table 1, parameters a_2, k, and f_0 are not changed significantly under the different current. Therefore, a_2, k, and f_0 can be considered constant values, and their mean values a_2 = 3874.38, k = 0.31, and f_0 = 36.12 are selected. Parameters a_1 and a_3 increase linearly with the increase in current, and the relationship between the parameters and current is obtained through fitting, as shown in Figure 5.

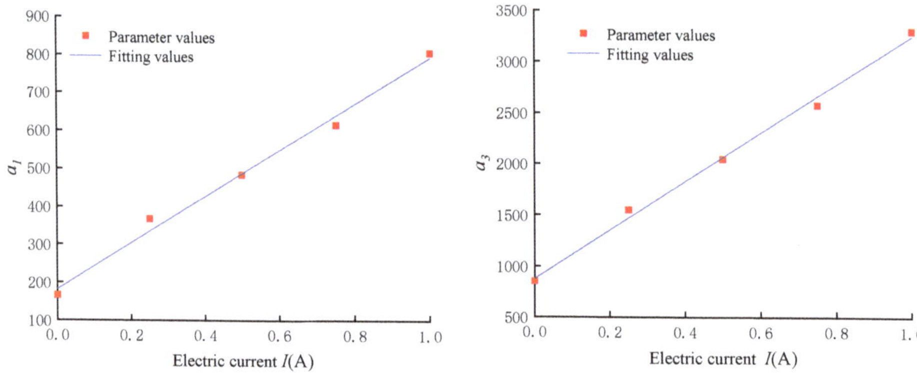

(**a**) (**b**)

Figure 5. The relation between a_1, a_3 and current. (**a**) Parameter a_1, (**b**) parameter a_3.

As shown in Figure 5, parameters a_1 and a_3 are linearly correlated with the current, which can be obtained by linear fitting as shown below:

$$\begin{cases} a_1 = m_1 I + n_1 = 611.04I + 181.63 \\ a_3 = m_2 I + n_2 = 2371.46I + 877.79 \end{cases} \tag{4}$$

The expression of the improved hyperbolic tangent model after parameter identification is obtained as follows:

$$F = (m_1 I + n_1)\tanh(a_2(\dot{x} + kx)) + (m_2 I + n_2)(\dot{x} + kx) + f_0 \tag{5}$$

According to the improved hyperbolic tangent model expressed in Equation (5) after parameter identification, the mechanical properties of the MR damper was simulated and compared with the test results, as shown in Figure 6. The force–displacement curve of the improved hyperbolic tangent model after parameter identification has a high degree of fit with the test curve. For the force–velocity curve, the nonlinear hysteretic characteristics of the MR dampers in low, and the high-speed regions are well described. Therefore, the hyperbolic tangent model can be applied to the subsequent semi-active control.

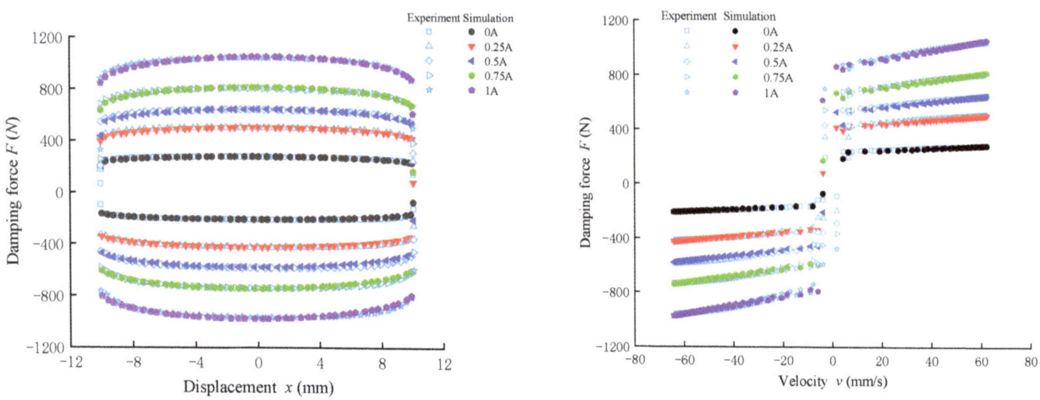

(**a**) (**b**)

Figure 6. Comparison of simulation results and experimental tests. (**a**) Force displacement, and (**b**) force velocity.

2.4. ANFIS Inverse Model of MR Damper

The ANFIS is a multilayer feedforward network that uses neural network learning algorithms and fuzzy reasoning to map an input space to an output space [29]. The ANFIS algorithm has a strong adaptive ability and self-learning ability, and it has a strong approximation ability to the nonlinear system. Therefore, ANFIS is suitable for the modeling of complex systems, such as MR dampers. Figure 7 shows the structure of ANFIS. There are five layers in total, and the function types of the nodes in each layer are the same.

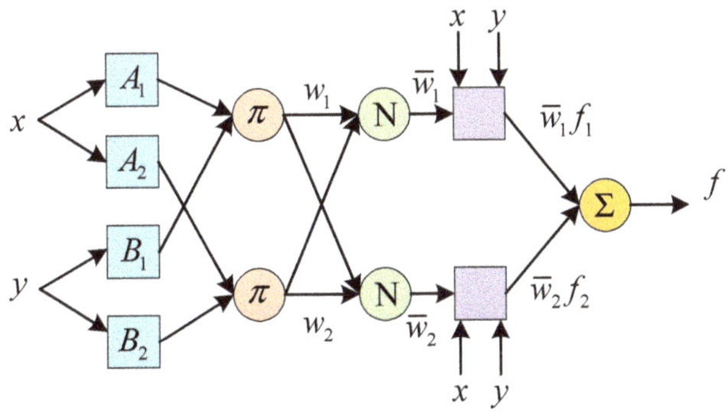

Figure 7. Structure diagram of ANFIS with two inputs.

Layer 1 is used to fuzzify the input signal, and its node output is defined as follows:

$$\begin{cases} o_{1,i} = \mu_{A_i}(x), i = 1,2\ or \\ o_{1,i} = \mu_{B_{i-2}}(x), i = 3,4 \end{cases} \tag{6}$$

where $o_{1,i}$ is the output of the ith node of layer 1, and x represents the input of the node, respectively. A_i (or B_{i-2}) represents fuzzy sets (negative big, negative small, positive big, positive small, etc.). The membership function of the fuzzy set is as follows:

$$\mu_{A_i}(x) = \frac{1}{1+\left(\frac{x-c_i}{a_i}\right)^{2b_i}}, \mu_{B_{i-2}}(x) = \frac{1}{1+\left(\frac{y-c_i}{a_i}\right)^{2b_i}} \tag{7}$$

The above equation is bell-shaped functions (the maximum is 1, the minimum is 0), where $\{a_i, b_i, c_i\}$ is the premise parameter used to adjust the shape of the function.

Layer 2 is represented by π, whose function is to calculate the applicability of each rule w_i. The specific operation is to take the product of the membership degree of each input signal as the applicability of this rule, as shown in the equation below.

$$o_{2,i} = w_i = \mu_{A_i}(x) \times \mu_{B_i}(y), i = 1,2 \tag{8}$$

Layer 3 is represented by circle N as shown in the Figure 7. Layer 3 is used to normalize the applicability of each rule, as shown in Equation (9).

$$o_{3,i} = \overline{w}_i = \frac{w_i}{w_1 + w_2}, i = 1,2 \tag{9}$$

The function of Layer 4 is to calculate the output of each rule. The node function is expressed as follows:

$$o_{4,i} = \overline{w}_i f_i = \overline{w}_i (p_i x + q_i y + r_i), i = 1,2 \tag{10}$$

where parameters $\{p_i, q_i, r_i\}$ represent an adjustable parameter set.

Layer 5 sums up all the inputs of the node and outputs the results to calculate the total output of the system.

$$o_{5,i} = \sum_{i=1}^{2} \overline{w}_i f_i = \frac{\sum_{i=1}^{2} w_i f_i}{\sum_{i=1}^{2} w_i} \tag{11}$$

ANFIS can obtain the required fuzzy rules and membership function after repeated training based on a large amount of data. Usually, the BP algorithm or hybrid algorithm is used to train the data.

The adaptive neuro fuzzy inference system (ANFIS) was used to establish the inverse model of the MR damper in this paper [30]. The training schematic diagram of the ANFIS inverse model of the MR damper is shown in Figure 8. In order to establish the inverse model of ANFIS, the velocity, displacement and expected damping force at the current moment as well as the velocity and desired damping force provided by the forward model at the previous moment were selected as the input, while the control current at the current moment was taken as the output.

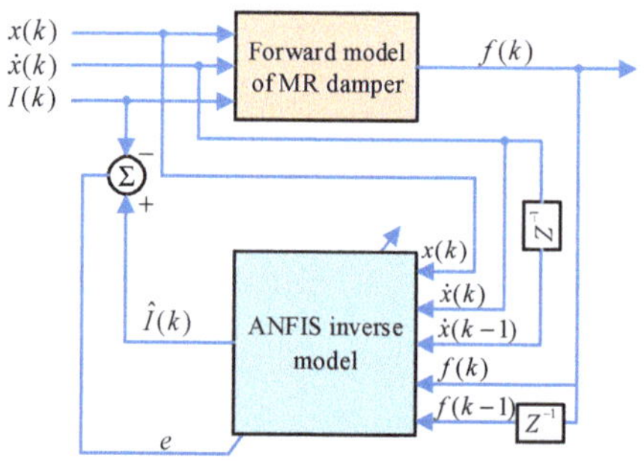

Figure 8. Schematic diagram of the ANFIS inverse dynamics model.

Two sets of data were needed to model a MR damper, using an adaptive neuro-fuzzy reasoning system. One set was used to train the inverse model of ANFIS, and the other set was used to test the accuracy of the model. A sinusoidal function with an amplitude of 20 mm was used for the displacement signal, and a sinusoidal function with an amplitude of 0~2 A was used for the control current signals. There were 1500 data points of which 1000 points were used for training and 500 points were used for validation.

Figures 9 and 10 show the comparison between the experimental and predicted values as well as the error of the ANFIS inverse dynamics model of the MR damper, respectively. The results illustrate that the predictive control current of the ANFIS inverse model can track the target control current well, and verify the effectiveness of the model.

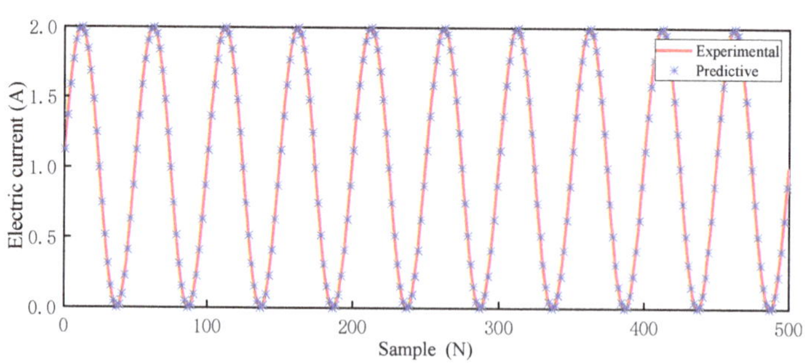

Figure 9. Verification of training data for ANFIS inverse dynamics model.

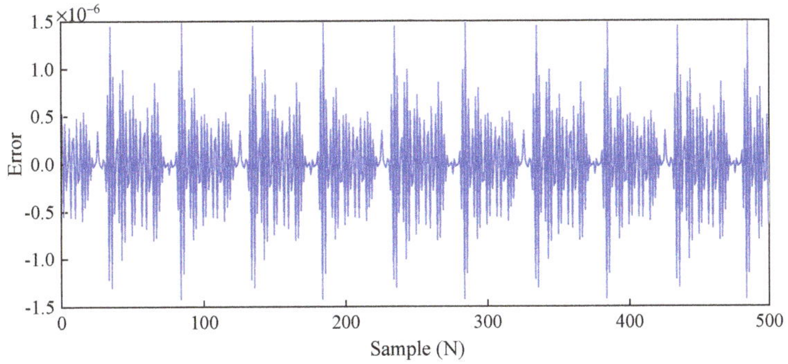

Figure 10. Error of ANFIS inverse dynamics model.

3. Modeling of Quarter Vehicle MR Semi-Active Air Suspension System

3.1. Elastic Model of Air Spring

The principle of the air suspension vehicle is that the air spring uses the air compressibility to play a supporting role. The vertical supporting force F_v produced by the air spring is the following:

$$F_v = PA \quad (12)$$

where P is the air pressure in the airbag, and A is the effective bearing area.

The air spring moves vertically in the process of working, and the air spring bag is stretched or compressed, due to the change in volume. This process exchanges heat with the outside world. Assuming that the time required for the movement of the air spring is short enough, it can be regarded as an adiabatic process [31]. The state equation of the gas at this time can be regarded as follows:

$$(P + P_a)V^m = (P_a + P_0)V_0^m \quad (13)$$

where V_0 is the volume of the air spring in the initial state, V is the working volume of the air spring, P_0 is the gas pressure in the initial state of the air spring, and P_a is the standard atmospheric pressure.

According to Equation (13), the pressure in the air bag can be obtained as follows:

$$P = (P_0 + P_a)(V_0/V)^m - P_a \quad (14)$$

Substituting Equations (13) and (14) to Equation (12), it can be derived as follows:

$$F = [(V_0/V)^m(P_0 + P_a) - P_a]A \quad (15)$$

When the air spring works, its volume and bearing area are related to the displacement, and thus, taking the derivative of Equation (15) with respect to the displacement, the stiffness k_a of the air spring is yielded, and it can be regarded as follows:

$$k_a = \frac{dF}{dx} = P\frac{dA}{dx} - A\frac{m}{V}(P + P_a)\frac{dV}{dx} \quad (16)$$

When the air spring is in an equilibrium state, there are the following relations:

$$P = P_0, A = A_0, V = V_0$$

Therefore, the air spring stiffness k_0 under an equilibrium state is obtained as follows:

$$k_0 = p_0\frac{dA}{dx} - A_0\frac{m}{V}(p + p_a)\frac{dV}{dx} \quad (17)$$

The natural frequency of the air spring is the following:

$$f_0 = \frac{1}{2\pi}\sqrt{\frac{k_0}{m}} \tag{18}$$

3.2. Modeling of Quarter Vehicle MR Semi-Active Air Suspension System

Passive suspension is mainly composed of a passive damper and spring with immutable stiffness, and its structure is relatively simple, as shown in Figure 11. The damping force and stiffness of passive suspension cannot be adjusted in real time with different road surface and vehicle parameters. Under a certain suspension relative speed, only a fixed suspension output damping force can be provided. Therefore, the vibration damping capacity of passive suspension decreases significantly, and its road adaptability is poor under a certain range of road surface conditions, especially on a bad road surface.

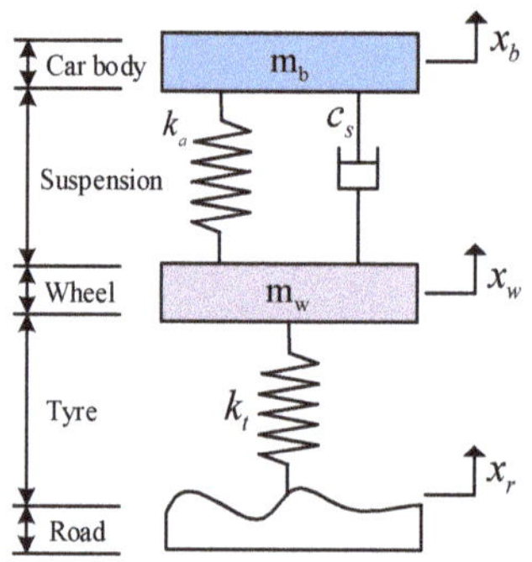

Figure 11. Model of quarter car passive suspension.

Figure 12 shows the model of a quarter vehicle MR semi-active air suspension system, which is composed of a car body (sprung mass), wheel (unsprung mass), tire stiffness, damping and suspension stiffness.

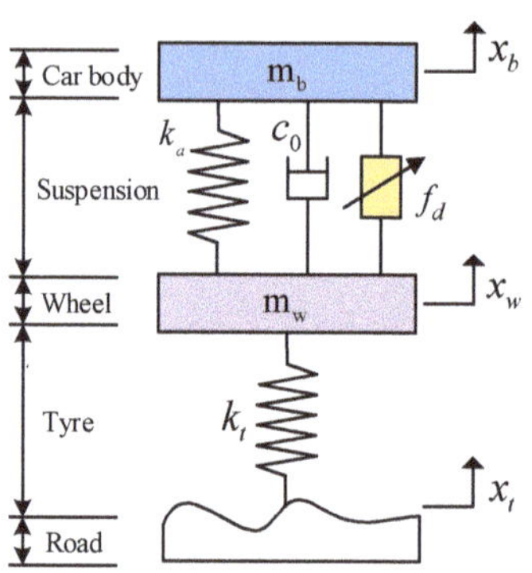

Figure 12. Model of quarter vehicle air suspension.

According to Newton's second law, the motion equation of the system is as follows:

$$\begin{cases} m_b\ddot{x}_b + k_a(x_b - x_w) + c_0(\dot{x}_b - \dot{x}_w) = f_d \\ m_w\ddot{x}_w + k_a(x_w - x_b) + c_0(\dot{x}_w - \dot{x}_b) + k_t(x_w - x_t) = -f_d \end{cases} \tag{19}$$

where m_b and m_w are the mass of the sprung body and the unsprung wheel, respectively, k_a is the stiffness of the air spring, k_t is the stiffness of the tire, x_b is the displacement of the mass of the body, x_w is the displacement of the wheel, x_t is the displacement of the road surface excitation, and f_d depicts the output damping force of the MR damper.

The system state variable can be denoted as follows:

$$x = (x_b, x_w, \dot{x}_b, \dot{x}_w) \tag{20}$$

The output variable can be depicted as follows:

$$y = (\ddot{x}_b, x_b - x_w, k_t(x_w - x_t)) \tag{21}$$

where $\ddot{x}_b$ is the acceleration of the car body, $x_b - x_w$ is the dynamic deflection of the suspension, and $k_t(x_w - x_t)$ is the dynamic load of the tire.

The state space expression of the suspension system is as follows:

$$\begin{cases} \dot{x} = Ax + Bu \\ y = Cx + Du \end{cases} \tag{22}$$

where

$$A = \begin{bmatrix} 0 & 0 & 1 & 0 \\ 0 & 0 & 0 & 1 \\ -\frac{k_a}{m_b} & \frac{k_a}{m_b} & -\frac{c_0}{m_b} & \frac{c_0}{m_b} \\ -\frac{k_a}{m_w} & -\frac{k_a+k_t}{m_w} & \frac{c_0}{m_w} & -\frac{c_0}{m_w} \end{bmatrix}, B = \begin{bmatrix} 0 & 0 \\ 0 & 0 \\ 0 & \frac{1}{m_b} \\ \frac{k_t}{m_w} & -\frac{1}{m_w} \end{bmatrix}$$

$$C = \begin{bmatrix} -\frac{k_a}{m_b} & \frac{k_a}{m_b} & -\frac{c_0}{m_b} & \frac{c_0}{m_b} \\ 1 & -1 & 0 & 0 \\ 0 & k_t & 1 & 0 \end{bmatrix}, D = \begin{bmatrix} 0 & \frac{1}{m_b} \\ 0 & 0 \\ -k_t & 0 \end{bmatrix}$$

4. Design of Sliding Mode Controller for MR Semi-Active Air Suspension

4.1. Reference Model of Sliding Mode Controller

Figure 13 is the reference model of sliding model controller based on the knowledge of optimal control theory proposed by Karnopp [32], which has excellent robustness and excellent control performance. According to Newton's second law, the dynamic differential equation is as follows:

$$\begin{cases} m_b\ddot{x}_{b1} + k(x_{b1} - x_{w1}) + c_{b1}\dot{x}_{b1} = 0 \\ m_w\ddot{x}_{w1} + k(x_{w1} - x_{b1}) + k_t(x_w - x_t) + c_{w1}\dot{x}_{w1} = 0 \end{cases} \tag{23}$$

where m_b and m_w represent sprung body mass and unsprung wheel mass, respectively; k and k_t denote the spring stiffness and tire stiffness, respectively; x_{b1}, x_{w1}, and x_{t1} are the body displacement, wheel displacement and displacement of the road surface excitation, respectively; and c_{b1} and c_{w1} are the damping coefficients.

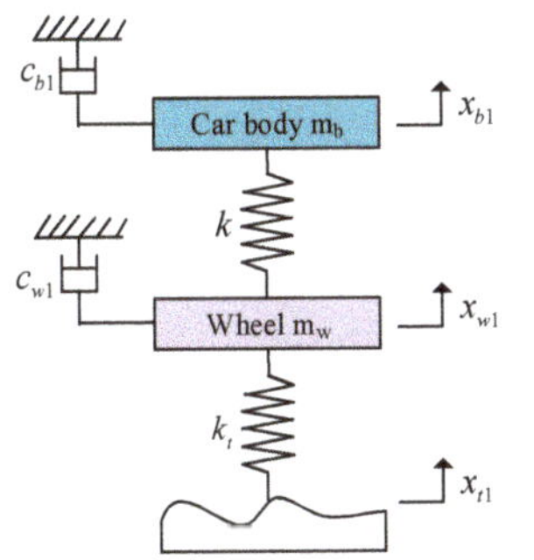

Figure 13. Reference model.

The state variables are selected as follows:

$$X_1 = [x_{b1}, x_{w1}, \dot{x}_{b1}, \dot{x}_{w1}]^T \tag{24}$$

Substituting Equation (24) into Equation (23), it can be obtained the following:

$$\dot{X}_1 = A_1 X_1 + E_1 \omega \tag{25}$$

where

$$A_1 = \begin{bmatrix} 0 & 0 & 1 & 0 \\ 0 & 0 & 0 & 1 \\ -\frac{k_t+k}{m_w} & \frac{k}{m_w} & -\frac{c_{b1}}{m_w} & 0 \\ \frac{k}{m_b} & -\frac{k}{m_b} & 0 & -\frac{c_{b1}}{m_b} \end{bmatrix}, \ E_1 = \begin{bmatrix} 0 \\ 0 \\ \frac{k_t}{m_w} \\ 0 \end{bmatrix}, \ \omega = x_{t1}$$

4.2. Design of Sliding Mode Controller

The design principle of the sliding mode controller (SMC) is to make the actual controlled system track the motion of the reference system and produce the sliding mode in the error dynamics system between the actual controlled system and the reference system. The error vector e of the air suspension system is defined as the velocity error, displacement error and integral of displacement error, and its matrix form can be expressed as follows:

$$e = \begin{bmatrix} \dot{x}_b - \dot{x}_{b1} & x_b - x_{b1} & \int (x_b - x_{b1}) \end{bmatrix}^T \tag{26}$$

$$\dot{e} = \begin{bmatrix} \ddot{x}_b - \ddot{x}_{b1} & \dot{x}_b - \dot{x}_{b1} & x_b - x_{b1} \end{bmatrix}^T \tag{27}$$

Therefore, the error dynamics equation is defined as follows:

$$\dot{e} = A(t)e + B(t)X + H(t)X_r + G(t)u \tag{28}$$

In the design of switching the surface of the sliding mode variable structure, the pole assignment method is generally used to set the function of the switching surface. The output value of the controller is the controllable damping force, and thus its switching surface can be expressed as follows:

$$s = cx[c_1, c_2, \cdots, c_n] \cdot [x_1, x_2, \cdots, x_n]^T \tag{29}$$

Parameter n is 3, so the sliding mode switching surface function is as follows:

$$s = ce \tag{30}$$

The state Equation (29) is simplified as follows:

$$\begin{bmatrix} s_1 \\ s_2 \end{bmatrix} = \begin{bmatrix} 0 & 1 \\ -c_1 & -c_2 \end{bmatrix} \begin{bmatrix} e_1 \\ e_2 \end{bmatrix} \tag{31}$$

All control points of sliding mode motion must have good dynamic quality and be asymptotically stable when entering the sliding mode switching surface. Equation (31) is the motion differential equation of the sliding mode, and its characteristic equation is as follows:

$$D(\lambda) = \lambda^2 + c_2\lambda + c_1 \tag{32}$$

The values of c_1 and c_2 can be obtained by making the characteristic root equal to the given pole.

The transfer function in standard form for a second-order system is as follows:

$$\Phi(s) = \frac{Y(s)}{U(s)} = \frac{\omega_n{}^2}{s^2 + 2\xi\omega_n s + \omega_n{}^2} \tag{33}$$

The two closed-loop poles of the system are $s_{1,2} = -\xi\omega_n \pm j\omega_n\sqrt{1-\xi^2}$. Only when the pole is on the left half of the complex plane does the control system have partial oscillations. The desired number of poles n is set as 3. Among them, the dynamic performance of the control system is determined by the dominant pole, and the far pole has little influence on it. It can be obtained that the characteristic root of the system is $-3.9 \pm 5.2j$, under the condition of $\sigma \leq 15\%$, $t_p \leq 0.7$, $\omega_n = 6.5$, and $\xi = 0.6$. Then, c = [42.25, 7.8, 1] can be obtained.

Therefore, the switching surface function of MR semi-active air suspension is as follows:

$$s = [\ (\dot{x}_b - \dot{x}_{b1}) \quad c_1(x_b - x_{b1}) \quad c_2(\int (x_b - x_{b1})\ )]^T \tag{34}$$

According to Equation (30), it can be obtained as follows:

$$\dot{s} = c\dot{e} = cA(t)e + cB(t)X + cH(t)X_r + cG(t)u \tag{35}$$

From the above, it can be developed as follows:

$$U = (cG)^{-1}[\dot{s} - (cA(t)e + cB(t)X + cH(t)X_r] \tag{36}$$

To minimize the error between the actual controlled model and the reference model, let $\dot{s}$ be 0, and Equation (36) can be deduced as follows:

$$U = (cG)^{-1}[(cA(t)e + cB(t)X + cH(t)X_r] \tag{37}$$

The equivalent output damping force of the sliding mode controller can be obtained as follows:

$$U_{eq} = (c_0 - c_1 m_b)(\dot{x}_b - \dot{x}_{b1}) + (k_a - c_2 m_w)(x_b - x_{b1}) - k_a(x_w - x_{w1}) - c_0(\dot{x}_w - \dot{x}_{w1}) \tag{38}$$

To improve sliding mode motion, the reaching law is defined as follows:

$$\dot{s} = -\varepsilon \cdot \text{sgn}(s),\ (\varepsilon > 0) \tag{39}$$

where sgn(s) is a sign function, and ε is the rate of control point approaching the switching surface, which affects the system stability. The smaller ε is, the smaller the approaching rate is, while the larger ε is, the greater the approaching rate is, and the more severe the system chattering is. In the sliding mode control, ε is 0.1 according to the experience and debugging results. The system stability was verified according to the Lyapunov stability criterion. If $V(x)$ is positive definite and $\dot{V}(x)$ is negative definite, the system is stable. Select the Lyapunov function as follows:

$$V(x) = \frac{1}{2}s^T s \tag{40}$$

Then it can be derived as follows:

$$\dot{V}(x) = s^T\dot{s} = s^T(-\varepsilon \text{sgn}(s)) \leq 0 \tag{41}$$

According to the Lyapunov stability criterion, it can be inferred that the system is stable. Therefore, the final sliding mode control law of the system is as follows:

$$U = U_{eq} + U_{sw} = U_{eq} + [cG]^{-1}\dot{s} = U_{eq} + \varepsilon m_s \text{sgn}(s) \tag{42}$$

where m_s is the coefficient related to s and $\dot{s}$.

Finally, the variable damping force by the sliding mode controller is as follows:

$$U_d = \begin{cases} U_{eq} + U_{sw} & [U_{eq} + U_{sw}](\dot{x}_b - \dot{x}_{b1}) \geq 0 \\ 0 & [U_{eq} + U_{sw}](\dot{x}_b - \dot{x}_{b1}) \leq 0 \end{cases} \tag{43}$$

4.3. Design of Fuzzy Sliding Mode Controller

The structure diagram of FSMC is shown in Figure 14. In order to make the controlled system track the motion of the reference system, the sliding mode switching surface function s is designed according to the error between the controlled system and the reference system. The equivalent output damping force U_{eq} of the sliding mode controller can be obtained. At the same time, ε and the switching control force U_{sw} are obtained by the fuzzy controller. Finally, the force required for the damper (predictive damping force) can be obtained. According to the inverse model of the damper, the control current is obtained, and the actual control force is generated by controlling the MR damper.

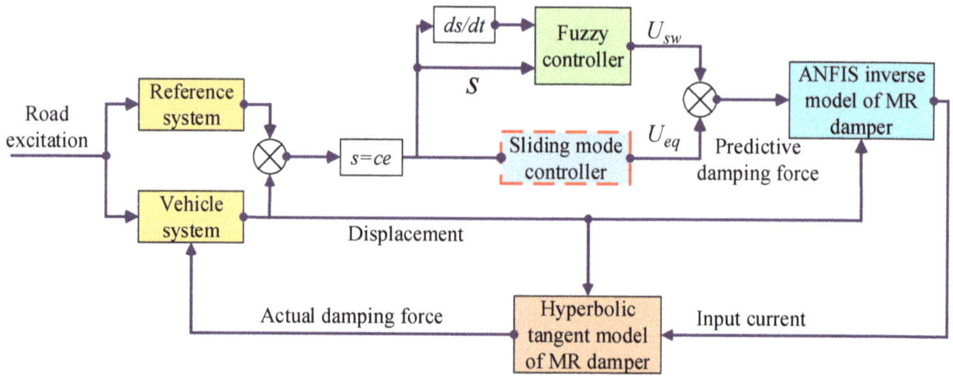

Figure 14. FSMC structure diagram of MR semi-active air suspension.

In the fuzzy system, s and $\dot{s}$ are defined as inputs of the fuzzy controller, and the output value of the fuzzy controller is defined as ε of the sliding mode controller. As shown in Figure 15, the fuzzy domain of the fuzzy input and output is set as $\{-3, -2, -1, 0, 1, 2, 3\}$. Language variables describing input and output variables are divided into seven levels: NB (negative big), NM (negative medium), NS (negative small), ZE (zero), PS (positive small), PM (positive medium), and PB (positive big). The triangular membership function is selected as the membership function of fuzzy input S, SC and output E, respectively.

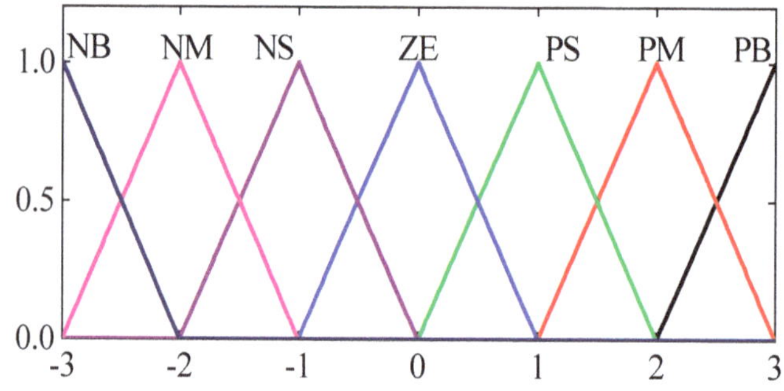

Figure 15. Membership function of input and output variables.

If the controller satisfies the conditions of $s\dot{s} < 0$, the fuzzy control rules shown in Table 2 are established. The control target can be expressed as follows:

$$U = -(cG)^{-1}c[A(t)e + G(t)X + H(t)X_r] + \varepsilon_{fuzzy} m_s \text{sgn}(s) \tag{44}$$

where ε_{fuzzy} is the output of the fuzzy sliding mode controller.

Table 2. Fuzzy control rules.

S \ E SC	NB	NM	NS	ZE	PS	PM	PB
NB	NB	NB	NM	NM	NS	NS	ZE
NM	NB	NB	NM	NS	NS	NS	ZE
NS	NM	NM	NS	ZE	ZE	ZE	ZE
ZE	NM	NM	NS	ZE	ZE	PS	PS
PS	NS	NS	ZE	ZE	PS	PS	PM

5. Simulation Analysis of Fuzzy Sliding Mode Control

5.1. Random Road Excitation Response

In order to verify the effectiveness of the FSMC method, the random road excitation was selected as the road input for simulation analysis. The white noise signal after the forming filter was usually used as the time domain signal of the random road surface. The real road surface of vehicle driving can be regarded as follows:

$$\dot{q}(t) = -2\pi f_1 q(t) + 2\pi n_0 \sqrt{G_q(n_0) v} w(t) \tag{45}$$

where f_1 is the lower cutoff frequency of the road filter, $q(t)$ donates the road surface excitation, $G_q(n_0)$ is the road roughness coefficient, $w(t)$ is the white noise with a covariance of 1, and v is the vehicle speed.

In the simulation analysis, MATLAB software was used for numerical simulation of the system, in which the sampling period is 0.001 s. Figure 16 shows the random input waveform of the road surface, using the filtering white noise generation method, where f_1 is 0.01 Hz, the C-level road roughness coefficient $G_q(n_0)$ is 256×10^{-6} m^3, and the vehicle velocity v is 20 m/s. Table 3 shows the parameters of semi-active air suspension system required for simulation.

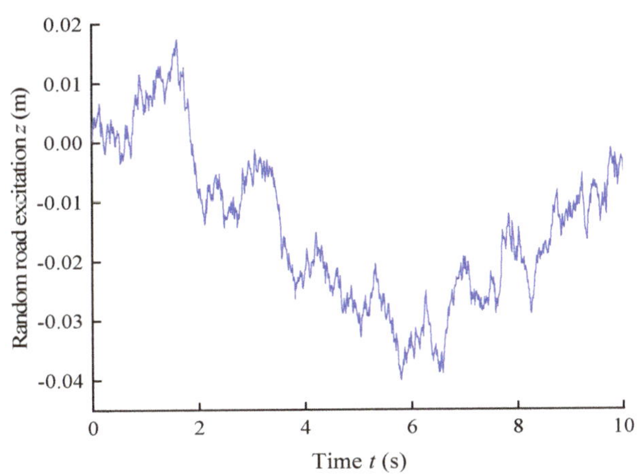

Figure 16. Random input waveform of road surface.

Figure 17 shows the time domain of the 2-DOF quarter vehicle semi-active air suspension controlled by the FSMC method. The random road excitations of passive suspension, SMC control and FSMC control suspension are the same, shown in Figure 16. It can be seen that the semi-active air suspension controlled by FSMC switching has significantly improved the damping performance, compared with the passive suspension and the semi active air suspension controlled by SMC.

Table 3. Model parameters of two-degree-of-freedom quarter vehicle passive air suspension system.

Parameters	Symbol	Unit	Values
Sprung mass	m_b	kg	400
Unsprung mass	m_w	kg	40
Tire damping	K_t	N/m	170,000
Damping coefficient	C_{b1}	N/(m/s)	3550
Damping coefficient	C_{w1}	N/(m/s)	3550
Suspension stiffness	K	N/m	17,000
Initial volume of air spring	V_0	m^3	0.0078
Initial area of air spring	A_0	m^2	0.0381
Initial pressure of air spring	P_0	MPa	0.382

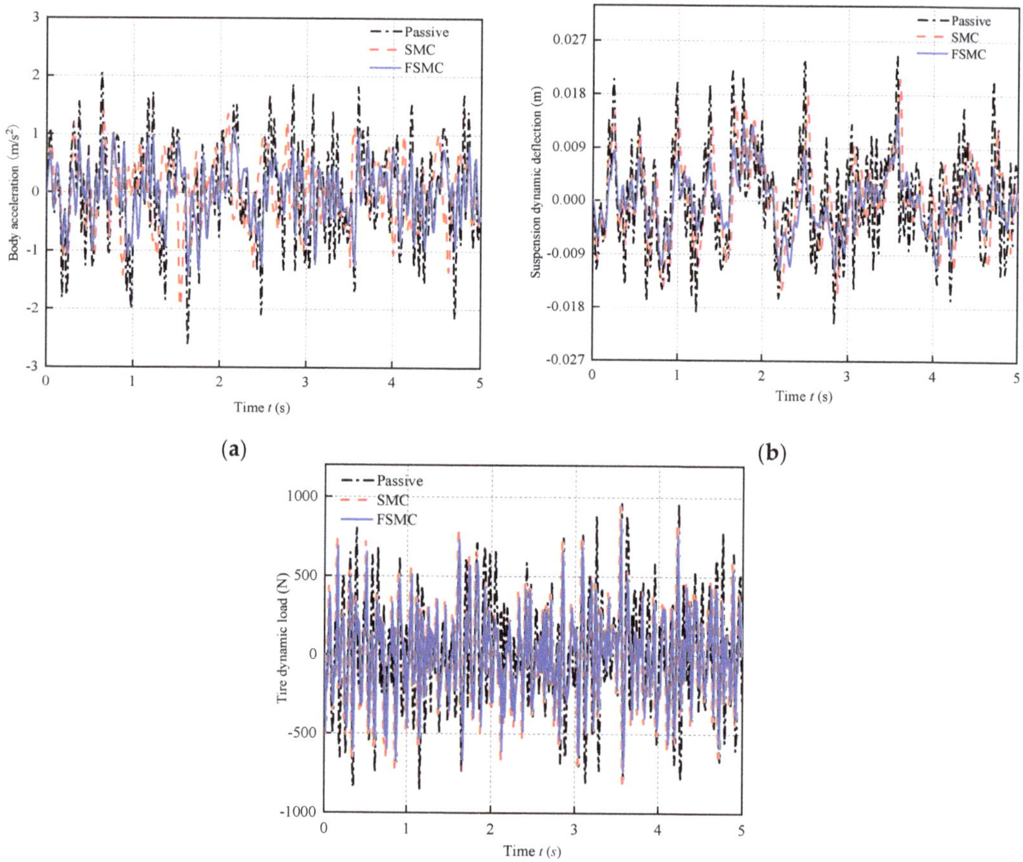

Figure 17. Time domain simulation curve of air suspension system under random road excitation. (**a**) Body acceleration, (**b**) suspension dynamic deflection, (**c**) tire dynamic load.

Table 4 shows the root mean square (RMS) values of the air suspension performance indexes under a random road surface. Compared with passive suspension, the body acceleration under SMC control and FSMC control decreases by 15.9% and 26%, respectively, the suspension dynamic travel decreases by 16.9% and 29.4%, respectively, and the wheel dynamic load decreases by 7.4% and 8.4%, respectively.

Table 4. RMS values of suspension system indexes under random road excitation.

Control Strategies	Body Acceleration	Suspension Stroke	Tire Dynamic Load
Passive suspension	0.857	7.73	307.5
SMC control	0.721	6.42	284.6
FSMC control	0.634	5.46	282.1

Figure 18 shows the power spectral density (PSD) under the input of random road surface. In the high frequency range, the power spectral density of FSMC control and sliding mode control is similar to that of the passive system in terms of body acceleration, suspension dynamic travel and tire dynamic load, due to frequency conversion interference. However, in the low frequency domain where the body is sensitive to the resonance of the car body, the 2-DOF quarter car semi-active air suspension controlled by FSMC is obviously superior to the controlled suspension and the sliding mode control suspension, the occupant's comfort is enhanced, and the resonance of the car body is effectively prevented.

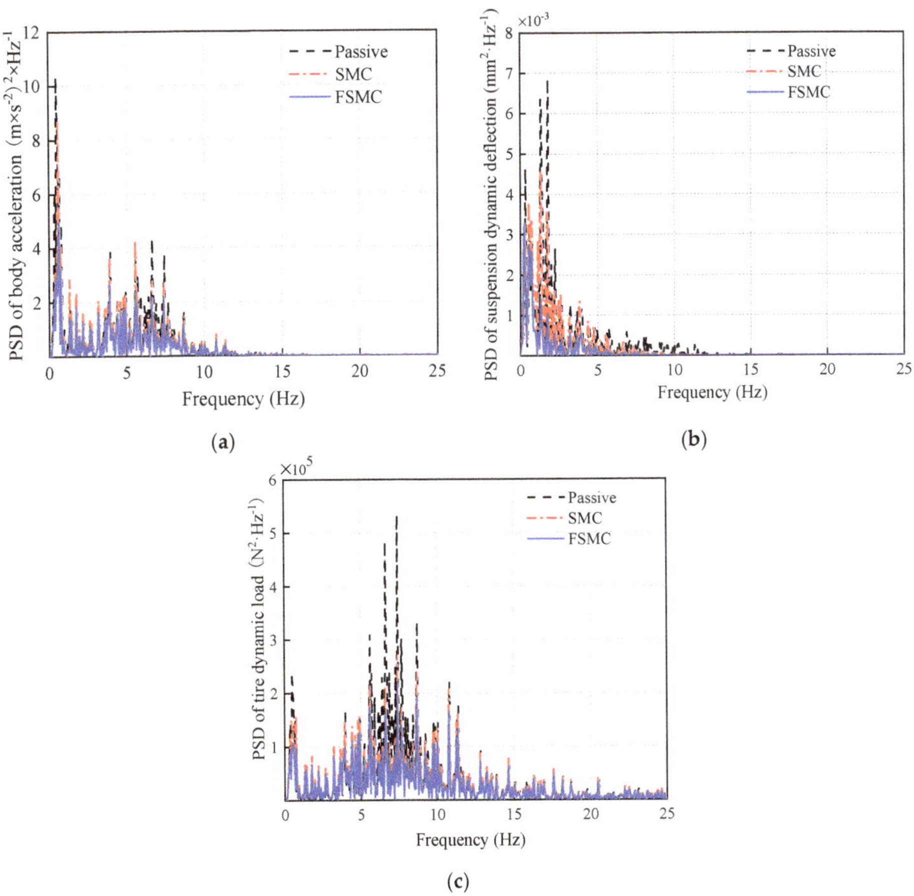

Figure 18. Simulation curve of power spectral density of air suspension system under random road excitation. (**a**) Body acceleration, (**b**) suspension dynamic deflection, (**c**) tire dynamic load.

5.2. Impacted Pavement Excitation Response

Vehicle driving is not only the existence of random continuous vibration; sometimes, there will be low-lying, protrusion sections, which bring strong impact to the vehicle. Therefore, in order to fully reflect the vibration reduction ability of the vehicle suspension system, the impact road surface model was established to replace the impact vibration of the vehicle on the road surface. Figure 19 shows the simulation curve of impact road surface input, and its mathematical model is as follows:

$$x_0(t) = \begin{cases} \frac{H}{2}\left[1 - \cos\left(\frac{2\pi u}{W_0} t\right)\right] & \frac{W_0}{u} \le t \le \frac{2W_0}{u} \\ 0 & other \end{cases} \quad (46)$$

where $x_0(t)$ represents the input displacement of the impact road, H is the height of the impact road with a value of 0.07 m, W_0 represents the width of the impact road with a value of 0.08 m, u donates the vehicle speed, and the value is 3.08 km/h.

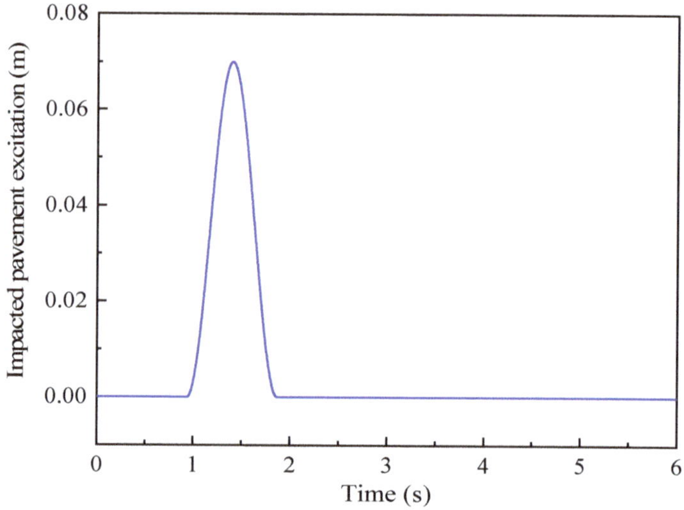

Figure 19. Impacted pavement input simulation curve.

The simulation curve of the 2-DOF quarter vehicle semi-active air suspension controlled by FSMC impacting the road surface is shown in Figure 20. It reflects that the peak value of the FSMC-controlled semi-active air suspension is significantly reduced, compared with that of the passive and SMC-controlled semi-active air suspension under the same impact road excitation.

Compared with passive suspension, the peak reduction percentages of body acceleration, suspension dynamic travel and wheel dynamic load of semi-active suspension under SMC control and FSMC control are shown in Figure 21, in which the body acceleration under SMC control and FSMC control are 13.9% and 27.5%, respectively, lower than that of passive suspension. The suspension dynamic travel is reduced by 12.8% and 26.5%, respectively, while the wheel dynamic load is reduced by 5% and 9.6%, respectively.

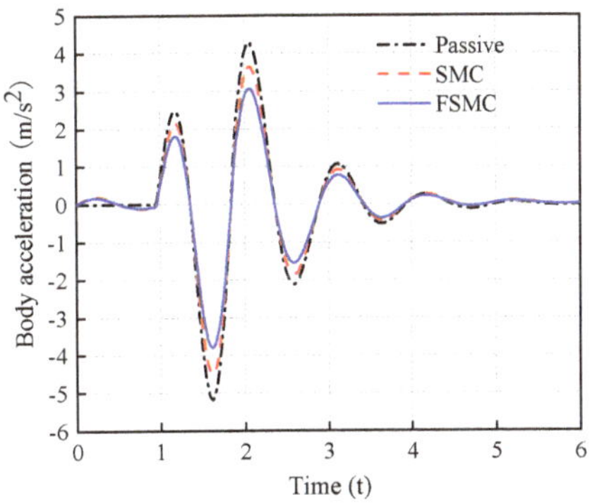

(a)

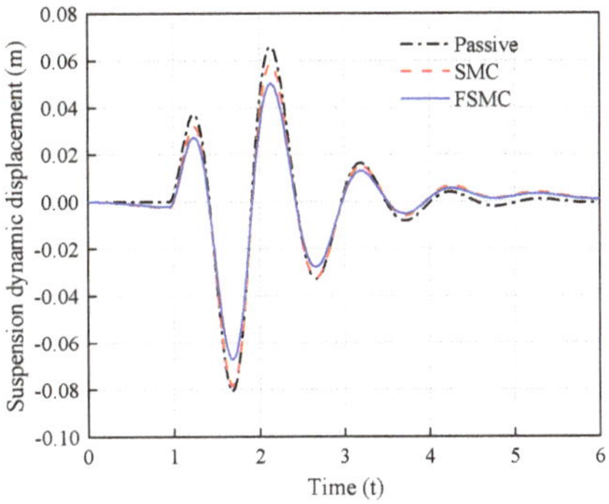

(b)

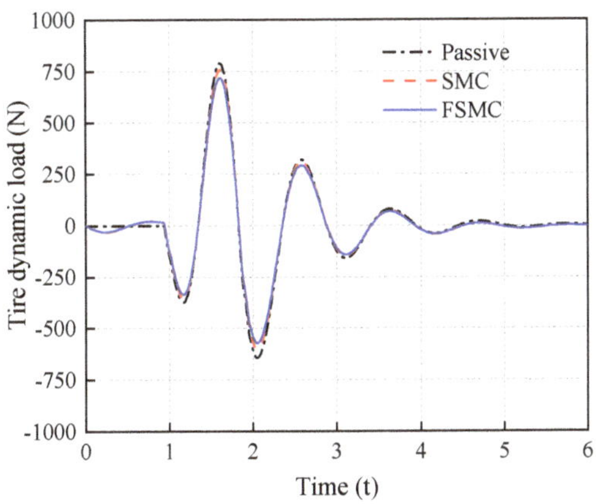

(c)

Figure 20. Simulation curve of air suspension system under impact road excitation. (**a**) Body acceleration, (**b**) suspension dynamic displacement, (**c**) tire dynamic load.

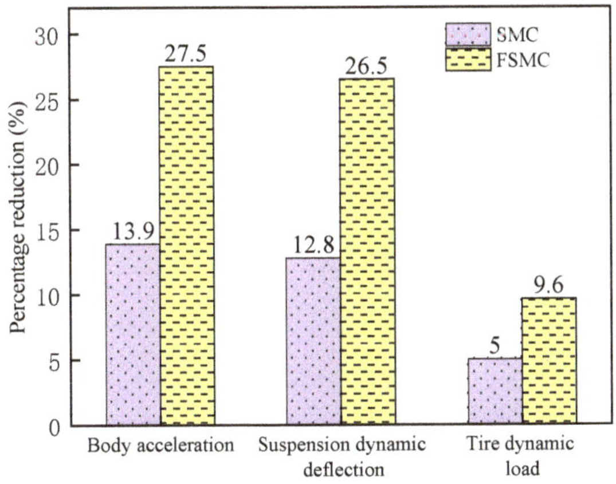

Figure 21. Percentage reduction of peak response under impacted pavement surface.

6. Conclusions

Based on the experimental data, the parameters of the improved hyperbolic tangent model were identified by the genetic algorithm, and the magnetic parameter identification results were consistent with the experimental data. The identified parameters of the improved hyperbolic tangent model could be used in semi-active control. The predictive control current of ANFIS inverse model can track the target control current well, which verifies the effectiveness of the model.

The semi-active air suspension system model of quarter vehicle equipped with MR damper was established, and the feasibility of the FSMC strategy was verified by taking the random road excitation and impact road excitation as input. Compared with the passive suspension, the vehicle acceleration under SMC control and FSMC control is reduced by 15.9% and 26%, respectively; the suspension dynamic travel is reduced by 16.9% and 29.4%, respectively; and the wheel dynamic load is reduced by 7.4% and 8.4%, respectively, under random road excitation. With the impact of road excitation, the peak acceleration of the vehicle based on SMC control and FSMC control decreases by 13.9% and 27.5%, respectively; the peak dynamic travel of suspension decreases by 12.8% and 26.5%, respectively; and the peak dynamic load of tire decreases by 5% and 9.6%, respectively, compared with the passive suspension. It can be concluded that the control strategy proposed in this article has a better damping effect and can improve the driving comfort and stability of the vehicle.

Author Contributions: G.L. contributed to the design of the suspension control algorithm and edited the paper; Z.R. carried out the simulation analysis and wrote the first draft; R.G. designed the suspension system and provided the data; G.H. contributed the control idea and revised the paper. All authors have read and agreed to the published version of the manuscript.

Funding: This research was funded by the National Natural Science Foundation of China (No. 52165004), Key R&D projects of Jiangxi Province of China (No. 20192BBEL50012).

Institutional Review Board Statement: Not applicable.

Informed Consent Statement: Not applicable.

Conflicts of Interest: The authors declare no conflict of interest.

References

1. Nguyen, S.D.; Bao, D.L.; Ngo, V.H. Fractional-order sliding-mode controller for semi-active vehicle MRD suspensions. *Nonlinear Dyn.* **2020**, *101*, 795–821. [CrossRef]
2. He, F.; Zhao, J.; Wang, H.Y. Study on control strategies of semi-active air suspension for heavy vehicles based on road-friendliness. *Adv. Mater. Res.* **2012**, *452*, 328–333. [CrossRef]
3. Zhang, Z.; Wang, J.; Wu, W.; Huang, C. Semi-active control of air suspension with auxiliary chamber subject to parameter uncertainties and time-delay. *Int. J. Robust. Nonlinear Control* **2020**, *30*, 7130–7149. [CrossRef]
4. Gokul, P.S.; Malar, M.K. A contemporary adaptive air suspension using LQR control for passenger vehicles. *ISA Trans.* **2019**, *93*, 244–254.
5. Fan, S.Y.; Chuang, C.W.; Feng, C.C. Apply Novel Grey Model Integral Variable Structure Control to Air-Ball-Suspension System. *Asian. J. Control* **2016**, *18*, 1359–1364. [CrossRef]
6. Elsaady, W.; Oyadiji, S.O.; Nasser, A. Magnetic circuit analysis and fluid flow modeling of an MR damper with enhanced magnetic characteristics. *IEEE Trans. Magn.* **2020**, *56*, 1–20. [CrossRef]
7. Hu, G.; Yi, F.; Liu, H.; Zeng, L. Performance analysis of a novel magnetorheological damper with displacement self-sensing and energy harvesting capability. *J. Vib. Eng. Technol.* **2021**, *9*, 85–103. [CrossRef]
8. Hu, G.; Yi, F.; Tong, W.; Yu, L. Development and evaluation of a MR damper with enhanced effective gap lengths. *IEEE Access* **2020**, *8*, 156347–156361. [CrossRef]
9. Sun, S.S.; Ning, D.H.; Yang, J.; Du, H.; Zhang, S.W.; Li, W.H. A seat suspension with a rotary magnetorheological damper for heavy duty vehicles. *Smart Mater. Struct.* **2016**, *25*, 105032. [CrossRef]
10. Wang, D.H.; Liao, W.H. Magnetorheological fluid dampers: A review of parametric modelling. *Smart Mater. Struct.* **2011**, *20*, 023001. [CrossRef]
11. Mughal, U.N. State of the art review of semi active control for magnetorheological dampers. In Proceedings of the ICNPAA 2016 World Congress, La Rochelle, France, 4–8 July 2016; AIP Publishing LLC: New York, NY, USA, 2016; Volume 1798, p. 020101.
12. Khalid, M.; Yusof, R.; Joshani, M.; Selamat, H.; Joshani, M. Nonlinear identification of a magneto-rheological damper based on dynamic neural networks. *Comput.-Aided Civ. Inf.* **2014**, *29*, 221–233. [CrossRef]

13. Kwok, N.M.; Ha, Q.P.; Nguyen, T.H.; Li, J.; Samali, B. A novel hysteretic model for MR fluid dampers and parameter identification using particle swarm optimization. *Sens. Actuators A Phys.* **2006**, *132*, 441–451. [CrossRef]
14. Guo, S.; Yang, S.; Pan, C. Dynamic modeling of MR damper behaviors. *J. Intell. Mat. Syst. Struct.* **2006**, *17*, 3–14. [CrossRef]
15. Choi, S.B.; Choi, J.H.; Lee, Y.S.; Han, M.S. Vibration control of an er seat suspension for a commercial vehicle. *J. Dyn. Syst. Meas. Control* **2003**, *125*, 60–68. [CrossRef]
16. Pang, H.; Shang, Y.; Yang, J. An adaptive sliding mode–based fault-tolerant control design for half-vehicle active suspensions using T–S fuzzy approach. *J. Vib. Control* **2020**, *26*, 1411–1424. [CrossRef]
17. Wang, G.; Chadli, M.; Basin, M.V. Practical terminal sliding mode control of nonlinear uncertain active suspension systems with adaptive disturbance observer. *IEEE/ASME Trans. Mechatron.* **2020**, *26*, 789–797. [CrossRef]
18. Fei, J.; Xin, M. Robust adaptive sliding mode controller for semi-active vehicle suspension system. *Int. J. Innov. Comput. Inf. Control* **2012**, *8*, 691–700.
19. Du, M.; Zhao, D.; Yang, B.; Wang, L. Terminal sliding mode control for full vehicle active suspension systems. *J. Mech. Sci. Technol.* **2018**, *32*, 2851–2866. [CrossRef]
20. Nguyen, S.D.; Lam, B.D.; Choi, S.B. Smart dampers-based vibration control-Part 2: Fractional-order sliding control for vehicle suspension system. *Mech. Syst. Signal. Pract.* **2021**, *148*, 107145. [CrossRef]
21. Zheng, L.; Li, Y.N.; Baz, A. Fuzzy-sliding mode control of a full car semi-active suspension systems with MR dampers. *Smart Struct. Syst.* **2009**, *5*, 261–277. [CrossRef]
22. Wang, R.; Liu, W.; Ding, R.; Meng, X.; Sun, Z.; Yang, L.; Sun, D. Switching control of semi-active suspension based on road profile estimation. *Veh. Syst. Dyn.* **2021**, 1–21. [CrossRef]
23. Pang, H.; Liu, F.; Xu, Z. Variable universe fuzzy control for vehicle semi-active suspension system with MR damper combining fuzzy neural network and particle swarm optimization. *Neurocomputing* **2018**, *306*, 130–140. [CrossRef]
24. Hu, G.; Liu, Q.; Ding, R.; Li, G. Vibration control of semi-active suspension system with magnetorheological damper based on hyperbolic tangent model. *Adv. Mech. Eng.* **2017**, *9*, 1687814017694581. [CrossRef]
25. Huang, Y.; Na, J.; Wu, X.; Gao, G.; Guo, Y. Robust adaptive control for vehicle active suspension systems with uncertain dynamics. *Trans. Inst. Meas. Control* **2018**, *40*, 1237–1249. [CrossRef]
26. Ming, L.; Yibin, L.; Xuewen, R.; Shuaishuai, Z.; Yanfang, Y. Semi-active suspension control based on deep reinforcement learning. *IEEE Access* **2020**, *8*, 9978–9986. [CrossRef]
27. Ye, H.; Zheng, L. Comparative study of semi-active suspension based on LQR control and H_2/H_∞ multi-objective control. In Proceedings of the 2019 Chinese Automation Congress (CAC), Hangzhou, China, 22–24 November 2019; pp. 3901–3906.
28. Rath, J.J.; Veluvolu, K.C.; Defoort, M. Adaptive super-twisting observer for estimation of random road excitation profile in automotive suspension systems. *Sci. World J.* **2014**, *2014*, 203416. [CrossRef]
29. Chang, F.J.; Chang, Y.T. Adaptive neuro-fuzzy inference system for prediction of water level in reservoir. *Adv. Water Resour.* **2006**, *29*, 1–10. [CrossRef]
30. Kim, H.; Lee, H.; Kim, H. Asynchronous and synchronous load leveling compensation algorithm in airspring suspension. In Proceedings of the 2007 International Conference on Control, Automation and Systems, Seoul, Korea, 17–20 October 2007; pp. 367–372.
31. Liu, H.; Lee, J.C. Model development and experimental research on an air spring with auxiliary reservoir. *Int. J. Automot. Technol.* **2011**, *12*, 839–847. [CrossRef]
32. Karnopp, D.; Crosby, M.J.; Harwood, R.A. Vibration control using semi-active force generators. *J. Eng. Ind.* **1974**, *96*, 619–626. [CrossRef]

applied sciences

MDPI

Article

Effect of Volume Fraction of Fine Sand on Magnetorheological Response and Blocking Mechanisms of Cementitious Mixtures Containing Fe_3O_4 Nanoparticles

Chizya Chibulu , **Mert Yücel Yardimci and Geert De Schutter** *

Magnel-Vandepitte Laboratory, Department of Structural Engineering and Building Materials, Ghent University, 9052 Ghent, Belgium

* Correspondence: geert.deschutter@ugent.be

Citation: Chibulu, C.; Yardimci, M.Y.; De Schutter, G. Effect of Volume Fraction of Fine Sand on Magnetorheological Response and Blocking Mechanisms of Cementitious Mixtures Containing Fe_3O_4 Nanoparticles. *Appl. Sci.* **2022**, *12*, 10104. https://doi.org/10.3390/app121910104

Academic Editors: Antonio Concilio, Salvatore Ameduri, Ignazio Dimino, Vikram G Kamble and Rosario Pecora

Received: 11 August 2022
Accepted: 28 September 2022
Published: 8 October 2022

Abstract: Active rheology control (ARC) or active stiffening control (ASC) is a concept with which the conflicting rheological requirements during different stages of concrete casting can be reconciled. For instance, formwork leakage could be reduced by actively controlling structuration at the formwork joints, without having the negative impact of increased structuration during pumping and form filling. Using the concepts of magnetorheology, an active control methodology was thus recently developed by the authors to study the control of formwork leakages under pressure. This was performed using a small-scale laboratory test setup, using cementitious pastes containing magnetisable particles. To upscale from paste to mortar, the effect of volume fraction of sand on the magnetorheological (MR) response and blocking mechanisms of mixtures containing Fe_3O_4 nanoparticles is thus investigated in the current study. The MR response is determined using storage modulus tests, and the impact of ASC for leakage reduction is investigated by measuring the flow rate. Experimental results show that increasing the sand volume beyond a threshold causes a reduction in mobility of the magnetic particles, and thus lowers the MR effect. Despite this reduction in the MR effect at high sand volume, the increased particle interactions induce clogging and filtration effects, drastically lowering the flow rate. Applying the ASC method refines the voids in the clog, thereby eliminating the filtration effect. It is concluded that ASC can be used on mortar, with the expectation that there would be a reduction in the magnetorheological effect with increasing volume of fine aggregates.

Keywords: active stiffening control; magnetic field; Fe_3O_4 nanoparticles; blocking; packing density; formwork leakage; flow rate

1. Introduction

Active rheology control (ARC) or active stiffening control (ASC) is a developing concept that offers a solution to the conflicting rheological requirements during the different stages of the casting process [1–4]. During pumping, for example, the concrete should be fluid enough to facilitate the pumping process; however, once the concrete is in the formwork, high fluidity gives rise to increased formwork pressure and formwork leakage. The ability to adjust the fresh concrete properties to match the desired behaviour could thus be of advantage. The concrete can be fluid while flowing in the pumping pipes, and then actively stiffened while in the formwork to prevent formwork leakage. ARC or ASC can be achieved by applying an external signal to the concrete that triggers a change in the material behaviour [3]. Based on this principle of ARC, an innovative active control methodology has recently been proposed in which a magnetic field is applied to the cementitious material to invoke a change in the rheological behaviour due to the presence of 'active' ingredients in the mixture. The active ingredients used were magnetisable particles in the form of Fe_3O_4 nanoparticles [5–8] and fly ash [9,10]. Although magnetorheological control is a fairly new concept for cementitious mediums, this is a technology already widely used in other industrial applications such as brakes, hydraulic valves, and shock absorbers [11].

The interest in magnetorheological (MR) fluids for industrial applications arises from the rapid and reversible field-dependent rheological changes in the fluid [12]. When a magnetic field is applied, the magnetic particles acquire dipole moments and align with the magnetic field to form chains or clusters, giving the fluid an additional yield stress component. To initiate flow, these chains or clusters must be deformed or broken.

Building out from the principles of MR fluids, a magnetic-field-based active stiffening control methodology was developed to study the efficacy of ASC for formwork leakages of flowable pastes under high pressure. This was performed using a small-scale laboratory test setup in which a pressurised column with a narrow opening at the bottom was filled with cementitious material to simulate formwork leakage under pressure [9,13]. Using the ASC methodology, reductions in both the flow rate and net mass of leaked material were observed for pressures lower than 30 kPa. This was attributed to the formation of magnetic clusters due to the application of a magnetic field across the outflow width. Although the formation of clusters was less effective at higher pressures, the study showed that clusters formed under a pressure of 10 kPa were able to withstand pressures of up to 100 kPa [9]. Following these promising results obtained using cementitious pastes, the current study is aimed at studying the effects of fine aggregates on the ASC methodology, in a first step to upscaling to mortar phase.

Mortar can be considered as a concentrated suspension, with rigid particles of different sizes suspended in the bulk fluid. In such a system, there are strong interactions between particles due to their close proximity. The concentration, shape, size distribution, and surface properties of the particles all affect the strength of these interactions. In general, there is a net attraction between particles which leads to flocculation [14]. Convergence of such a fluid towards a bottleneck, such as in the case of formwork leakage through a gap, is likely to cause blockage of flow. In general, clogging of granular material is a complex phenomenon that is dependent on the outlet size and type, the physical properties of the particles, and the particle packing [15]. In the case of mortar, however, the presence of the cementitious pastes in which the grains are embedded also influences the contact network between particles [16]. The more paste available, the lower the frequency of collisions between coarser particles, and thus the lower the particle interlocking effect [17]. Additionally, due to the heterogenous nature of cementitious mortars, this blocking effect can induce filtration of the liquid phase through the granular skeleton, as observed in extrusion flows, for example [18], leading to a local change in properties at the flow confinement.

The presence of aggregates clearly has an influence on the nature of the flow, particularly when flowing through a small gap. It can also be expected that the presence of aggregates would influence the magnetorheological response of the fluid, since the magnetically induced particle interactions are strongly dependent on the distance between the magnetic particles within the matrix of the fluid [19]. As such, the current study investigates the effect of sand volume fraction on the magnetorheological response and active stiffening control methodology previously introduced by the authors.

2. Experimental Programme

2.1. Materials

The basic materials used in this study include Portland cement (CEM 52.5 N), spherical Fe_3O_4 nanoparticles (98% purity), a polycarboxylate ether superplasticiser (PCE), a cellulose-based viscosity-modifying admixture (VMA), and fine sand. The cement had a median particle size of approximately 7.2 μm and specific gravity of 3.2. The Fe_3O_4 nanoparticles used in the study had a particle size of 100 nm and specific gravity of 4.95, as stated by the manufacturer. The sand was a very fine quartz sand with median particle size of 170 μm and specific gravity of 2.65. A very fine sand was selected for testing to remain within the limit of maximum particle sizes permissible to conduct rheology experiments within the confined geometry of the plate–plate rheometer. The gap between the shearing plates should be at least 5–10 times the maximum particle size in the mixture [20]. Table 1 shows the chemical composition and Figure 1 shows the particle size distribution of the cement powder and sand.

Table 1. Chemical composition of cement and sand (wt. %).

Element	SiO_2	Al_2O_3	Fe_2O_3	CaO	MgO	K_2O	SO_3	Others
Cement	19.4	6.04	4.12	61.5	1.25	5.35	0.48	1.86
Sand	99.0	0.06	0.70	0.02	-	0.30	-	0.06

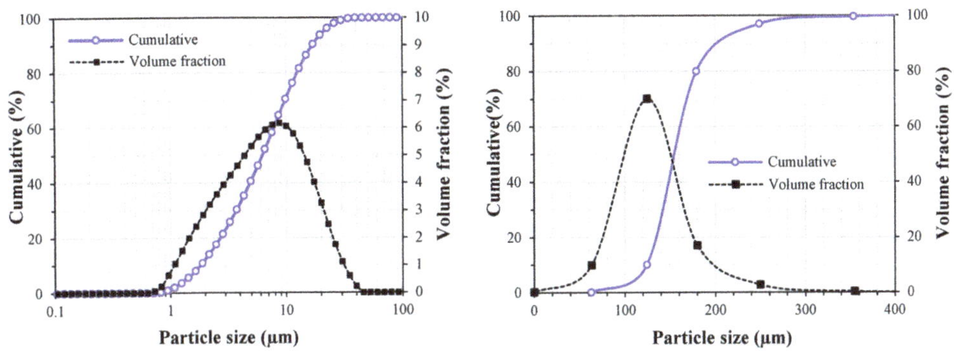

Figure 1. Particle size distribution of (**left**) cement powder; (**right**) sand.

The magnetic properties of the cement and magnetic nanoparticles (MNPs) were measured by a vibrating sample magnetometer (VSM) at 25 °C. The magnetisation curves of the materials are plotted in Figure 2, and the magnetic parameters are summarised in Table 2.

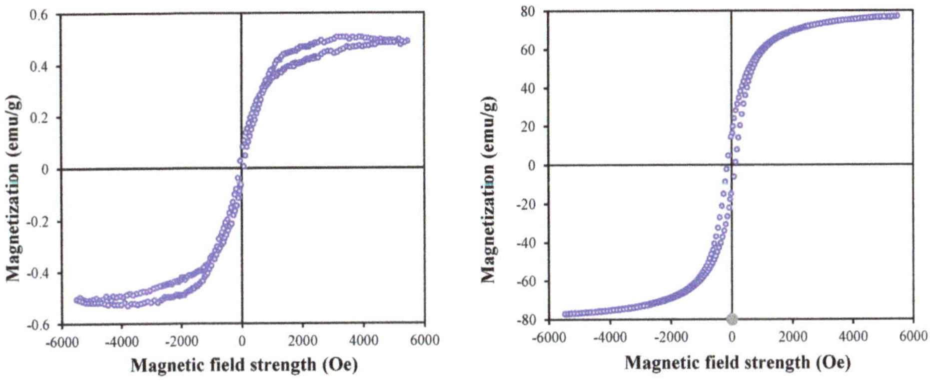

Figure 2. Magnetisation curves of (**left**) Portland cement, (**right**) MNPs.

Table 2. Magnetic properties of cement and Fe3O4 nanoparticles (MNP) used in the study.

Material	Saturation Magnetisation (M_s, emu/g)	Remnant Magnetisation (M_r, emu/g)	Coercive Field (H_c, Oe)
Cement	0.59	0.05	70.72
MNP	77.56	10.23	108.27

Five mixtures were prepared to study the effect of volume concentration of fine aggregates on the magnetorheological response and confined flow under magnetic field.

The water to cement ratio of the paste and the volume of MNPs used was kept constant. The quantity of superplasticiser, and in the instance of the reference paste the quantity of VMA, was adjusted to attain a mini slump spread diameter of 255 ± 5 mm. The sand amount was added to the paste in 10% volumetric replacements of the total volume of paste. Mixtures M0, M1, M2, M3, and M4 thus have 0%, 10%, 20%, 30%, and 40% total volume of sand. This is visually represented in Figure 3, and the mixture proportions of the pastes are shown in Table 3.

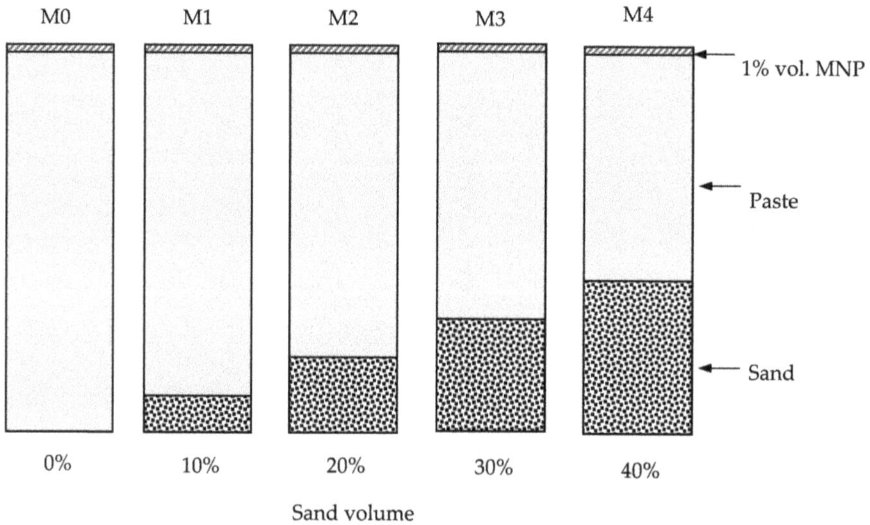

Figure 3. Schematic representation of mixture proportions.

Table 3. Mixture proportions and fresh properties.

Mix. No	W/C (-)	PCE (wt. % Cement)	VMA (wt. % Cement)	Sand (% Total Volume)	NP (% Total Volume)	Mini Slump (±5 mm)	Wet Density (g/cm^3)
M0	0.55	-	0.025	0	1	255	1.91
M1	0.55	0.05	0	10	1	255	2.01
M2	0.55	0.15	0	20	1	255	2.03
M3	0.55	0.4	0	30	1	255	2.13
M4	0.55	1.0	0	40	1	255	2.16

2.2. *Methodology*

2.2.1. Particle Packing Density

The packing density of solids in a granular system is a basic parameter that characterises the properties of the system. Since we can consider concrete or mortar as a granular system, the particle packing would thus also significantly influence the properties of these mixtures [21]. The packing density, minimum void ratio, and excess water ratio were determined using the centrifugal consolidation method [22,23]. The mixtures were placed in standard 50 mL centrifuge tubes and centrifuged at 3500 rpm for 10 min. When the test-tubes are centrifuged, the particles in the mixture are compacted, thereby reducing the amount of water needed to fill the voids in between the compacted particle matrix. After centrifuging, the excess amount of water thus forms a layer at the top of the compacted mixture. This excess water layer was removed using a pipette and measured, and the amount of water remaining in the compacted sample could be determined. The excess water ratio W_e could then be calculated according to Equation (1).

$$W_e = \frac{V_{w(excess)}}{V_s} \tag{1}$$

The packing density, φ_{max} was then calculated as shown in Equation (2):

$$\varphi_{max} = \frac{V_s}{V_s + V_w} \tag{2}$$

where V_s and V_w are the volume of total solids and the volume of the water remaining in the consolidated mass, respectively. The water remaining in the consolidated material was taken as the mass of the initial water in the mixture (mixing water), minus the excess water removed, assuming the particles did not absorb any water. The minimum void ratio V_{min} defined as the volumetric ratio of voids to solids can be determined from the packing density and is calculated as shown in Equation (3). Generally, the higher the packing density, the smaller the volume of voids needing to be filled [24].

$$V_{min} = (1 - \varphi_{max}) / \varphi_{max} \tag{3}$$

2.2.2. Oscillatory Time Sweep Test

The oscillatory time sweep test was performed to evaluate the structural build-up with and without magnetic field application by using a rotational parallel-plate rheometer (Anton Paar MCR-102, Graz, Austria). Tests with magnetic field application were conducted with a magnetic field strength of approximately 0.65 T, to match the magnetic field applied in the pressure flow test. The magnetic field was applied at the start of the time sweep test. The strain amplitude and frequency during the time sweep tests were 0.001% (within the linear viscoelastic region) and 2 Hz, respectively. To determine the linear viscoelastic region (LVER), strain sweep tests were conducted without a magnetic field from 0.0001% to 10% and constant frequency of 2 Hz prior to the time sweep tests. The critical strain as shown in Figure 4 was approximately 0.004%, and thus a strain amplitude of 0.001% could be selected for the time sweep test.

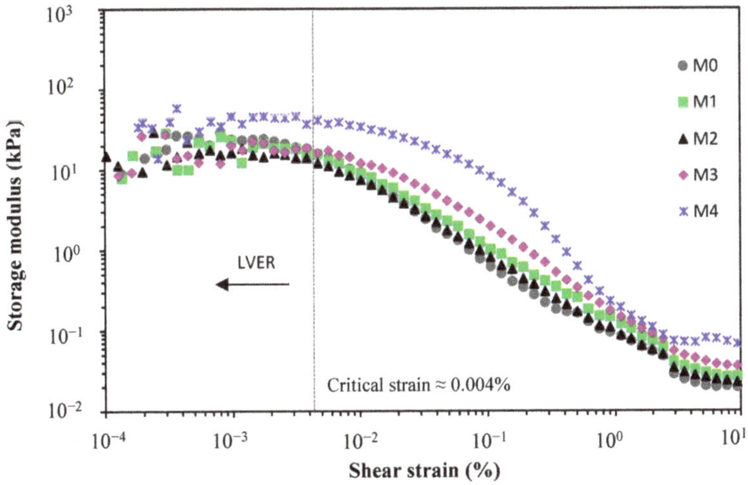

Figure 4. Strain sweep test results showing LVER of the mixtures.

2.2.3. Pressure-Driven Flow Test

A small-scale test setup that was previously developed [9,13] was used to study the flow rate of the mixtures through a confined geometry under the impact of applied pressure to imitate formwork leakage. The test setup includes a vertical pressurised cylinder with a flow outlet positioned at the bottom of the cylinder. The flow outlet used in this study had a rectangular cross-section, with dimensions of 2 mm × 20 mm, and height of approximately 7 mm. To apply a magnetic field, the bottom plate of the cylinder was replaced by a plate with similar geometry, but instead with static magnets (see Figure 5). This configuration creates a magnetic field across the flow width, similar to MR fluids in the valve mode [12,25]. This allows for the deposition of the magnetic particles at the walls, thereby pinching the flow width together and reducing the flow rate like an MR fluid in the magnetic gradient pinch mode [26].

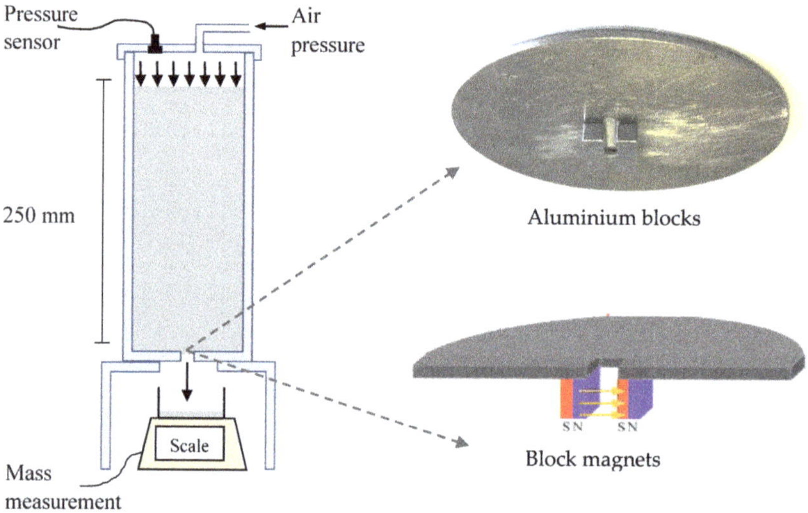

Figure 5. Test setup to study pressure-driven flow of cementitious paste [9].

Previous experiments showed a threshold pressure at approximately 30 kPa at which the active stiffening control was no longer sufficient to cause blockage of flow. The overhead pressure was thus kept constant at 30 kPa for each test and two tests were performed for each mixture to check repetition.

3. Results and Discussion

3.1. Particle Packing Density and Excess Water Ratio

The packing densities of the particles, which is defined as the ratio of the volume of solid particles to the bulk volume occupied by the particles [27], are given in the first column of Table 4 and plotted against the volume fraction of sand in Figure 6. As expected, the packing density, φ_{max}, increases with increasing sand volume fraction. This is due to the occupying effect which follows when the volumetric fraction of coarser particles is less than the optimum value [27,28]. When the larger particles (sand grains) are added to smaller particles (cement and MNP), the larger particles occupy the space within the bulk and porous volume, thereby increasing the packing density. The minimum void ratio (V_{min}) and excess water ratio (W_e) were are also determined and are given in Table 4, and W_e is plotted against the volume fraction of sand in Figure 6.

Table 4. Particle packing and excess water ratio.

Sample	φ_{max}	V_{min}	W_e
M0	0.47	1.13	0.59
M1	0.54	0.85	0.46
M2	0.59	0.69	0.33
M3	0.64	0.57	0.22
M4	0.67	0.49	0.12

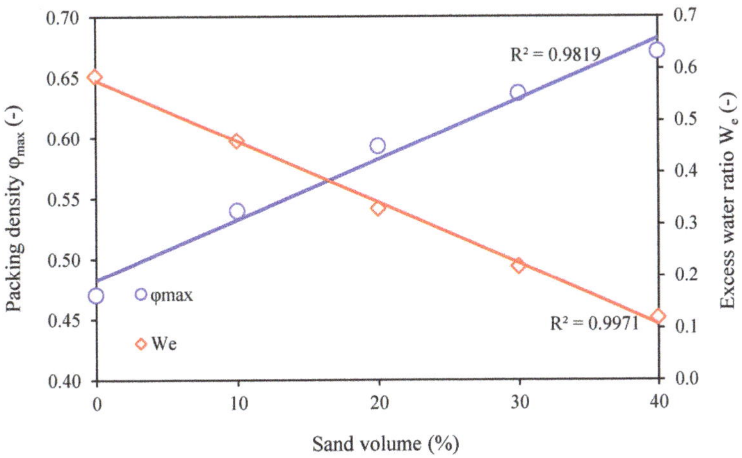

Figure 6. Packing density and excess water ratio as a function of sand volume fraction.

In a state of maximum packing density, the coarser particles are tightly packed together, with just enough fine particles available to fill the voids between the coarser particles. Since there are no excess fine particles separating the coarse particles from each other, the coarser particles would slide against each other, resulting in large interparticle interactions and interlocking. Conversely, in the state of maximum mass flow rate, the coarse particles are separated from each other by the excess fines since there are more than enough fines necessary to fill the voids. The fine particles act like ball bearings, lowering the particle interlocking action and the smooth movement of coarse particles within the fluid [17]. Extending this theory to cementitious pastes, the excess amount of water in the paste is responsible for dispersing the particles and lubricating the cement paste [27]. Increasing the sand volume in the mixture from 0 to 40% therefore moves the mixtures from a state of mass flow rate towards maximum packing density, which is evident by the increasing φ_{max} and decreasing W_e. This could arguably also lead to a reduction in mobility of the MNPs to form clusters due to the obstruction caused by the increasing presence of larger solid particles in the mixture as illustrated in Figure 7. The magnetorheological response of the mixtures was therefore investigated by measuring the storage modulus of the mixtures with and without magnetic field application and is discussed in the following section.

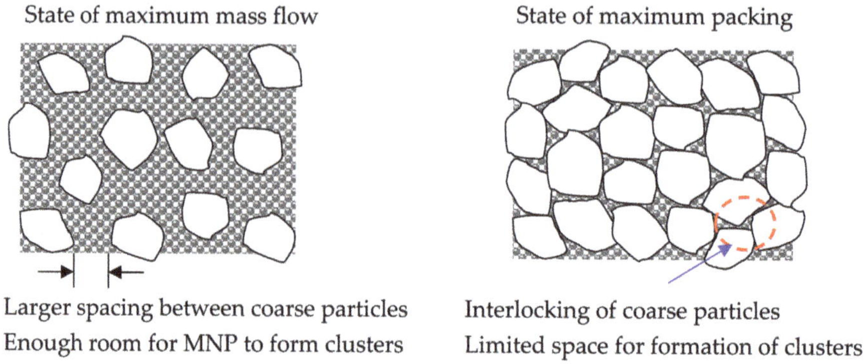

Figure 7. Effect of aggregate content (after [17]).

3.2. Storage Modulus

The effect of sand volume fraction on the magnetorheological response determined by measuring the storage modulus of the mixtures with and without magnetic field application is shown in Figures 8 and 9. In the absence of the external magnetic field, the mixtures all have a storage modulus in a similar range after 300 s (G'_{300s} < 150 kPa). When a magnetic field of 0.65 T is applied, the initial rate of increase in storage modulus (and hence the structural build-up) and final storage modulus after 300s of all the mixtures increases significantly to varying degrees. This is consistent with previous studies in which the storage modulus of cementitious pastes containing MNPs was measured with/out a magnetic field. The increase in storage modulus is due to the increased structuration caused by the formation of magnetic clusters within the paste. The magnitude of increase in the storage modulus under a magnetic field provides an indication of the degree of magneto responsiveness and ability to form clusters. Generally, a larger relative increase in storage modulus indicates a higher magneto response [5,7,8,29].

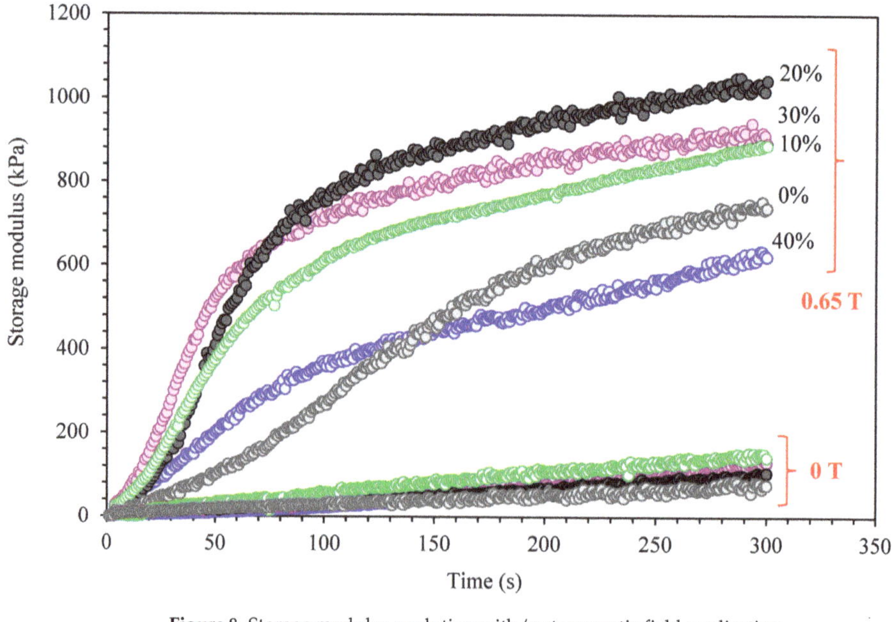

Figure 8. Storage modulus evolution with/out magnetic field application.

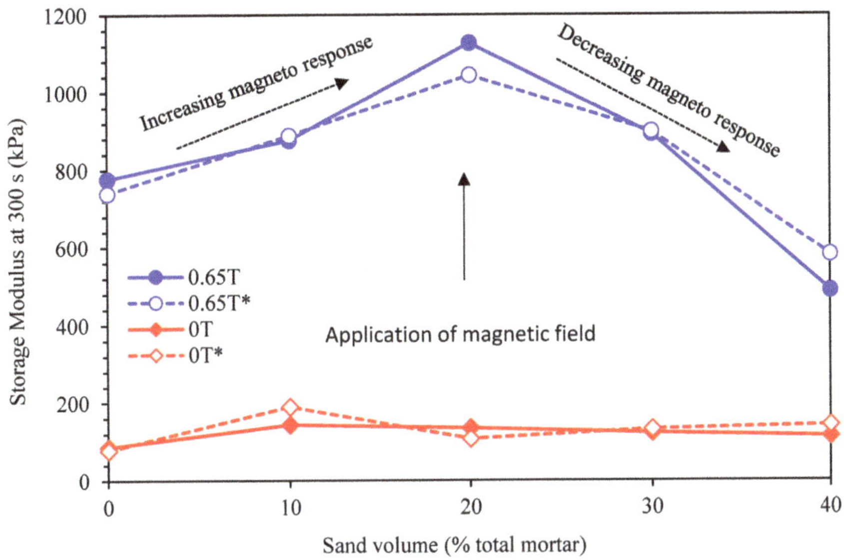

Figure 9. Effect of sand volume fraction on magneto response and increase in storage modulus at 300 s. * Repetition of test on a different batch.

Increasing the volume percentage of sand particles from 0 to 20% had the effect of increasing the magnetorheological response of the mixture, as shown in the storage modulus evolution in Figure 8 and final storage modulus after 300 s in Figure 9. It is hypothesised that in up to 20% volume of sand, the presence of sand grains in the mixture does not have a negative impact on the ability of the magnetic particles to form chains or clusters. Due to the low packing density, the sand grains are loosely packed within the cementitious paste, giving the magnetic particles more freedom to move and form clusters. As the percentage volume of sand is increased from 20 to 40%, it is theorised that the presence of the sand grains begins to inhibit the movement of the magnetic particles, thereby lowering the magnetorheological response. At 40% sand volume, the magnetorheological response is the lowest. The result thus suggests that there is a threshold volume percentage of fine aggregate that can be added to the paste after which the magnetorheological effects begin to decline due to the presence of aggregates. In practice, it is more desirable to formulate the mixture composition for maximum packing density to improve the strength and durability indexes of the concrete [30]. It can thus be hypothesised that for more practical mixture compositions containing fine aggregates of even larger sizes than used in this study, the magnetorheological response would not be as high as its paste counterparts. Nevertheless, despite the M4 having the lowest magneto response, the magnitude of increase in storage modulus is still noteworthy. The storage modulus increased by a factor of 4.2. Moreover, in comparison to the time sweep tests that are performed in nearly static conditions, formwork leakage flows occur in dynamic conditions, of which the particle concentration also plays a key role in the blocking mechanisms. The higher sand volume could have a positive effect on the blocking mechanisms despite the lower magneto response. This is discussed in the following section.

3.3. Flow Rate

The mass of material flowing out of the test cylinder was measured continuously during testing. The loss of mass from the test cylinder relative to the initial mass in the 250 mm height of fluid (M_{loss}) could thus be determined. The M_{loss} as a function of time under 30 kPa overhead pressure is plotted in Figure 10. Despite having different packing densities, the M0, M1, M2, and M3 mixtures exhibited similar flow behaviour, i.e., mass flow, with similar flow rates when no magnetic field was applied. The slope of the M_{loss} curve remained relatively constant until the end of the test when the test cylinder was almost completely emptied, except for the material in the dead zones. The M4 mixture, however, deviated from mass flow and exhibited significant filtration effects from the onset of flow. Only the liquid fraction was lost from the cylinder, and therefore the percentage mass loss was extremely low. However, the filtration process continued until most of the liquid fraction within the vicinity of the gap was lost.

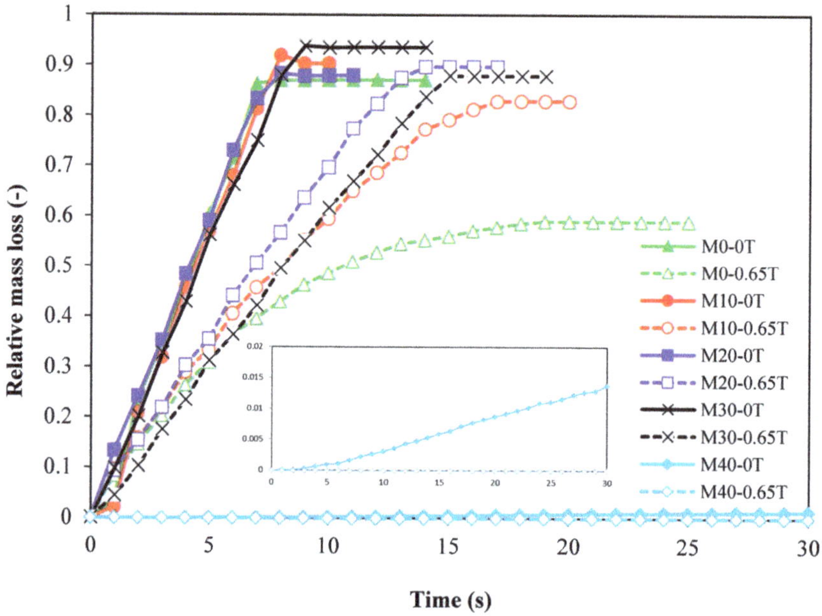

Figure 10. Relative mass loss as a function of time without ASC (solid lines) and with ASC (dashed lines).

On the contrary, when a magnetic field of 0.65 T was applied, the flow rates of all the mixtures decreased. The M1, M2, and M3 mixtures only show a reduction in slope, with no changes to the net mass of outflow at the end of the test. The M0 mixture, however, not only showed a reduction in the initial slope, but the slope continued to decrease until it reached a plateau, indicating that complete blockage occurred. This resulted in an approximately 40% reduction in mass loss. For the M40 mixture, the slope was effectively reduced to zero and the filtration effect was not detectable. The reduction in slope is due to the formation of magnetic clusters at the outflow region [9]. As the mixture flows through the magnetic field, the magnetisable particles are deposited onto the walls of the outflow channel that pinch the flow width inwards, like an MR fluid in the magnetic gradient pinch mode [26].

The mass flow rate $\dot{m}$ can be obtained by computing the slope of the M_{loss} curves. For comparative reasons, the mass flow rate was converted into a volumetric flow rate since the mixtures had different densities. The linear slope indicates a constant flow rate, with little to no blocking, and thus it was assumed that the density of the material flowing out of the cylinder remained unchanged. In the cases where blocking was observed, the flow rate was only computed during the initial flow stage when the slope of the M_{loss} curve was approximately constant [9]. In the case of filtration, since it was predominantly the liquid fraction that was lost from the sample, the density of material was assumed to be that of water. The computed flow rates with and without ASC are plotted in Figure 11 and show the effect of percentage sand volume on the magnetically induced flow reduction.

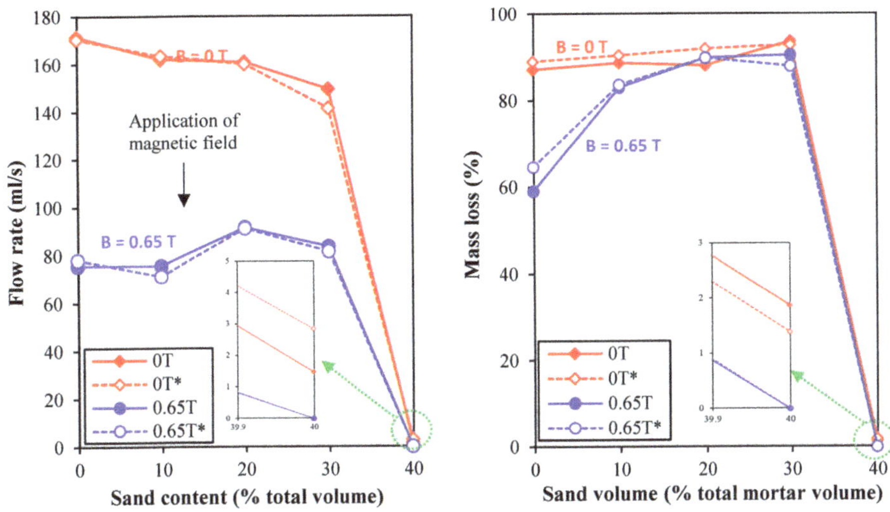

Figure 11. Effect of sand content on (**left**) mass flow rate, (**right**) mass loss (%). * Repetition of test on a different batch.

Our results showed that when there are no sand grains in the mixture, the packing density is lower, the excess water is higher, and the MNPs have more freedom to move and form clusters. As the clusters deposit on the walls of the outflow, they accumulate and lead to complete blockage of the flow. When the sand volume is increased up to 30%, the packing density is increased, and the excess water decreased. However, it is hypothesised that the percentage concentration of sand in the mixture is still sufficiently low, such that the interparticle interactions between the sand particles are still below the level at which large interlocking actions can be observed. There is still sufficient room for the aggregates to move freely within the paste. However, this is theorised to have a negative effect on the formation of clusters, in that the movement of the aggregates within the paste can lead to the breakage of magnetic clusters formed during the shearing action caused by the applied overhead pressure. This could explain why the M1, M2, and M3 pastes did not block completely despite having a higher or similar magnetic response in the storage modulus results in comparison to the M0 mixture. For the M4 mixture, however, it is assumed from the higher packing density and low excess water that there is sufficient interparticle interaction between the sand grains to cause interlocking effects, particularly as the fluid is flowing through the outflow region. The increased friction between particles leads to a jamming of the flow and leads to filtration effects. This is consistent with the theory that the probability of a granular blocking event increases when the particles are closer to each other, which becomes more probable when the volume fraction of particles is increased. In fact, if the mixture is stable such that there is no dynamic segregation that would locally increase the volume at the flow confinement, granular blocking during concrete casting is said to only fractionally be dependent on the number and size of the coarsest particles passing through the confinement, the volume fraction, and the size of the confinement (e.g., space between reinforcement bars), and not necessarily the rheological behaviour of the suspending fluid [31]. Considering that no visible segregation or bleeding was observed in the mixtures used in this study, it can be assumed that the blocking was caused by the increase in volume concentration of sand coupled with the flow confinement. Due to the non-homogenous nature of the mixtures, the blocking induced the filtration of the interstitial fluid through the granular skeleton. The applied overhead pressure enhances the friction between the grains, increasing the stresses and pressure gradients, which in turn increases the fluid filtration [32].

When a magnetic field was applied, however, the magnetic particles form clusters that further refine the voids in the clog created by the agglomeration of sand particles at the outflow. Due to this pore refinement, the loss of interstitial fluid is prevented, thereby preventing the formation of honeycombs in the finished surface, as shown in Figure 12, which shows the hardened surface of the M4 mixture after the tests with and without the ASC method. A previous study shows that if the fluid intrinsically exhibits blocking, the application of the magnetic field further enhances this blocking behaviour [9]. Thus, although the magneto response is decreased due to increasing sand content, this is overwhelmingly compensated for by the increase in particle interactions which lead to blocking.

Figure 12. Hardened surface after the test: (**left**) no magnetic field application; (**right**) magnetic field of 0.65 T applied.

4. Conclusions

Effect of volume fraction of fine sand on the magnetorheological response and blocking mechanisms of cementitious mixtures containing Fe_3O_4 nanoparticles was investigated for the application of formwork leakage control. From the preceding results and discussions, the following conclusions can be drawn:

(1) The magnetorheological response, measured using the oscillatory time sweep tests, increases with increasing volume fraction of sand until a threshold value after which the presence of fine aggregates begins to limit the movement of the MNPs.
(2) Using ASC, the flow rate is significantly reduced for all mixtures due to the formation of magnetic clusters at the outflow. However, the flow reduction is reduced when the sand concentration is low in comparison to the reference paste.
(3) The filtration of the interstitial fluid caused by clogging effects at higher aggregate content can be reduced or eliminated when the ASC methodology is applied due to the pore refinement resulting from the formation of magnetic clusters.

In summary, the ASC methodology can be used on mortar, with the expectation that there would be a reduction in the magnetorheological effect with increasing volume of fine aggregates. Regardless, there is still great potential for the ASC method to be used for formwork leakage control due to the counteracting blocking effect caused by the increased particle interactions, which are further enhanced when the ASC method is applied. Further studies are needed to reveal the limits of the active stiffening control method to reduce or prevent formwork leakage.

Author Contributions: Data curation, C.C.; Formal analysis, C.C.; Funding acquisition, G.D.S.; Investigation, C.C.; Supervision, M.Y.Y. and G.D.S.; Visualization, C.C.; Writing—original draft, C.C.; Writing—review & editing, M.Y.Y. and G.D.S. All authors have read and agreed to the published version of the manuscript.

Funding: This research was funded by the European Research Council (ERC) under the European Union's Horizon 2020 research and innovation program (grant agreement No. 693755).

Data Availability Statement: The data presented in this study are available on request from the corresponding author.

Acknowledgments: This paper is a deliverable of the ERC Advanced Grant project 'SmartCast'. The authors gratefully acknowledge the financial support from ERC.

Conflicts of Interest: The authors declare no conflict of interest.

References

1. de Schutter, G. *Smart Casting of Concrete Structures by Active Control of Rheology (SmartCast)*; Grant Project 693755, 2016–2021; European Commission: Luxembourg, 2016.
2. de Schutter, G. SmartCast: Smart Casting of Concrete Structures. 17 October 2017. Available online: https://youtu.be/ZgVwNkyvJFg (accessed on 17 October 2017).
3. De Schutter, G.; Lesage, K. Active control of properties of concrete: A (p)review. *Mater. Struct.* **2018**, *51*, 123. [CrossRef] [PubMed]
4. de Schutter, G.; Lesage, K.; el Cheikh, K.; de Schryver, R.; Muhammad, M.; Chibulu, C. Introduction to the concept of active rheology control in case of pumping of cementitious materials. In Proceedings of the 2nd International RILEM Conference Rheology and Processing of Construction Materials, Dresden, Germany, 8–11 September 2019.
5. Jiao, D.; El Cheikh, K.; Shi, C.; Lesage, K.; De Schutter, G. Structural build-up of cementitious paste with nano-Fe3O4 under time-varying magnetic fields. *Cem. Concr. Res.* **2019**, *124*, 105857. [CrossRef]
6. Jiao, D.; Lesage, K.; Yardimci, M.; El Cheikh, K.; Shi, C.; De Schutter, G. Rheological Properties of Cement Paste with Nano-Fe_3O_4 under Magnetic Field: Flow Curve and Nanoparticle Agglomeration. *Materials* **2020**, *13*, 5164. [CrossRef] [PubMed]
7. Jiao, D.; Lesage, K.; Yardimci, M.Y.; El Cheikh, K.; Shi, C.; De Schutter, G. Structural evolution of cement paste with nano-Fe3O4 under magnetic field—Effect of concentration and particle size of nano-Fe3O4. *Cem. Concr. Compos.* **2021**, *120*, 104036. [CrossRef]
8. Jiao, D., Lesage, K.; Yardimci, M.Y.; EL Cheikh, K.; Shi, C.; De Schutter, G. Quantitative assessment of the influence of external magnetic field on clustering of nano Fe3O4 particles in cementitious paste. *Cem. Concr. Res.* **2021**, *142*, 106345. [CrossRef]
9. Chibulu, C.; Yardimci, M.Y.; Jiao, D.; De Schryver, R.; Lesage, K.; De Schutter, G. Active stiffening control by magnetically induced blocking in confined flow of fly ash pastes. *Constr. Build. Mater.* **2021**, *313*, 125485. [CrossRef]
10. Jiao, D.; Lesage, K.; Yardimci, M.Y.; Shi, C.; De Schutter, G. Possibilities of fly ash as responsive additive in magneto-rheology control of cementitious materials. *Constr. Build. Mater.* **2021**, *296*, 123656. [CrossRef]
11. Carlson, D.J. Mr Fluids and Devices in The Real World. *Int. J. Mod. Phys. B* **2005**, *19*, 1463–1470. [CrossRef]
12. Gedik, E.; Kurt, H.; Recebli, Z.; Balan, C. Two-dimensional CFD simulation of magnetorheological fluid between two fixed parallel plates applied external magnetic field. *Comput. Fluids* **2012**, *63*, 128–134. [CrossRef]
13. Chibulu, C.; el Cheikh, K.; Yardimci, M.Y.; de Schutter, G. Experimental Study of Formwork Tightness as a Function of Rheological Properties of SCC. In *Rheology and Processing of Construction Materials*; Mechtcherine, V., Khayat, K., Secrieru, E., Eds.; RheoCon 2019, SCC 2019; RILEM Bookseries; Springer: Cham, Switzerland, 2019; Volume 23, pp. 476–481.
14. Banfill, P.F.G. The rheology of fresh cement and concrete—A review. In Proceedings of the 11th International Cement Chemistry Congress, Durban, South Africa, 11–16 May 2003.
15. Xiao, Y.; Han, Y.; Jia, F.; Liu, H.; Li, G.; Chen, P.; Meng, X.; Bai, S. Research on clogging mechanisms of bulk materials flowing trough a bottleneck. *Powder Technol.* **2020**, *381*, 381–391. [CrossRef]
16. El Cheikh, K.; Rémond, S.; Khalil, N.; Aouad, G. Numerical and experimental studies of aggregate blocking in mortar extrusion. *Constr. Build. Mater.* **2017**, *145*, 452–463. [CrossRef]
17. Fung, W.; Kwan, A. Effect of particle interlock on flow of aggregate through opening. *Powder Technol.* **2014**, *253*, 198–206. [CrossRef]
18. Perrot, A.; Rangeard, D.; Mélinge, Y.; Estellé, P.; Lanos, C. Extrusion Criterion for Firm Cement-Based Materials. *Appl. Rheol.* **2009**, *19*, 53042.

19. Samal, S.; Škodová, M.; Blanco, I. Effects of Filler Distribution on Magnetorheological Silicon-Based Composites. *Materials* **2019**, *12*, 3017. [CrossRef]
20. Ferraris, C.; Martys, N.S. Concrete Rheometers. In *Understanding the Rheology of Concrete*; Woodhead Publishing Limited: Cambridge, UK, 2012; pp. 63–82.
21. Fung, W.W.S.; Kwan, A.K.H.; Wong, H.H.C. Wet packing of crushed rock fine aggregate. *Mater. Struct.* **2009**, *42*, 631–643. [CrossRef]
22. Fennis, S.A.A.M.; Walraven, J.C.; Nijland, T.G. Measuring the packing density to lower the cement content in concrete. In *Tailor Made Concrete Structures*; Taylor & Francis Group: London, UK, 2008; pp. 419–424.
23. Dai, X.; Aydin, S.; Yardimci, M.Y.; Qiang, R.; Lesage, K.; De Schutter, G. Rheology, early-age hydration and microstructure of alkali-activated GGBFS-Fly ash-limestone mixtures. *Cem. Concr. Compos.* **2021**, *124*, 104244. [CrossRef]
24. Raj, N.; Patil, S.G.; Bhattacharjee, B. Concrete Mix Design by Packing Density Method. *J. Mech. Civ. Eng. (IOSR-JMCE)* **2014**, *11*, 34–46. [CrossRef]
25. Hajalilou, A.; Mazlan, S.A.; Lavvafi, H.; Shameli, K. *Field Responsive Fluids as Smart Materials*; Springer Nature: Singapore, 2016.
26. Goncalves, F.D.; Carlson, J.D. An alternate operation mode for MR fluids—Magnetic gradient pinch. *J. Physics Conf. Ser.* **2009**, *149*, 01250. [CrossRef]
27. Wong, H.H.C.; Kwan, A.K.H. Packing density of cementitious materials: Measurement and modelling. *Mag. Concr. Res.* **2008**, *60*, 165–175. [CrossRef]
28. Wong, V.; Chan, K.W.; Kwan, A.K.H. Applying Theories of Particle Packing and Rheology to Concrete for Sustainable Development. *Organ. Technol. Manag. Constr. Int. J.* **2013**, *5*, 844–851. [CrossRef]
29. Jiao, D.; Lesage, K.; Yardimci, M.Y.; el Cheikh, K.; Shi, C.; de Schutter, G. Rheological behavior of cement paste with nano-Fe3O4 under magnetic field: Magneto-rheological responses and conceptual calculations. *Cem. Concr. Compos.* **2021**, *120*, 104035. [CrossRef]
30. Ghoddousi, P.; Javid, A.A.S.; Sobhani, J. Effects of particle packing density on the stability and rheology of self-consolidating concrete containing mineral admixtures. *Constr. Build. Mater.* **2014**, *53*, 102–109. [CrossRef]
31. Roussel, N.; Nguyen, T.; Yazoghli, O.; Coussot, P. Passing ability of fresh concrete: A probabilistic approach. *Cem. Concr. Res.* **2009**, *39*, 227–232. [CrossRef]
32. Perrot, A.; Lanos, C.; Melinge, Y.; Estellé, P. Mortar physical properties evolution in extrusion flow. *Rheol. Acta* **2007**, *46*, 1065–1073. [CrossRef]

applied sciences

MDPI

Article

Magnetically-Induced Pressure Generation in Magnetorheological Fluids under the Influence of Magnetic Fields

Purwadi Joko Widodo †, Eko Prasetya Budiana †, Ubaidillah Ubaidillah † and Fitrian Imaduddin *,†

Mechanical Engineering Department, Universitas Sebelas Maret, Jl. Ir Sutami No 36A, Surakarta 57126, Indonesia; purwadijokow@staff.uns.ac.id (P.J.W.); ekoprasetya@staff.uns.ac.id (E.P.B.); ubaidillah_ft@staff.uns.ac.id (U.U.)
* Correspondence: fitrian@ft.uns.ac.id
† These authors contributed equally to this work.

Abstract: This study aims to observe the magnitude of the Magnetorheological Fluids (MRFs) pressure due to the application of a magnetic field. This was accomplished by placing the MRFs in a U-shaped tube, then applying a magnetic field generated by a magnetic coil. A finite element simulation for the magnetic field was carried out to estimate the magnetic field strength generated by the coil variable to the current input given in the simulated apparatus. Changes in MRFs pressure were recorded using a data logger to better observe the fluid pressure phenomena occurring in the MRFs with respect to current input variations. The results showed that the magnetic field influences the MRFs fluid pressure proportionally. The slope is not constant as the magnetic field effect to the fluid pressure gets stronger when the current input is higher. However, there are also an adverse effect of heat generated in the coil in higher current, which results in coil performance degradation and reduces the magnetic field strength.

Keywords: magnetorheological fluids; pressure; magnetic fileds; U-shaped tube

Citation: Widodo, P.J.; Budiana, E.P.; Ubaidillah, U.; Imaduddin, F. Magnetically-Induced Pressure Generation in Magnetorheological Fluids under the Influence of Magnetic Fields. *Appl. Sci.* **2021**, *11*, 9807. https://doi.org/10.3390/app11219807

Academic Editors: Rosario Pecora, Antonio Concilio, Salvatore Ameduri, Ignazio Dimino, Vikram G Kamble and José M. Franco

Received: 22 July 2021
Accepted: 12 October 2021
Published: 20 October 2021

1. Introduction

As a smart material, Magnetorheological Fluids (MRFs) are discussed by many researchers due to their favourable characteristics for advanced actuator applications. Existing literature on MRFs include studies on the material's synthesis and actuator design [1,2], behaviour modelling and control [3,4], and its applications [5]. As such, this long history of MRFs development has made it one of the foremost choices for advanced actuator development in automation technologies. Although MRFs are affected by changes in material viscosity when exposed to a magnetic field, these changes can be quickly controlled and are reversible by generating or negating the influence of the magnetic field which, in turn, influences the properties of the magnetorheological material [6].

MRFs change from a viscous liquid to a solid when exposed to an external magnetic field. This is due to changes in the MRFs structure wherein a dipole torque along the external magnetic field lines causes the ferromagnetic particles in the MRFs to form a chain-like structure. When the external magnetic field is removed, the MRFs returns to its original state within milliseconds as the particles disperse due to thermal motion. When the strength of the magnetic field increases, the pressure on the ferromagnetic particle chain structure increases, causing the MRFs' viscosity to increase. In such cases, the yield stress that can be applied under steady conditions is proportional to the strength of the magnetic field applied. However, the yield strength varies according to the magnetization properties of the metal particles used in the MRFs which are a function of the concentration and properties of the metal particles [7].

In terms of MRFs operation modes, most studies have investigated the ability of MRFs to form a chain-like structure (barrier) when a magnetic field is applied and concluded that the formation of this barrier increases the shear stress of the material. They also note that changes in shear stress alter its ability to withstand external forces. Due to its ability to form a chain-like structure during operation, MRFs is utilized in several operating modes using certain control techniques. As such, MRFs can be operated in many ways depending on the requirements of an application.

Various every day devices, such as shock absorbers [8,9], brakes [10,11], clutches [12], earthquake dampers [13,14], control valves [15,16], mounts [17–19], thermal energy transfers [20], biomedical applications [21], precision polishing [22], and sound propagation devices [23,24], are made possible due to the properties of these smart materials. Based on the studies above, it is evident that MRFs are mostly used in energy dissipating devices and not in an active energy actuating device due to its operational modes.

For practical reasons, devices that use MRFs can be categorized into four operating modes: (1) shear-mode, (2) squeeze-mode, (3) valve-mode, and (4) gradient-pinch-mode [25]. These operating modes rely on value changes in the shear stress of the material when a magnetic field is applied [26]. Most studies conclude that microstructures similar to the formed chain increase the value of the force against the applied external force [27]. This change in force elicits a change in the shear strength of the MRFs, which increases the shear resistance of the material and its ability to withstand external forces. This feature is utilized in most applications that apply MRFs as a smart material.

On the other hand, only a few studies investigate MRFs as liquid possessing ferromagnetic properties, which could perform active actuation. Departing from the knowledge that the magnetic field induction is capable of causing the particle in MRFs to move in a specific direction, these movements hypothetically should be able to generate internal fluid pressure. It is, therefore, interesting to investigate the potential of MRFs in this respect. This study will observe the significance of these movements to cause an increase in the fluid pressure and analyze the relationship between the magnetic field and the fluid pressure. The results will be an essential finding that leads to discovering a new MRFs working mode. This study is started by designing a theoretical U-shaped tube apparatus in the Finite Element Method Magnetics (FEMM) software package. The simulation was conducted to calculate the respective magnetic field strength for each variation of the current input. The experiment was then conducted by arranging the U-shaped tube apparatus and installing the pressure transducer to measure the generated pressure of the MRFs for every time the current to the electromagnetic coil is changed.

2. Materials and Methods

In a Poiseuille flow channel configuration, it is known that the magnetic field would generate a pressure difference and regulate the pressure drop, as occurs in the valve operating mode [28]. However, in a Poiseuille flow arrangement, the pressure difference is generated because the existing fluid flow is obstructed and thus, the measured pressure difference is identified as flow pressure drop occurs due to obstruction. The valve mode of the MRFs, for example, examined the magnitude of the pressure drop as a proportional contribution of the magnetic field strength and the yield stress of the fluid. On the other hand, the pressure measurement arranged in this paper is intended to observe the possibility of the magnetic field to generate pressure to the MRFs when it is at rest. Therefore the nature of the measurement is different from the measurement conducted at the valve mode arrangement.

In this study, the observation will be conducted using a U-shaped tube arrangement. The U-shaped tube method was chosen due to its simple pressure measurement process [29], as unequal pressure in the legs of the tube will result in different liquid levels, which indicate a pressure difference in the channel of the U-shaped tube. The U-shaped tube would be initially set at rest, with balanced pressure between both legs without a magnetic field. Once the liquid in both legs is level, the magnetic field is slowly applied by increasing the current input to the coil. The liquid levels are observed to see whether there were changes in the liquid levels in both legs when the magnetic field was applied. Figure 1 showed the reaction of the MRFs in the U-shaped tube when an external magnetic field was applied, and there is a clear difference between the liquid levels of both legs, which indicated a change in pressure. The level change, in this case, is caused by the magnetic induction to the magnetic particles immersed in the MRFs that provides magnetic force attraction to the particles until it each reach equilibrium position.

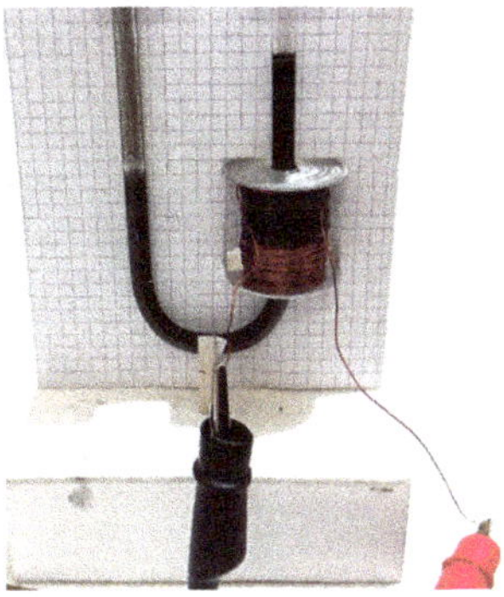

Figure 1. Different MRFs levels in both legs indicating a pressure change.

In order to quantify the observation, the magnetic field need to be estimated because it is not possible to insert the Gaussmeter probe in the U-shaped tube arrangement. The magnetic field estimation is conducted using FEMM an open-source software, 2-D finite element software widely used in magnetic analysis and simulations [11,30–33]. The FEMM simulation was carried out to simulate the magnetic circuits that exist in the theoretical model of the U-shaped tube apparatus. The material properties involved in the magnetic circuit are also used in the calculation, including the B-H curve values of the MRFs. To verify the accuracy of FEMM in estimating magnetic field strength a Gaussmeter test was also carried out on the upper surface of the coil on the axis of the coil hole, as shown in Figure 2a where measurements were possible. Figure 2b shows the observation point and magnetic field contour of the FEMM simulation. The observation point is 1 mm above the coil line due to the thickness of the bobbin that has been used. From the test result with Gaussmeter and FEMM Simulation, a comparison of results is obtained as shown in Figure 3. Based on the data obtained, that the simulation results are close to the measurement results, with an average error of 3%. Similar methods have been used in previous studies, such as in [11,30,33].

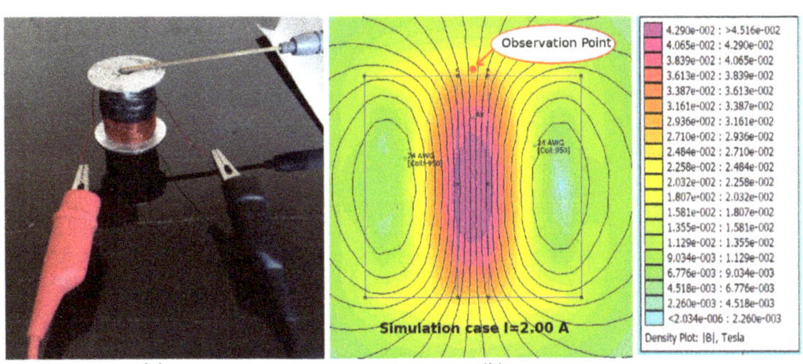

(a) (b)

Figure 2. FEMM verification setup, (**a**) measurement setup (**b**) FEMM Simulation contour plot.

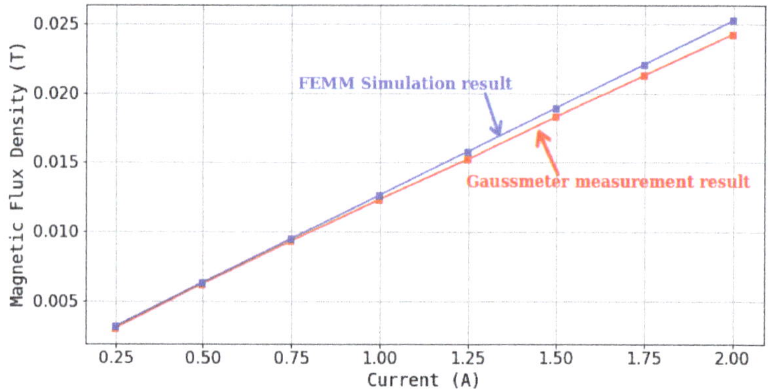

Figure 3. Comparison Results of Gaussmeter Measurement and FEMM Simulation.

An experiment was then conducted with the research apparatus; a U-shaped tube was designed and made out of glass (Figures 4 and 5). With the experimental apparatus model as shown in Figure 4, the volume of air trapped in the channel between MRFs and transducer is 1079.375 mm^3.

The control variable of this study was the direct current (DC) supply, with a voltage of 12 volts and variable currents to generate varying magnetic field strengths for the MRFs. The current input variations were varied from 0.25 A to 2.00 A with 0.25 A intervals. An Arduino microcontroller unit (MCU), a laptop, and CuteCom graphical serial device terminal served as a data logger. An MPX5010DP differential pressure sensor (sensitivity 450 mV/Kpa ~1 mV = 2.22 Pa) was used as the pressure data reader with a typical zero offset of 200 mV [34].

The MRF-122EG was poured into the U-shaped tube before the DC current is applied to the magnetic coil. The current input is alternated between 0.25 A, 0.50 A, 0.75 A, 1.00 A, 1.25 A, 1.50 A, 1.75 A, And 2.00 A every 10 s. The changes in pressure were monitored using the pressure sensor and recorded using the data logger at a sampling speed of 10 milliseconds (ms).

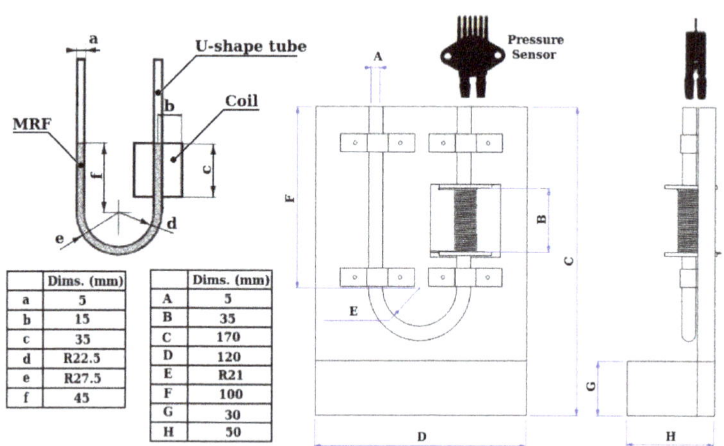

	Dims. (mm)
a	5
b	15
c	35
d	R22.5
e	R27.5
f	45

	Dims. (mm)
A	5
B	35
C	170
D	120
E	R21
F	100
G	30
H	50

Figure 4. The U-shaped tube and other testing equipment.

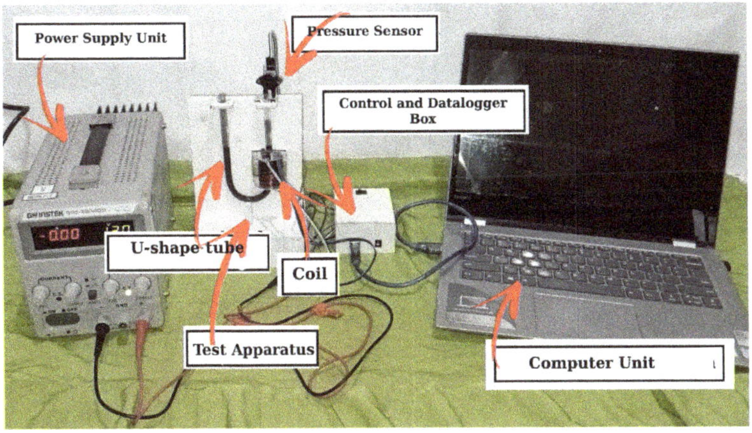

Figure 5. The research apparatus.

3. Results

3.1. Simulation Results

The results of the FEMM simulation are shown in Figure 6a. The figure illustrates the magnetic flux generated around the electromagnetic coil, and the path coincides with the MRFs. According to the figure, it can be seen that the flux line coincides with the MRFs in a parallel direction, which is different from the common flux-fluid arrangement in the valve mode. As seen in Figure 6b, the observation line was used to point to the sampling section of the apparatus, in which the magnetic flux density data value will be collected and plotted from the simulation. According to the contour plot in Figure 6b, the magnetic flux density appears to be uniform along the mid-section of the y-axis.

The correlation curve between the applied current and the magnetic flux density at observasion line is shown in Figure 7. The figure shows that the magnetic flux density is linearly correlated with the current input with a slope of about 0.125 T/A.

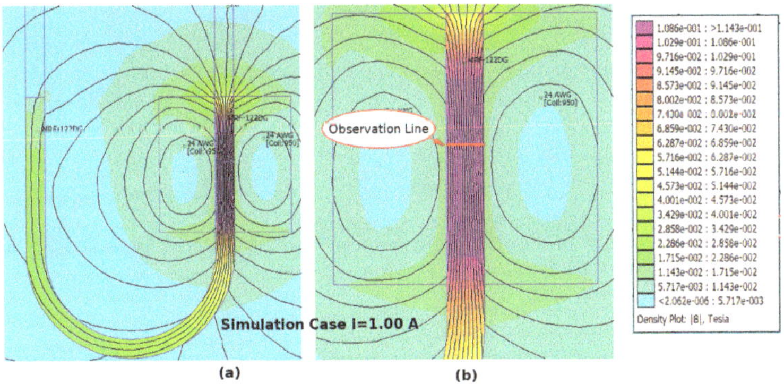

Figure 6. (**a**) Results of the FEMM software simulation (current input = 1 A); (**b**) The observation line that was used in the simulation.

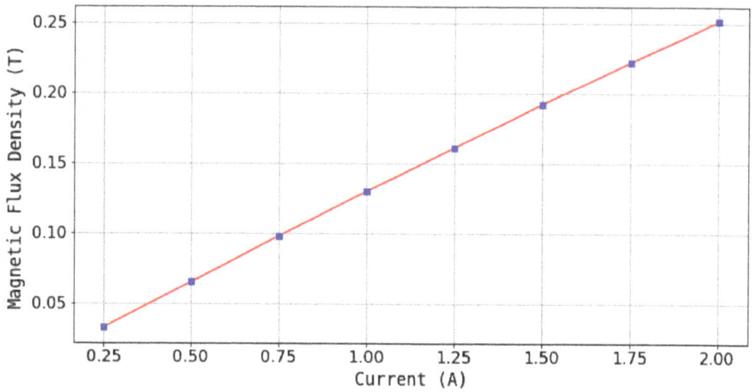

Figure 7. A graph of the correlation between current and magnetic flux density in the FEMM software simulation.

3.2. Experiment Results

The experimental results generally showed a significant effect of the magnetic field, which was represented by the amount of current input, to the change of MRFs pressure in the testing channel. For instance, Figure 8 shows the pressure values obtained at a current input of 1.00 A. The current input (yellow curved line) is seen in two states; the demagnetized state (Current = 0 A) and the magnetized state (Current = on); which results in changes of MRFs pressure. The MRFs pressure (blue curved line) is at its lowest when it is under a demagnetized state. But when the current input is turned on, the MRFs pressure increases proportionally to the input current.

As seen in Figure 9 the increase in MRFs pressure is consistent as the input current increases. At a lower current input, the change in MRFs pressure was relatively small as it only rose about 13.06 Pa to 18.39 Pa at 0.25 A current excitation. However, at a higher current input, the slope of MRFs pressure to current input is increased. For example, the average MRFs pressure was measured at about 267.80 Pa at 1.00 A of current excitation. Similar trends are still observed at input currents of 1.25 A, 1.50 A, 1.75 A, and 2.00 A with average changes in MRFs pressure of 517.16 Pa, 709.20 Pa, 904.64 Pa, 918.55 Pa, and 1020.01 Pa, respectively. The detail of overall changes in MRFs pressure during experiments is presented in Table 1.

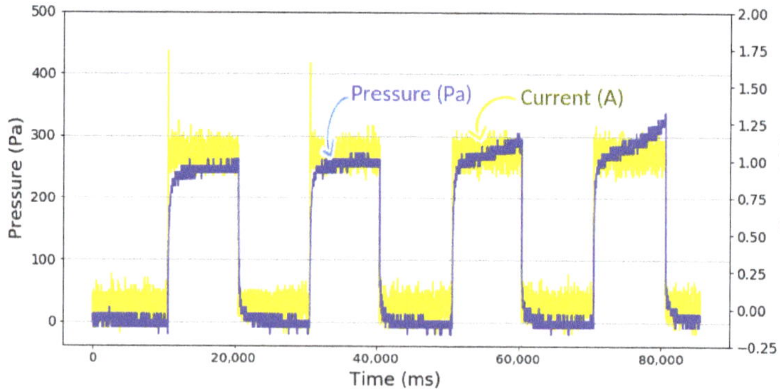

Figure 8. Experiment results of MRFs pressure at an input current of 1.00 A.

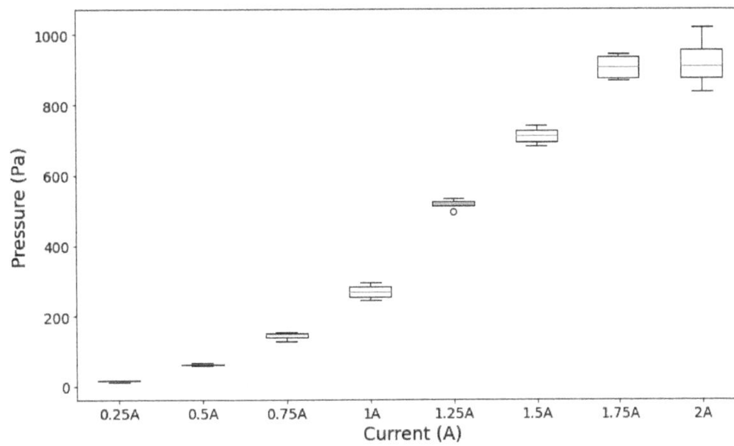

Figure 9. A box plot of MRFs pressure (Pascal) and input current (Ampere).

Table 1. Experiment Data.

Current Input (A)	0.25	0.50	0.75	1.00	1.25	1.50	1.75	2.00
$\triangle P_{min}$ (Pa)	13.06	58.98	129.17	244.49	496.04	679.79	867.36	835.50
$\triangle P_{max}$ (Pa)	18.39	65.91	155.64	292.53	533.77	739.02	942.50	1020.01
$\triangle P_{avg}$ (Pa)	15.53	62.80	144.77	267.80	517.16	709.20	904.64	918.55
$\triangle P_{median}$ (Pa)	15.34	63.16	147.14	267.09	519.42	708.99	904.34	909.34
Std Deviation	2.68	3.12	11.52	21.29	15.63	25.85	38.37	78.84

Note : $\triangle P_{min}$ is the minimum pressure change that occurs during magnetization at the observation cycle at each given current, $\triangle P_{max}$ is the maximum pressure change that occurs during magnetization at the observation cycle at each given current, $\triangle P_{avg}$ is the average pressure change that occurs during magnetization at the observation cycle, Std Deviation is the standard deviation of the measured pressure.

4. Discussion

According to the Figure 8, the observations were carried out for several cycles where each cycle consist of 10 s demagnetized state and 10 s magnetized state. From the figure, it was seen that the pressure was relatively stable in the initial cycle, while the pressure tended to increase in the subsequent cycles. In Figure 8, it can be seen that there is an increase in the pressure slope in the continuation cycle, due to the accumulation of magnetic particles in the previous cycle. This case occurs because, during demagnetization, the magnetic particles cannot be completely re-distributed in the fluid.

It is evident that increases in the magnetic field have caused changes in the MRFs pressure. This is because the pressure sensor detects increases in the barometric pressure of the air above the MRFs. As seen in the boxplot shown in Figure 9, the MRFs pressure readings obtained during the experiment has only small deviations, which indicates the measurement consistency. The most apparent deviation in MRFs pressure occurred at an input current of 2.00 A. In general, the data was normally distributed except for the input current of 1.25 A, where the sole outlier is measured.

According to the results, a sudden change of slope was detected when current excitation reached 2.00 A. This condition is slightly different from the other region, which might be the result of the heat dissipated from the coil when excited with a high current. During the experiment, the temperature of the coil bobbin was found to exceed 60 °C when excited with a 2.00 A current. The heat generated will slightly increase the wire resistance and decreased coil performance which was indicated by an observed decrease in the current input to the coil after few cycles of measurement. A similar case was also reported in the previous work about thermal modeling on heated injectors [35]. During the first cycle, the change in MRFs pressure was measured at 1020.01 Pa as the input current reached

1.88 A. In the subsequent cycles, the change in MRFs pressure is decreased because the input current could only reach 1.85 A, 1.84 A, and 1.83 A.

Figure 10 depicts the graph obtained from the experiment at an input current of 1.00 A. It shows that the MRFs pressure had a relatively fast response time and reacted quickly to the input current. Response time is defined as the period that a sensor requires to reach 90% of the total signal response [36]. It took approximately 530 ms to get more values up to 90% of the maximum value. Meanwhile, it took 1810 ms; or less than 2 s, to get to the maximum value based on the data logger. It is evident that there is a trend for bigger changes in MRFs pressure when the strength of the magnetic field increased, which, in this case, was represented by the increased input current.

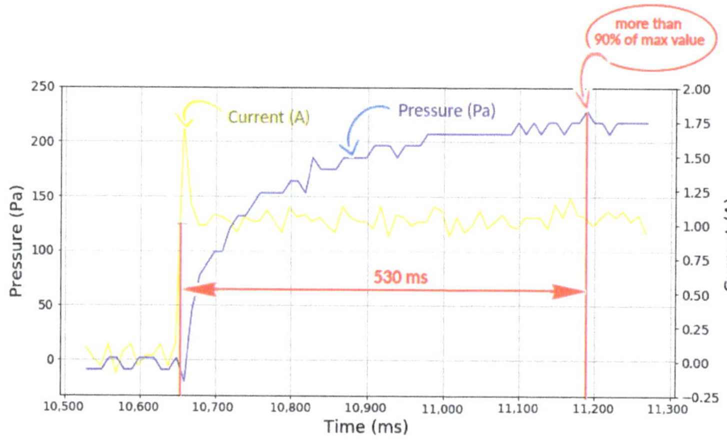

Figure 10. Changes in MRFs pressure at an input current of 1.00 A.

5. Conclusions

The experimental assessment to observe the effect of magnetic field induction on the fluid pressure of magnetorheological fluids (MRFs) has been conducted. The results showed that the MRFs fluid pressure could proportionally be increased with the increase of the magnetic field. The indication was observed visually with the fluid level change in the U-shaped tube and detected by the pressure sensor. Around 900 Pa of fluid pressure was recorded when the MRFs were subjected to 2.00 A current input to the electromagnet. According to the magnetic simulation, such an arrangement was estimated to produce about 0.25 T. The change of fluid pressure is positively correlated to the change of the magnetic field strentgh. Overall, the observation results indicate a strong relationship between MRFs fluid pressure and the magnetic field induction, which could pave the way to discovering a new MRFs operational mode.

Author Contributions: Conceptualization: P.J.W., U.U. and F.I.; Data curation: P.J.W.; Methodology: P.J.W., U.U. and F.I.; Supervision: E.P.B., U.U. and F.I.; Writing—original draft: P.J.W.; Writing—review & editing: P.J.W., E.P.B., U.U. and F.I. All authors have read and agreed to the published version of the manuscript.

Funding: Sebelas Maret University Doctoral Program Scholarship and Penelitian Unggulan UNS.

Institutional Review Board Statement: Not applicable.

Informed Consent Statement: Not applicable.

Data Availability Statement: The data presented in this study are available on request from the corresponding author. The data are not publicly available due to privacy.

Conflicts of Interest: The authors declare no conflict of interest.

Abbreviations

The following abbreviations are used in this manuscript:

AWG	American Wire Gauge
DC	Direct Current
FEMM	Finite Element Method Magnetics
MCU	Microcontroller Unit
MRFs	Magnetorheological Fluids
PVC	Polyvinyl Chloride

References

1. Utami, D.; Ubaidillah; Mazlan, S.A.; Imaduddin, F.; Nordin, N.A.; Bahiuddin, I.; Abdul Aziz, S.A.; Mohamad, N.; Choi, S.-B. Material Characterization of a Magnetorheological Fluid Subjected to Long-Term Operation in Damper. *Materials* **2018**, *11*, 2195. [CrossRef]
2. Imaduddin, F.; Mazlan, S.; Zamzuri, H. A design and modelling review of rotary magnetorheological damper. *Mater. Des.* **2013**, *51*, 575–591. [CrossRef]
3. Bahiuddin, I.; Mazlan, S.A.; Shapiai, M.I.; Choi, S.-B.; Imaduddin, F.; Rahman, M.A.A.; Ariff, M.H.M. A new constitutive model of a magneto-rheological fluid actuator using an extreme learning machine method. *Sens. Actuators A Phys.* **2018**, *281*, 209–221. [CrossRef]
4. Ubaidillah; Hudha, K.; Akmar, F.; Kadir, A. Modelling, characterisation and force tracking control of a magnetorheological damper under harmonic excitation. *Int. J. Model. Identif. Control* **2011**, *13*, 9–21. [CrossRef]
5. Ubaidillah; Lenggana, B.; Son, L.; Imaduddin, F.; Widodo, P.; Harjana, H.; Doewes, R. A new magnetorheological fluids damper for unmanned aerial vehicles. *J. Adv. Res. Fluid Mech. Therm.* **2020**, *73*, 35–45. [CrossRef]
6. Imaduddin, F.; Mazlan, S.A.; Ubaidillah; Idris, M.H.; Bahiuddin, I. Characterization and modeling of a new magnetorheological damper with meandering type valve using neuro-fuzzy. *J. King Saud Univ.-Sci.* **2017**, *29*, 468–477. [CrossRef]
7. Gadekar, P.; Kanthale, V.S.; Khaire, N.D. Magnetorheological Fluid and its Applications. *Int. J. Curr. Eng. Technol.* **2017**, *7*, 32–37.
8. Idris, M.H.; Imaduddin, F.; Ubaidillah; Mazlan, S.A.; Choi, S.-B. A concentric design of a bypass magnetorheological fluid damper with a serpentine flux valve. *Actuators* **2020**, *9*, 16. [CrossRef]
9. Milecki, A.; Hauke, M. Application of magnetorheological fluid in industrial shock absorbers. *Mech. Syst. Signal Process.* **2012**, *28*, 528–541. [CrossRef]
10. Hu, G.; Wu, L.; Li, L. Torque characteristics analysis of a magnetorheological brake with double brake disc. *Actuators* **2021**, *10*, 23. [CrossRef]
11. Nya'Ubit, I.R.; Priyandoko, G.; Imaduddin, F.; Adiputra, D.; Ubaidillah. Torque characterization of t-shaped magnetorheological brake featuring serpentine magnetic flux. *J. Adv. Res. Fluid Mech. Therm.* **2020**, *78*, 85–97.
12. Latha, H.; Pantangi, U.S.; Seetharamaiah, N. Design and manufacturing aspects of magneto-rheological fluid (mrf) clutch. *Mater. Today Proc.* **2017**, *4*, 1525–1534. [CrossRef]
13. Jamalpour, R.; Nekooei, M.; Moghadam, A. Seismic response reduction of steel mrf using sma equipped innovated low-damage column foundation connection. *Civ. Eng. J.* **2017**, *3*, 1–14. [CrossRef]
14. Dyke, S.J.; Spencer, B.F.; Sain, M.K.; Carlson, J.D. Seismic Response Reduction Using Magnetorheological Dampers. *IFAC Proc. Vol.* **1996**, *29*, 5530–5535. [CrossRef]
15. Satria, R.R.; Ubaidillah, U.; Imaduddin, F. Analytical approach of a pure flow mode serpentine path rotary magnetorheological damper. *Actuators* **2020**, *9*, 56. [CrossRef]
16. Abd Fatah, A.Y.; Mazlan, S.; Koga, T.; Zamzuri, H.; Imaduddin, F. Design of magnetorheological valve using serpentine flux path method. *Int. J. Appl. Electromagn. Mech.* **2016**, *50*, 29–44. [CrossRef]
17. Sarkar, C.; Hirani, H.; Sasane, A. Magnetorheological smart automotive engine mount. *Int. J. Curr. Eng. Technol.* **2015**, *5*, 419–428.
18. Zeinali, M.; Mazlan, S.A.; Choi, S.-B.; Imaduddin, F.; Hamdan, L.H. Influence of piston and magnetic coils on the field-dependent damping performance of a mixed-mode magnetorheological damper. *Smart Mater. Struct.* **2016**, *25*, 055010. [CrossRef]
19. Sapinski, B. Simulation of an MR squeeze-mode damper for an automotive engine mount. In Proceedings of the 17th International Carpathian Control Conference (ICCC), IEEExplore, High Tatras, Slovakia, 29 May–1 June 2016; pp. 641–644.
20. Yildirim, G.; Genc, S. Experimental study on heat transfer of the magnetorheological fluids. *Smart Mater. Struct.* **2013**, *22*, 085001. [CrossRef]
21. Adiputra, D.; Rahman, M.A.A.; Bahiuddin, I.; Ubaidillah; Imaduddin, F.; Nazmi, N. Sensor number optimization using neural network for ankle foot orthosis equipped with magnetorheological brake. *Open Eng.* **2021**, *11*, 91–101. [CrossRef]
22. Bhavsar, P.; Unune, D.R. Magnetorheological Polishing Tool for Nano-Finishing of Biomaterials. In Proceeding of the 10th International Conference on Precesion, Meso, Micro and Nano Engineering, Chennai, India, 7–9 December 2017; pp. 6–10.
23. Purnomo, E.D.; Ubaidillah; Imaduddin, F.; Yahya, I.; Mazlan, S. Preliminary experimental evaluation of a novel loudspeaker featuring magnetorheological fluid surround absorber. *Indones. J. Electr. Eng. Comput.* **2020**, *17*, 922–928. [CrossRef]

24. Dyana Kumara, K.A.; Ubaidillah, U.; Priyandoko, I.Y.G.; Wibowo, W. Magnetostatic Simulation in a Novel Magnetorheological Elastomer Based Loudspeaker Surround. In Proceedings of the 2019 6th International Conference on Electric Vehicular Technology (ICEVT), Bali, Indonesia, 18–21 November 2019; pp. 295–299.
25. Ahamed, R.; Choi, S.-B.; Ferdaus, M.M. A state of art on magneto-rheological materials and their potential applications. *J. Intell. Mater. Syst. Struct.* **2018**, *29*, 2051–2095. [CrossRef]
26. Rahman, M.; Chao, O.; Julai, S.; Ferdaus, M.M.; Ahamed, R. A review of advances in magnetorheological dampers: Their design optimization and applications. *J. Zhejiang Univ. Sci. A* **2017**, *18*, 991–1010. [CrossRef]
27. de Vicente, J.; Klingenberg, D.J.; Hidalgo-Alvarez, R. Magnetorheological fluids: A review. *Soft Matter* **2011**, *7*, 3701–3710. [CrossRef]
28. Ostadfar, A. Chapter 1—Fluid Mechanics and Biofluids Principles. In *Biofluid Mechanics*; Ostadfar, A., Ed.; Academic Press: London, UK, 2016; pp. 1–60.
29. Vijayaraghavan, G.K.; Sundaravalli, S. Pressure and Pressure Measuring Devices. In *Fluid Mechanics and Machinery*; Lakshmi Publications: Chennai, India, 2018; ISBN 978-93-831030-8-9
30. Ubaidillah; Wirawan, J.W.; Lenggana, B.W.; Purnomo, E.D.; Widyarso, W.; Mazlan, S.A. Design and Performance Analysis of Magnetorheological Valve for Upside-Down Damper. *J. Adv. Res. Fluid Mech. Therm.* **2019**, *63*, 164–173.
31. Yatchev, I.; Gueorgiev, V.; Ivanov, R.; Hinov, K. Simulation of the dynamic behaviour of a permanent magnet linear actuator. *Facta Univ.-Ser. Electron. Energy* **2010**, *23*, 37–43. [CrossRef]
32. Białek, M.; Nowak, P.; Lindner, T.; Wyrwał, D. FEMM Examination of the Gripper with Magnetorheological Cushion. In Proceedings of the 2020 International Conference Mechatronic Systems and Materials (MSM), Bialystok, Poland, 1–3 July 2020; pp. 1–5.
33. Ubaidillah; Lenggana, B.W. Finite Element Magnetic Method for Magnetorheological Based Actuators. In *Finite Element Methods and Their Applications*; IntechOpen: London, UK, 2020; pp. 1–18. [CrossRef]
34. Freescale Semiconductors Inc. *Technical Data Sheet: MPXV5010DP, 0 to 10 kPa, Differential, Gauge, and Absolute, Integrated, Pressure Sensors*; Freescale Semiconductors Inc.: Austin, TX, USA, 2012.
35. Volpato, O. *Thermal Modeling of Heated Injectors*; SAE Technical Papers 2011-36-0087; SAE: Warrendale, PA, USA, 2011. [CrossRef]
36. Arafat, M.M.; Dinan, B.; Akbar, S.A.; Haseeb, A.S.M.A. Gas sensors based on one dimensional nanostructured metal-oxides: A review. *Sensors* **2012**, *12*, 7207–7258. [CrossRef]

applied sciences

MDPI

Review

Review: A Survey on Configurations and Performance of Flow-Mode MR Valves

Janusz Gołdasz [1,*], Bogdan Sapiński [2], Michal Kubík [3], Ondřej Macháček [3], Wojciech Bańkosz [1], Thomas Sattel [4] and Aditya Suryadi Tan [4]

[1] Faculty of Electrical and Computer Engineering, Cracow University of Technology, ul. Warszawska 24, 31-155 Kraków, Poland; wojciech.bankosz@doktorant.pk.edu.pl
[2] Faculty of Mechanical Engineering and Robotics, AGH University of Science and Technology, al. A. Mickiewicza 30, 30-059 Kraków, Poland; deep@agh.edu.pl
[3] Faculty of Mechanical Engineering, Brno University of Technology, Technická 2896/2, 61669 Brno, Czech Republic; michal.kubik@vutbr.cz (M.K.); ondrej.machacek@vutbr.cz (O.M.)
[4] Department of Mechanical Engineering, Technical University of Ilmenau, Ehrenbergstraße 29, 98693 Ilmenau, Germany; thomas.sattel@tu-ilmenau.de (T.S.); aditya-suryadi.tan@tu-ilmenau.de (A.S.T.)
* Correspondence: jgoldasz@pk.edu.pl

Abstract: Magnetorheological (MR) actuators are semi-active devices controlled by magnetic stimuli. The technology has been commercialized in the automotive industry or high-quality optical finishing applications. It harnesses the rheology of smart fluids to result in the unique application of the material. By a wide margin, the most common example of an MR actuator is a flow-mode single-tube housing with a control valve (electromagnet with a fixed-size air gap filled with the MR fluid) operating in a semi-active vibration control environment. The analysis of the prior art shows that the developed configurations of MR valves vary in size, complexity, the ability to generate adequate levels of pressure, and the interactions with the MR fluid's rheology resulting in various performance envelopes. Moreover, miscellaneous testing procedures make a direct valve-to-valve comparison difficult. Therefore, in this paper we present a detailed and systematic review of MR control valves, provide classification criteria, highlight the operating principle, and then attempt to categorize the valves into groups sharing similarities in the design and performance envelope(s). Moreover, a simple performance metric based on the shear stress calculation is proposed, too, for evaluating the performance of particular valving prototypes. In the review, we discuss the key configurations, highlight their strengths and weaknesses and explore various opportunities for tuning their performance range. The review provides complementary information for the engineers and researchers with a keen interest in MR applications, in general. It is an organized and and critical study targeted at improvements in the categorization and description of MR devices.

Keywords: magnetorheological actuator; magnetorherological valve; magnetorheological fluid; design; dynamic range; performance

Citation: Gołdasz, J.; Sapiński, B.; Kubík, M.; Macháček, O.; Bańkosz, W.; Sattel, T.; Tan, A.S. Review: A Survey on Configurations and Performance of Flow-Mode MR Valves. *Appl. Sci.* **2022**, *12*, 6260. https://doi.org/10.3390/app12126260

Academic Editors: Antonio Concilio, Rosario Pecora, Ignazio Dimino, Salvatore Ameduri and Vikram G Kamble

Received: 20 May 2022
Accepted: 15 June 2022
Published: 20 June 2022

1. Introduction

Smart materials are capable of sensing external stimuli, i.e., temperature, light, electric field, magnetic field, stress, and reacting to them in a coordinated fashion. In many application fields, e.g., vibration suppression, medical systems, noise control, the implementation of a specific smart material has led to spectacular results, e.g., smart material-based actuators, drive systems, motors, sensors, etc. For instance, magnetorheological (MR) fluids (suspensions of micron-sized particles in a non-conductive carrier liquid) undergo a dramatic transition from that of a Newtonian fluid to a pseudo-solid. As discovered originally by J. Rabinov [1], the material develops a yield stress when exposed to magnetic field. The so called MR effect is stable, reversible fast and of sufficient magnitude to be utilized in real-world applications [2]. The range of prototype MR fluid-based devices is vast incorporating automotive suspension dampers and powertrain mounts, washing machine dampers, rotary clutches and brakes, exoskeletons, optical finishing systems, haptics, military gun recoil systems, etc. [3]. In particular, the technology has been successfully commercialized in automotive suspension systems and high-quality optics [4,5].

Regrettably, their electrical counterparts, electrorheological (ER) fluids, have failed to deliver commercial applications due to serious deficiencies, e.g., sensitivity to metallic impurities, lower yield stress and a very narrow operating temperature range [6]. Their yield stress is on the order of magnitude lower than that of MR fluids forcing the use of long flow channels and tight gaps, and it is limited by the maximum (breakdown) electric field strength. Their operating temperature range is narrow (usually 0–80 °C), and the response time varies with temperature [7]; the operating temperature range of MR fluids well exceeds that range. Moreover, their sensitivity to metallic contamination forces an extremely strict control process during the assembly, and the leakage current across the gap between the electrodes increases with temperature, resulting in high power consumption by these devices. Thus, despite the large interest from the industry in the early 1990s, a successful application of ER fluids has yet to materialize.

The magnetorheology is based on a similar principle as their ER counterparts. Again, the actuators are unique devices in many aspects. Their operating principle is based upon the modification of specific material properties (yield stress) by external magnetic fields in order to drive changes in the force (or torque) output of these actuators. In essence, the technology is valveless. As a result, the devices are structurally simple. At the device-level, they consist of three fundamental components or modules: housing, fluid, and solenoid valve (electromagnet). An external low-voltage supply is also needed for powering the actuator. This is in contrast to valve-based semi-active solenoid actuators in which flow restriction variations are achieved by manipulating the flow geometry—an advantage over the conventional semi-active actuators, and the MR valves do not incorporate small moving and precise components. The technology is not, however, rid of drawbacks. Despite the extensive research over the last 30 years, fluid stability, fluid tribology, increased weight, etc., are engineering challenges that the technology needs to cope with.

By a fashion, the actuators can be operated in at least one of the operating modes: flow (valve), shear (rotary), squeeze or gradient pinch [8,9]. The flow-mode seems to have been best utilized in vibration control applications, the shear-mode in rotary dampers and clutches, whereas the squeeze-mode in small-stroke mounts. The gradient pinch mode is an interesting variation of flow-mode featuring highly non-uniform and deliberate distributions of magnetic flux in the MR valve's flow channels [10]. Hybrid MR actuators take advantage of at least the two operating modes for increased performance or specific performance characteristics [11,12].

As a reminder, the basic set of application-specific performance requirements for a typical MR actuator should include [8]: (1) zero or minimum (safe) force (torque) in the off-state (deactivated) condition, (2) maximum damping force at the highest current (magnetic flux) condition, (3) power supply, (4) power consumption, (5) response time, (6) temperature operating range, heat dissipation. Other specifications may be related to friction, packaging, etc., not to mention manufacturability and cost.

An analysis of prior art in the field reveals that the most common area of application of MR actuators is vibration control [6]. Moreover, the application field has been dominated by single-mode (flow-mode) devices, which is even more evident while analyzing , e.g., the commercial MR actuators. The automotive MR dampers are simple single-tube & flow mode devices with dual coil valves. Similar observations can be made while inspecting civil engineering applications [13]. Therefore, the emphasis of this review is on flow-mode MR actuators powered from external power supplies and their valves in particular. Energy-harvesting MR actuators are beyond the scope of this review [14]. Our motivation behind the topic review is simple. Although several reviews have already been published on the subject [15–20], little attention has been paid to the generic performance of MR actuators. One important shortcoming of the previous review works is the lack of comprehensive and critical insight on the structures of MR actuators and the relationship with their performance. An even closer inspection of the prior art shows that very few MR configurations have a chance to stand up to the real-world requirements both in terms of the dynamic range, power draw, or the response time. Mostly, the previous studies summarize the research work performed with little attention to their design details and parameters and how they contribute to the damping force. Therefore, the purpose of this review study is to fill in the gap. To summarize, this study provides a systematic and organized review of existing configurations of flow-mode MR valves, highlights the contribution of specific features to the valve's output, and delivers a simple metric to evaluate their performance based on the ability to yield appropriate forces (pressures).

To highlight the selection criteria for the review, the analyzed set of references spans 30 years (1992–2022). Due to the reasons above, the analyzed references observe only the flow-mode prototypes for use in vibration control systems. The study includes research papers, as well as patents and patent applications of commercial enterprises. We cover feasibility and design studies, valve geometry optimization examples, presentations of novel features, etc. Modeling papers fall into the review scope here provided that they serve the sole purpose of demonstrating the unique design features and their contribution to the performance envelope enhancement and shaping.

The work is structured as follows: Section 2 provides the classification criteria based on the detailed analysis of various MR valve structures complemented by a fundamental analysis of the valve's operating principle and performance metrics in Section 3. In Section 4, we analyze the selected structures, their key features and performance characteris-tics. The analysis is complemented by a qualitative demonstration of the proposed performance metric. Finally, conclusions are drawn in Section 5.

We proceed with the review by providing the classification criteria based on the detailed analysis of various MR valve structures, highlight the valve's operating principles along with fundamental equations and performance metrics. The considerations are then the basis for a detailed analysis of selected configurations of MR valves. The analysis is complemented by a qualitative demonstration of the proposed performance metric. Finally, conclusions are drawn at the end of the review paper.

2. Method—Selection of Criteria

In this section, we carry out a detailed analysis of the structure of a flow-mode MR actuator, then propose criteria for the actuator and valve classification based on the outcome of this analysis. Briefly, consider a typical MR actuator featuring a single cylinder tube housing and a control valve assembly—see Figure 1. The valve incorporates one dual-coil electromagnet with one (homogenous) annular channel. The assembly is attached to a rod. The electrical connection to the coils in series is through the thru-hole in the rod. The pressurized gas chamber (not shown) is for volume compensation due to rod motion and fluid expansion due to temperature.

Let us briefly recall how the actuator works. In the non-energized condition, the current is not supplied to the coil(s), and no magnetic flux is generated in the annular channel. The flow losses in this condition are minimal and depend on the geometry of the flow channel and material properties of the fluid (viscosity, density) as well as the piston speed. As soon as the serial coils are energized, the magnetic fluxes that are induced in the electromagnet structure travel through the core into the annular gap, enter the outer ring, and then back into the core via the flow channel. These elements form the primary path for the magnetic flux. As the housing is usually ferromagnetic, portions of the fluxes enter the housing, too, and travel back to the rod via the MR fluid, thus, bypassing the annulus. As a result the fluid is energized in the regions above the active sections—the fluid's magnetic field dependent yield stress increases and the outcome is the resistance-to-flow build-up. The flow through the annulus can be initiated only when the pressure drop across the piston exceeds the resulting breakaway pressure due to the yield stress increase. Note that the fluid in this condition behaves in a rather complex fashion. Simply speaking, as soon as it enters the active zones its rheology changes from the near-Newtonian one to that of a pseudo-solid (Bingham) type only to return to the original condition in the non-active regions over the coil. That is the opposite of the situation in their electrical (ER) counterparts where the entire surface of the flow channel is energized, and thus contributes to the field-dependent losses as a whole.

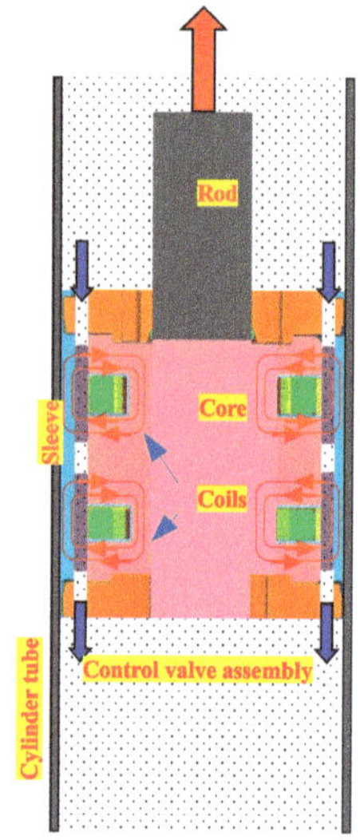

Figure 1. Example: MR actuator assembly.

The exemplary actuator incorporates several components which can be the basis for the classification of particular actuator configurations regardless of the operating mode, namely, the housing and the valve. Analyzing the valve's structure in detail, it is apparent that a set of classification criteria can be provided based on the design of the following elements: solenoid, solenoid's position relative to the housing, coil assembly, electrical connections (electrical interface to the power supply), and flow channel(s). The criteria can be extended to incorporate the optional bypass and the sensory (displacement, velocity, magnetic flux) or fail-safe add-ons.

In detail, a control valve incorporates at least one electromagnet assembly. Its sole purpose in the structure is to induce the magnetic field of sufficient strength for activating the fluid in the area over the magnetic poles and should meet particular performance requirements in terms of flux dynamic range, flux linearity vs the inducing current (ideal scenario) and response time. Next, considering the position of the electromagnet relative to the housing, the electromagnets can be placed inside the housing or on the outside of the tube. The external assemblies are usually stationary (integrated into the housing), whereas the internal electromagnets are usually attached to a moving rod although stationary configurations can be observed in the literature, too. Particular electrical connections (for supplying the voltage to the coil terminals) follow the method of attachment of the electromagnet to the rod and its relative position to the housing. The electromagnets incorporate at least one coil assembly wound transversely or radially. The air gap in the electromagnet acts as a flow channel of constant height with stationary surfaces constituting the flow path (as in flow-mode or pinch mode). A valve may feature several parallel flow channels, thus splitting the flow into several paths of equal or different resistance.

A hydraulic bypass is located in parallel to the main flow path(s); its main contribution is in shaping the valve's performance at near-zero flow rates. With ideal bypasses their function is to degrade the pressure drop down to zero at the zero flow rate. This is usually achieved by diverting a portion of the fluid flow to a non-energized flow path. A conventional digressive valve is a rare specimen in MR valves [21], and its efficiency can be questioned. Sensory mechanisms are often integrated into the housing or the control valve, or implemented in the control logic (flux control) [22].

Based on the analysis of the prior art, it is evident the simplest benchmark is a single-tube MR actuator with one control valve (having a single transversely wound coil assembly, one annular-type primary flow channel, and one bypass path—optional) attached to the moving rod and no sensory features. Although necessary, the pressurized gas container's location is of secondary importance. The function of the electromagnet is to ensure a uniform (or near-uniform) distribution of magnetic flux in the air gap (radially) and maximum (longitudinally). The flux distribution in the annulus is intrinsically nonhomogenous due to radial effects, however, for simplicity's sake, we assume the uniform flux distribution in the channel—the flat-plate approximation results in little error [23].

With the exception of the bypass, all these features can be observed in, e.g., [8,24]; see [25] for the description of an MR valve with a bypass path. Therefore, all the other configurations will be analyzed concerning the benchmark structure. The output of the actuator is usually represented in the form of the static force-velocity plot at various levels of the input current (magnetic flux, flux density)—see Figure 2, or pressure vs flow rate characteristics (more common for a valve). With the actuator such data are obtained by exciting the actuator with constant velocity or sine displacement inputs or a test rig and at fixed levels of the magnetic field in the annulus, then recording the peak forces corresponding to particular velocity amplitudes. For comparison, MR valve characteristics are acquired on a flow bench where the response of the valve (pressure drop) is measured as a function of flow rate and the magnetic field. In the example shown, the bypass presence translates into the high slope of the force-velocity curves below approximately 0.25 m/s. Above the threshold velocity, the flow through the valve is dominated by the annular gap resistance. Fluid properties aside, the forces in the first operating regime (below the knee point) can be altered mainly by modifying the bypass geometry, whereas the actuator's performance in the second regime depends on the primary flow channel geometry (annulus) as well as the electromagnet's capability to energize the fluid.

The actuator decomposition elements, the basis for comparison and classification of various configurations of MR actuators and the valves (which deviate from the benchmark design) are then highlighted in Figures 3 and 4, as well as Table 1. To clarify, the information concerns only flow-mode valves with (homogenous) annular flow paths. The structures with composite flow channels (usually in the form of an annular gap in series with a radial one) are not accounted for here. In this review, we limit our considerations to homogenous annular flow channels for the ease of the valve-to-valve comparison; the valves with performance-enhancing composite flow paths merit a separate review.

One may distinguish between the factors which merely affect the force output of the actuators (performance) and those which only facilitate the ease of assembly or the connection to the external circuit. In this review, we focus on those both contributing to the performance of the actuator and associated with the valve.

Table 1. Flow-mode MR actuators: classification features include exemplary references.

Feature	Key Characteristics	Ref.
Fluid volume	single-tube dual-tube	[8,24,25] [26–29]
Volume compensator	internal external	[8,24,25] [29–33]
Compensator's pressure level	low high	[28,29] [8,24,25]
Electromagnets per valve	single multiple	[8,24,25] [30,34–36]
Valve's operating mode	single mixed	[8,24,25] [11,37–40]
Location—electromagnet	internal external	[8,24,25] [41]
Relative position—electromagnet	non-stationary stationary	[8,24,25] [28,34,36]
Attachment method—electromagnet	rod (non-stationary) housing (housing)	[8,24,25] [28,34,42]
Magnetic flux distribution	uniform, quasi-uniform non-uniform (radial,axial)	[8,24] [9,25,43–45]
Electrical circuit	uni-polar bi-polar	[46] [8,24,47]
Coil assemblies per electromagnet	single multiple	[8,24,25] [48–53]
Coil-to-coil connections	serial parallel or mixed	[48,51] [54]
Coil arrangement in the electromagnet	radial transverse	[50,55–58] [8,24,25]
Flow channel function	primary secondary (bypass) or hybrid	[8,24] [25,44,59–62]
Flow channel shape	annular, planar radial, helix	Most [39,40,63–65]
Number of flow channels	single multiple (parallel)	[8,24,25,66] [48,66–68]
Flow channel height	constant variable (longitudinally, radially)	[8,24,25] [43,52,69,70]

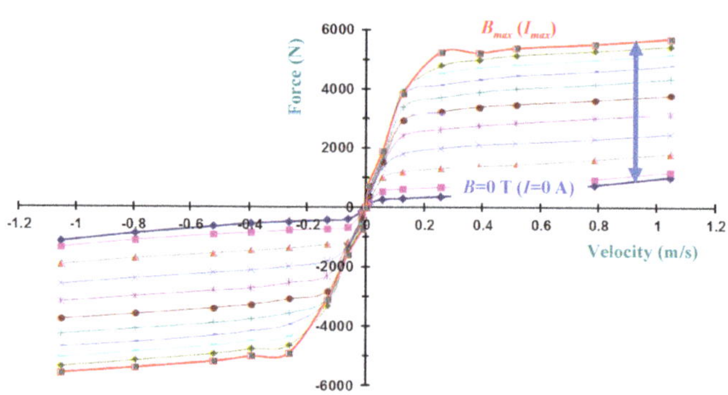

Figure 2. Example: static force–velocity characteristics (commercial actuator data courtesy of BWI Group); damping forces measured at fixed current levels from $I = 0$ to $I_{max} = 5$ A in 0.5 A steps.

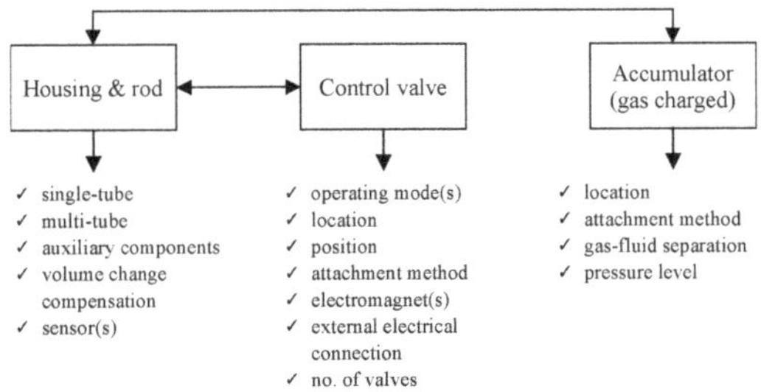

Figure 3. Layout: actuator decomposition.

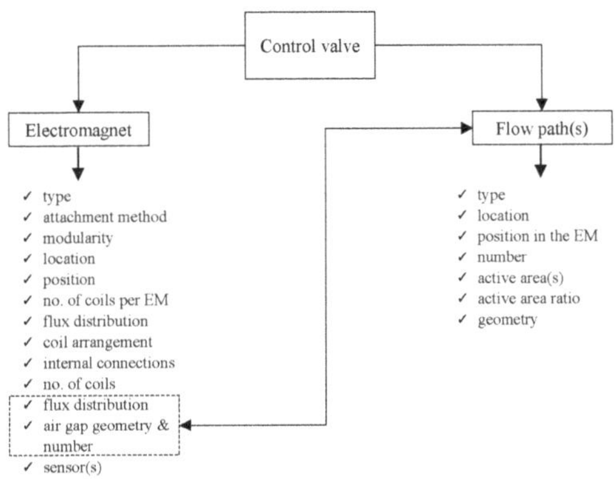

Figure 4. Layout: valve decomposition.

3. Fundamentals

Here, we discuss the MR valve's operating principle, highlight basic fundamental relationships, and then propose a simple shear stress metric for the valve's performance characterization and comparison. Let us first consider the simple MR valve (electromagnet—EM) in Figure 5. The electromagnet produces a magnetic field of required strength. It consists of a core of specific dimensions, air gap(s), coil winding(s), and a ring. Contrary to other electromagnets the ones used in flow-mode actuators contain no moving parts. The EM's fixed-height air gap(s) form the flow channels. The air gap length is L_g, and its size $h = (D_2 - D_c)/2$. The coil window dimensions are $W_c \times H_c$ (W_c—width, H_c—height), and it fits N_c number of wire turns. The outer diameter of the electromagnet is D_p. The current in the coil results in N_cI ampere turns. Considering the EM's circuit, the following simple relationship can be obtained.

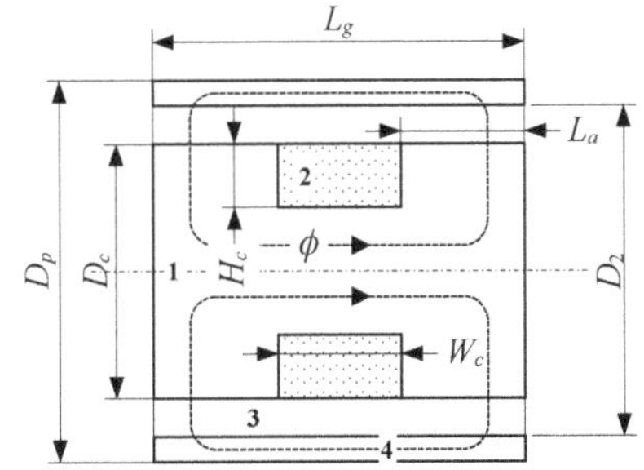

Figure 5. MR electromagnet (valve): 1—core, 2—coil, 3—annulus, 4—ring.

$$\sum_i H_i l_i = N_c I \tag{1}$$

As the annulus is the element of the largest reluctance, the above equation can be reduced as follows

$$2H_g h \approx N_c I \tag{2}$$

and $B_g = \mu_r(B_g)\mu_0 H_g$. Then,

$$2h\frac{B_g}{\mu_r(B_g)\mu_0} - N_c I = 0 \tag{3}$$

where μ_r—relative permeability of the fluid, μ_0—absolute permeability (vacuum), l_i—length of the ith magnetic circuit section with a constant cross-section area. By solving the equation, the average gap flux density B_g can be computed, H_g—gap magnetic field strength. Using Gauss law, the magnetic flux ϕ can be obtained

$$\phi = B_g A_m = B_g \pi D_m L_a = B_g \pi (D_c + h) L_a \tag{4}$$

where A_m—active surface (of one pole), $D_m = D_c + h$. The obtained value of B_g is yet to be verified for magnetic saturation; the valve should not saturate under any circumstances or increasing the current in the coil will result in little or no effect. Given the core and the coil dimensions, the circuital resistance is

$$R_c = \frac{\rho_c l_w}{A_w} = \frac{\rho_c 4 N_c (D_c - H_c)}{d_w^2} \tag{5}$$

and it is subject to the constraint $U_c = IR_c < U_{bat}$, U_c—voltage drop, U_{bat}—supply voltage, I—circuital (coil) current, ρ_c—copper resistivity, A_w—single wire cross-section area, l_w—wire length.

Let us then consider the simplest approximation of the Bingham plastic model usually used for modeling the output of MR (and ER) valves [71,72]

$$\Delta P = \Delta P(B_g, v) = \begin{cases} 12\frac{\mu L_g}{bh^3}Q & B = 0 \\ 4\frac{\tau_0(B_g)L_a}{h}\operatorname{sgn}(Q) + 12\frac{\mu L_g}{bh^3}Q & |B| \neq 0 \end{cases} \tag{6}$$

and the force $F_d = A_p \Delta P$, ΔP—pressure difference across the valve, $Q = A_p v$—flow rate, A_p—cross-section area (valve), b—gap width, v—velocity, τ_0—yield stress. Thus, the (absolute) dynamic range at the reference velocity v_{ref} is given as follows

$$\Delta F = F_d(B_{max}, v_{ref}) - F_d(0, v_{ref}) \tag{7}$$

or the turn-up ratio K_f is

$$K_F = \frac{F_d(B_{max}, v_{ref})}{F_d(0, v_{ref})} \tag{8}$$

Using the above simple or far more advanced formulas various optimization studies can then be performed in order to meet specific application criteria (range, force envelope, etc.) [73,74].

While studying the simple model of the electromagnet in the context of the other configurations such as multiple gap valves, it should be appreciated that certain designs of the MR valves may present a serious challenge if one is to achieve a particular force envelope. For example, in a valve having N parallel annular flow paths, the flux is required to travel across $2N$ air gaps, thus requiring an amplified number of ampere turns to meet the specified performance objectives. This may be difficult to realize given packaging constraints which are always present in real-life applications.

It is evident then, that there are a number of factors that influence the final form of the valve, namely, ampere turns, coil (wire) loading and configuration, material and fluid properties (permeability, yield stress, viscosity, density), packaging, active area ratio, flow channel geometry, placement and location in the electromagnet (valve), etc. They are to be considered simultaneously in the design process.

Given the above dynamic range metrics, the absolute dynamic range metric makes direct comparisons of various MR configurations difficult. They are usually prototyped for different applications, and have discernible performance targets, not to mention various and distinct fluid properties or electromagnet characteristics. Therefore, we propose a simple metric for performance evaluation of MR actuators which can be related to their capacity to generate adequate forces (pressures). For example, consider the annular gap illustrated in Figure 6. The wet area (at mid-gap) is then equal to $A_m = \pi D_m L_g$. Given the output force F_d, the shear stress τ can be calculated as follows

$$\tau = \frac{F_d}{A_m} \tag{9}$$

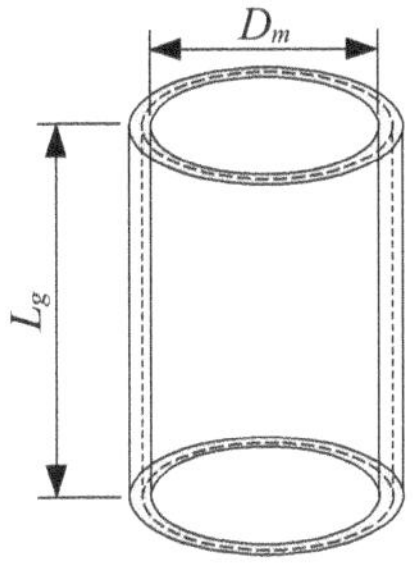

Figure 6. Shear gap, D_m—circumferential diameter, L_g—gap length.

The shear stress criterion was proposed based on the analysis of a similar metric that is commonly used for sizing the electrical actuators or motors [75]. In this area, it implies that the output torque (force) of an electrical motor or an actuator is proportional to the product of the rotor volume and the shear stress. In electrical motors the metric value varies from hundreds of pascals to roughly a hundred kilo pascals [76]. We show in the study that MR actuators well exceed that range.

The prior art analysis shows a variety of actuators of different configurations and sizes which makes a direct comparison difficult or even impossible. The above metric (calculated on a gap basis) is straightforward to obtain based on the information contained in research papers and optimal as it allows to relate the produced output force to the (shear) area. Moreover, as an alternative metric, the absolute shear stress range metric can be used instead

$$\Delta\tau = \frac{F_d(B_{max}, v_{ref}) - F_d(0, v_{ref})}{A_m} \tag{10}$$

The $\Delta\tau$ metric emphisizes the contributions of the active portions of the flow channel .

4. Analysis

In this section, we discuss the typical configurations of MR valves in relation to their performance envelopes. Specifically, the configurations are illustrated in Figures 7 and 8, and the performance envelopes they are capable of are highlighted simply in Figure 9. The range of analyzed structures covers the benchmark structure and its functional variations—bypass mechanisms, force asymmetry, etc. Structural means for maximizing the active area ratio and the dynamic range by varying the number of coils, flow channels, and coil's orientation relative to the core and the flow path are discussed here, too. The material is complemented by a brief analysis using the proposed shear stress metric.

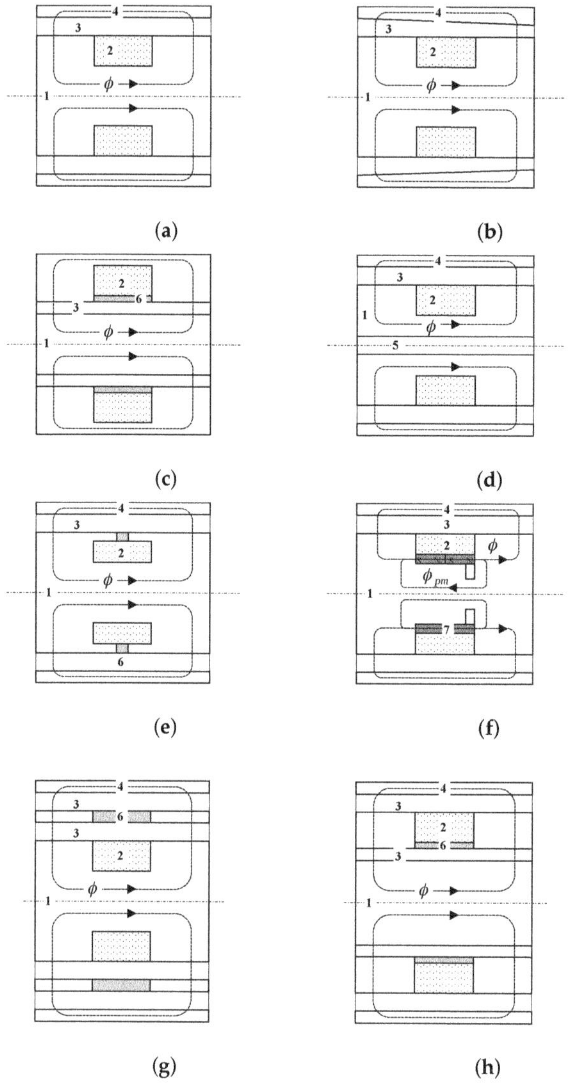

Figure 7. MR valves: configurations, 1—core, 2—coil(s), 3—primary flow path(s), 4—ring, 5—bypass, 6—(non-magnetic) spacer ring, 7—permanent magnet. (**a**) Benchmark valve [24,25]; (**b**) Variable height [69]; (**c**) Annulus location [77]; (**d**) Bypass valve [59,60,78,79]; (**e**) High active area ratio [80]; (**f**) Fail-safe valve [81]; (**g**) Dual annulus (concentric) [66]; (**h**) Dual annulus, adapted from [66,77,82].

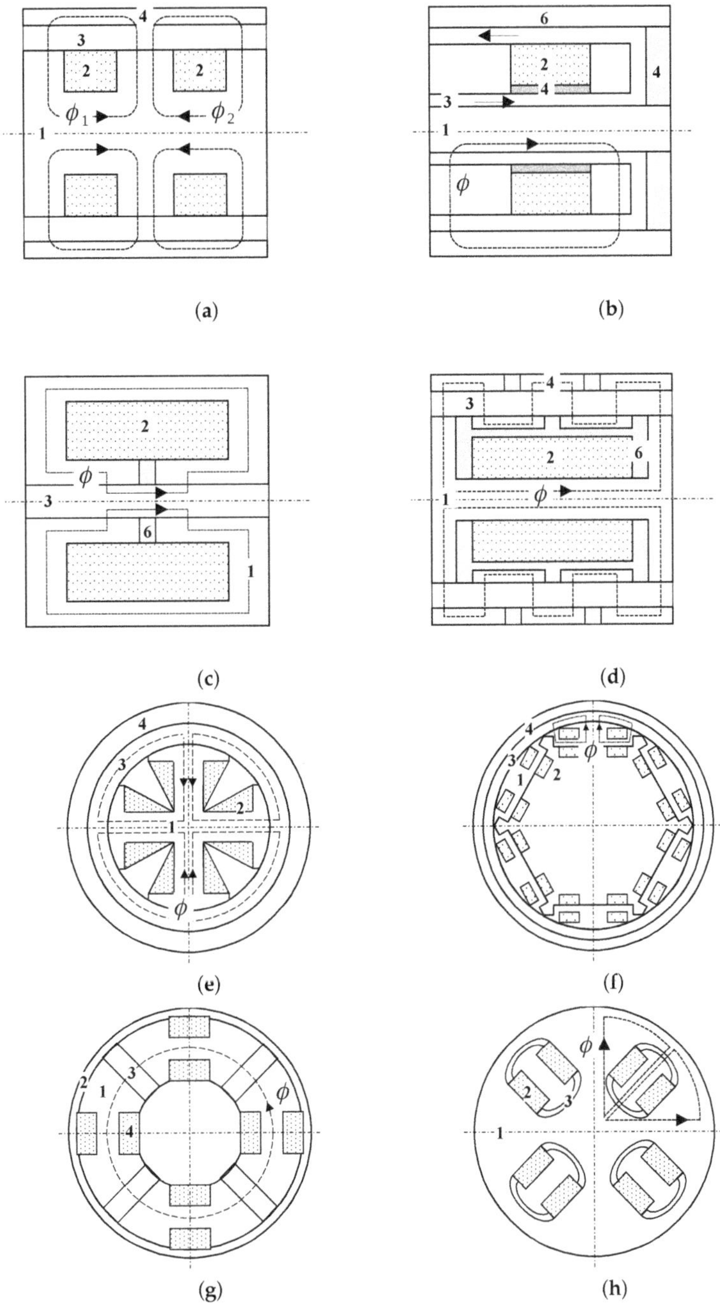

Figure 8. MR valves: configurations (continued), 1—core, 2—coil(s), 3—primary flow path(s), 4—ring, 6—(non-magnetic) spacer. (**a**) Multiple coils [51,52]; (**b**) Equivalent 'two-coil' valve [83]; (**c**) Pinch valve [9]; (**d**) Serpentine flux path [32,84,85]; (**e**) Multi-coils (radial) [50,55,56,86]; (**f**) Dual-coil multi-poles (radial) [87]; (**g**) Multi-pole valve (toroidal) [88,89]; (**h**) Multi-pole valve (toroidal)—adapted from [89].

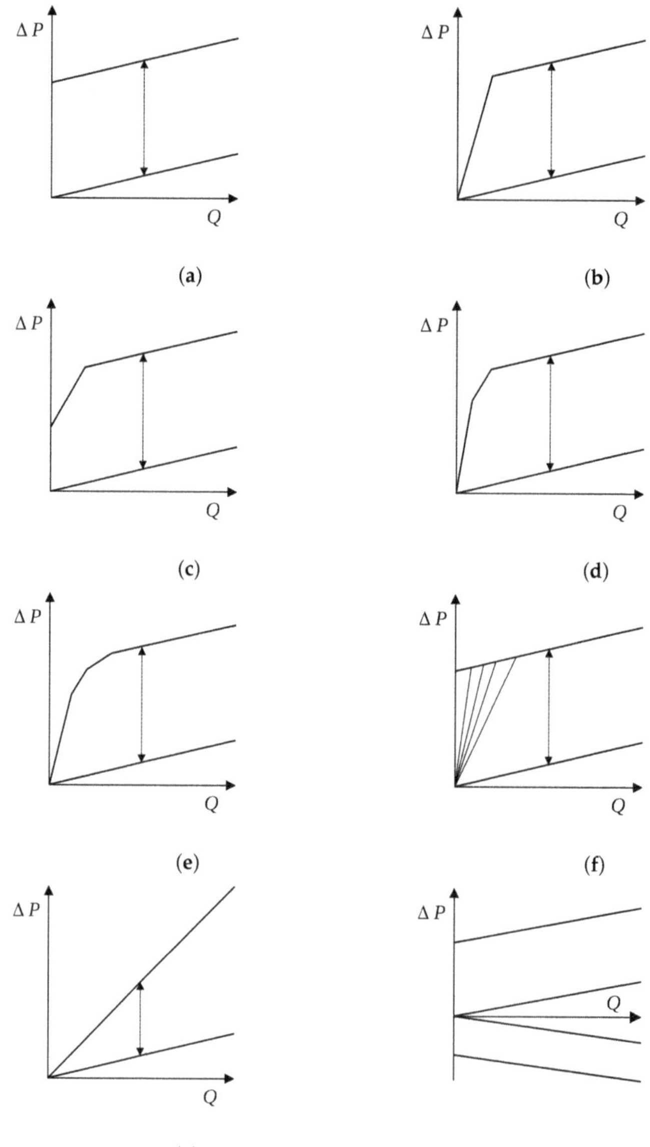

Figure 9. Steady-statepressure-flow rate $\Delta P - Q$ characteristics (envelopes) of MR valves: (**a**) benchmark, no bypass, (**b**) benchmark, hydraulic bypass, (**c**) benchmark, flux bypass, (**d**) benchmark & hybrid hydraulic and flux bypass, (**e**) parallel flow channels, (**f**) variable slope bypass—1, (**g**) variable slope bypass—2, (**h**) asymmetric valve.

There may be numerous reasons why a particular configuration may be preferred over another one within the context of a given application. Let us consider it briefly in the context of performance over the entire velocity (flow range). To begin with the benchmark valve, its single-slope envelope and its relatively high dynamic range may meet most semi-active vibration control application needs, however, the sharp transition at zero flow rate (see Figure 9a) eliminates such configuration from any practical use in vibration control.

This issue can be avoided with dual-slope damping force envelopes characterized by a high damping coefficient at low velocities (flow rates) and a small one at medium and high velocities. The solution is far more practical than the simple benchmark valve, however, the bypass or a combination of bypasses restricts the low-speed performance of the valve to a single value of the damping coefficient. This, in turn, may raise the need for a valve with the ability to vary the low-speed damping coefficient independently or in line with the flux in the main control circuit. Adding additional (energized) parallel flow paths (as well as magnetic circuits) allows shaping the upper and lower limits of the performance envelope according to the needs, however, at the expense of the simplicity and cost. Although the particular below-mentioned structures seem to have the ability to tune the damping force envelope almost to any needs, the valve's complexity increases tremendously. It seems that an optimized benchmark valve with the bypass path for the force roll-off at low-speed (flow rate) meets the requirements of most semi-active applications in the vibration control area. The observation is supported by our evaluation results using the shear stress metric below in the section.

4.1. Benchmark Valve(s)

Let us consider briefly the benchmark in Figure 7a. As already mentioned, it involves a transversely wounded coil onto a core. The core and the outer ring form the annulus for the fluid to pass through. At the same time, they constitute elements of the magnetic circuit of the valve. The current in the coil produces the magnetic flux in the structure. While passing through the annular gap, the flux energizes the fluid, thus developing a yield stress in the material. As a result, the fluid's resistance-to-flow is built up, and the output force (pressure drop) increases according to the induced flux level. Effectively, the changes in the rheology of the material translate into the relationship between the pressure drop ΔP and the flow rate Q illustrated in Figure 9a.

The benchmark configuration is structurally simple and compact, and it delivers satisfactory $\Delta P - Q$ envelopes as demonstrated in the analysis at the end of the section. It has been the basis for many of the other configurations developed over the last 30 years. Despite giving the engineer a sufficient amount of freedom in tuning the output of the valve according to the requirements, its application may result in numerous compromises. First of all, from the EM's perspective, saturating the core in the area below the coil can hardly be avoided. Second, the usual modifications of the core cross-section area below the coil window are often difficult due to packaging constraints. Last, the configuration provides little flexibility in the coil assembly placement. For instance, our analysis of the benchmark valve implementations in the available references has led us to conclude that most engineering efforts to avoid saturation in this region resulted in hardly efficient configurations with shallow and wide coil windows and short magnetic poles or substituting the core material.

In many analyzed configurations the active area ratio was below 1/2.

It is evident that the configuration in Figure 7a shares many similarities with the valves revealed in Figure 7b, as well as Figure 7c. The former has a unique feature of intrinsic asymmetric $\Delta P - Q$ envelopes in either direction of motion (achieved due to different flow losses) whereas the latter shows the annulus shifted below the coil window. Despite certain advantages [77], it can be argued that the valve in Figure 7c may be more liable to saturation. Moreover, the gap's smaller circumferential width (compared to the benchmark valve) may translate into steeper slopes of the $\Delta P - Q$ characteristics. This can be coped with by increasing the gap height (decreasing the yield stress in the annulus as a result, and thus degrading the pros of the particular valve's structure).

As a way of approximation, the magnetic field in the flow channel is radially uniform in the benchmark structure. However, it can be disrupted on purpose. Choi and Wereley [43] analyzed the simple structure with an eccentric gap. In this concept, the gap height varies continuously in the radial direction, and thus, the resulting yield stress. As a result, various breakaway pressures are achieved at each point of the gap.

4.2. Base Functionality: Benchmark Valve, Bypass

Next, let us analyze the valve in Figure 7d. The configuration incorporates a thru-hole in the structure. The bypass (usually located in the valve and rarely on the outside of it) is a feature to control pressure in the MR actuator by diverting a portion of the fluid flow into a secondary flow path. It is indispensable in suspension dampers in particular. There are several reasons for it. One characteristic feature of any flow-mode MR damper without the mechanism is the presence of high field-dependent intercept forces at the velocity of 0 m/s (of the force-velocity characteristics), which is a result of energizing the fluid in the control gap volume. Only upon exceeding the breakaway pressure can the flow through the main gap be initiated. This is typical of any controlled friction damper and unwanted in controlled chassis applications in passenger vehicles, for instance, due to stick-and-slip and severe harshness effects [61]. In the conventional shim-stack valves used in passive suspension dampers, the behavior can be achieved by preloading the shims in valving stacks. The bypass function in such valves is realized by thru-holes in pistons, orifices, or slots manufactured into the shims. The bypass is a simple means for altering the force-velocity characteristics in a passive damper or a semi-active MR damper so that the high intercept forces can be effectively degraded and shaped for optimal ride quality. It is theoretically possible to realize the control function with a non-bypass MR controllable damper, however, it would require the exact knowledge of the damper's relative velocity and ultra-fast controls.

MR bypasses are most often manufactured in the form of thru-holes (pure bypass) in the core assembly as in Figure 7d or cutouts on the inside/outside diameter of the sleeve (forming the primary path for the MR fluid flow) or the outer diameter of the core assembly (flux bypass) [59,60,78,79]. While the former resembles standard hydraulic bypasses as those used in any hydraulic suspension damper, for example, the latter takes advantage of the magnetic flux local distribution in the area of the cutout. As a result of the lower flux density distribution in the cutout region the yield stress is made non-uniform and degraded which effectively allows the fluid to pass through the gap at a lower pressure drop than in the other volume of the gap. The flux in the pure bypass is zero or near zero. As a result, the fluid's rheology within the bypass flow regime is virtually unaffected by magnetic flux variations in the electromagnet. The results in the envelope are simply illustrated in Figure 9b. For comparison, the flux bypass is located directly in the flux main path, and any changes in the flux level in the region translate into the breakaway pressure changes as in Figure 9c. Both features can be combined in one MR valve assembly to develop a dual-slope bypass for even more effective tuning of the damping force envelope at low velocities (flow rates) [79]—see Figure 9d.

The functionality of conventional bypasses is limited, though, as the slope of the curve (damping coefficient) below the breakaway force cannot be altered or it can be altered only in combination with the flux bypass. In simplified terms, they are characterized by a single and high damping coefficient. In general, the damping coefficient in the bypass region is independent of the magnetic field. For comparison, a variable slope bypass would allow altering the damping coefficient at low velocities. This would require shaping either the magnetic poles or the gaps(s) and/or designing the orifices in such a way to be controlled by the flux [9]. However, so far, the invention of the pinch mode has not yielded any viable variable slope contributions in the area despite the initial developments and claims by the inventors [9]. At the time of writing this review, no other experimental evidence with adequate performance had been developed to support the inventor's claim [45].

Rare contributions in this area have not met the expectations [45].

4.3. Add-Ons: Benchmark, Fail-Safe Features

With any MR valve the electrical short circuit or failure will prevent the valve from functioning. In this condition, the generated pressure drop by the valve will be at its minimum levels as in Figure 9a, for instance. The condition can be deemed unsafe. Arguably, fail-safe valves may be a remedy for this scenario [81,90,91]. In those valves, permanent magnets are used for generating the bias flux ϕ_{pm}, which, in turn, raises the induced pressure drop above the minimum levels with no need to energize the electromagnet. The positive flux $+\phi$ adds to the bias flux, and $-\phi$ subtracts from it. As such, the power draw can be reduced. The approach, however, is not rid of drawbacks. It requires sensing flux via hardware or software means (to achieve the true zero flux condition in the gap) and bi-directional power supplies. Long-term agglomeration of metallic particles may be an issue, too, thus making the use of dedicated fluid formulations necessary to delay the unwanted phenomenon.

4.4. Extended Performance: Benchmark Valve, High Active Area Ratio

In most MR valves, the active area ratio is rather mediocre, which is in contrast to a typical ER valve in which the active area ratio is equal to or approaches 100%. This can be obviated with a different topology (to be discussed in sections below) or by moving the coil assembly below the surface of the valve—see Figure 7e. This paves the way to increasing the active pole lengths. The valve is more difficult to assemble, and requires a modular structure. Moreover, the sections above the coil are easier to saturate and measures must be taken to prevent flux leakage between the neighboring poles. However, with this configuration almost the entire surface of the core may be used for energizing the fluid and the valve's dynamic range increased.

4.5. Extended Dynamic Range: Multiple Parallel Flow Channels

Adding a parallel flow path to the benchmark configuration is a simple way of increasing the dynamic range. It is achieved by diverting a large portion of the fluid flow into the other parallel flow channel. Moreover, splitting the flow promotes lower flow losses. This condition is, however, achieved by degrading the bottom curve of the $\Delta P - Q$ envelope rather than augmenting the maximum pressures, therefore, the benefits are rather doubtful. Moreover, it can be proven that the average flux density in the outer annulus is lower by the mean active surface area ratio. It forces the use of narrower gaps in the outer annulus, otherwise, the flow channel would be reduced to that of a huge bypass (leakage path). Despite obvious complexities and manufacturability issues, this configuration, allows for the dual-slope (or N slopes with N parallel flow channels) characteristics as in Figure 9e if needed. The original concept [66] shows the flow path above the coil window. In the view of [77], shifting one or both flow channels below the coil window (see Figure 7h) would be possible with all the consequences this particular configuration supplies.

4.6. Extended Dynamic Range: Multiple Serial Poles

Next, let us envision the valve in Figure 8a. The structure is a straightforward extension of the benchmark valve with the added flexibility of multiple coils. This allows splitting the flux path into several regions which can be independently tuned and activated (parallel connections only). Implementing the concept yields some improvement in dynamics, too. Note that the electromagnet coil's inductance L_c is proportional to the square of wire turns ($L_c \propto N_c^2$). Therefore, the sum of inductances of the two small coils is less than the inductance of a single large coil assembly. Moreover, the coils have the property of mutual inductance. Winding the coil onto the core in such a way that the current flows in the opposite directions further reduces the inductance. Given the same total resistance of the dual coil circuit (or the generic N-coil assembly) the concept reduces the time constant of the circuit L_c/R_c relative to the benchmark structure. Overall, the concept is more complex, requires additional machining work, and adds additional electrical connections, however, it yields a competitive advantage over the benchmark. The same principle can be realized by guiding the magnetic flux without the added complexity of two or more coil assemblies in the electromagnet. The concept in Figure 8b presented in [83] adds the extra active section to the annulus with just one coil assembly. Note, however, that the flow path's U-turn will augment the flow losses, and the flux will travel across four air gaps as in the parallel path structure. It is likely then that the dynamic range at higher flow rates will be severely compromised. In essence, the behavior of this particular category will be similar to the benchmark envelope in Figure 9a.

4.7. Extended Dynamic Range: Complex Flux Paths

The serpentine flux valve in Figure 8d is another attempt to increase the active area ratio [32]. The zig-zag flux structure is a simple modification of the benchmark valve. and the structure in Figure 7e with the extra sectioning of the outer ring. In the structure, the flux is carefully guided through the annulus crossing it back and forth several times. The inventors claim improvements in the effective active area ratio (and the increase in the dynamic range by a factor of 1.5 when compared against the benchmark valve), however, the structure is complex, and requires sectioning both the core and the outer ring, the use of non-magnetic elements in several locations in the two assemblies, and is rather difficult to assemble. Moreover, note that the flux leakage between the neighboring poles can hardly be avoided, thus reducing the valve's effectiveness. Without any bypass paths the particular valve is capable of delivering the characteristics as in Figure 7a.

4.8. Extended Dynamic Range: High Active Area Ratio Valves, Radial Coils

The concepts presented in Figure 8e,f fall into the category of the high active area ratio valves. With those configurations the ratio can be found to well exceed 90%. The concepts involve several coil assemblies distributed around the core with the wiring on radially protruding cores. The coils on the neighboring poles are wounded in the opposite directions. With this sort of feature the concept may be suitable for high force applications or for building compact actuators with an adequate performance envelope. In either case, the bottleneck area is the cross-section between the coil windows. The configurations are difficult to assemble and wind, but its high surface area ratio may provide a competitive advantage in specific application areas. Moreover, the flux distribution in the annulus is uniform along a single pole with a drop in the region between the adjacent poles, thus forming a natural flux bypass path for the fluid to pass through. In this plain form, its performance envelope may be typical of Figure 9c.

4.9. Extended Dynamic Range: Toroidal Cores with High Active Area Ratio, Multiple Flow Paths

In general, the category features several coil assemblies distributed on a toroidal core—see Figure 8g,h. The flow is then split into several parallel flow paths in the form of rectangular slits or semi-circular channels. The topology of the valve in Figure 8g enforces serial connections between the coils as energizing one coil will results in flux spread across all slits. The coils can be independently energized in the structure illustrated in Figure 8h. The effect may be the variable-slope envelope in Figure 9f as each section will deliver a distinct breakaway pressure drop. The two structures offer lots of freedom in tuning the valve's performance. Varying the flow path size, tuning particular coil will allow for a transition between the characteristics in Figure 9e,f. The complexity of this family of valves is on par with the configurations having the radially distributed core assemblies as described in the section above.

4.10. Variable Slope: Gradient Pinch Valve

Despite evident differences between the benchmark valve and the gradient pinch assembly in Figure 8c, the latter belongs to the flow-mode category. The flux in the structure is directed in the direction parallel to the fluid flow. This is contrary to all the other MR valves in which the flux passes the annulus in the direction perpendicular to the flow. The resulting flux is non-uniformly distributed both in the radial and the axial direction and concentrated mostly in the regions adjacent to the poles. As the resulting yield stress decreases towards the center of the channel, the valve will operate in a manner similar to the controlled orifice, although this is achieved mostly via material means and not the geometry. This is the recent and relatively unknown addition to the family of MR valves; its performance limits are unknown, not to mention less than satisfactory understanding of the pinch mode underlying mechanism. The original experiment of the inventors [9] has not been repeated so far, the existing studies are theoretical, and the experimental ones perform below the expectations. The concept, however, promises a distinct performance from the one which can be realized with the benchmark structure—see Figure 9g. The pinch mechanism is the subject of the authors' on-going research.

4.11. Shear Stress Metric Evaluation

In this section, we evaluated the shear stress metric for several configurations for which the geometry information and testing conditions can be reconstructed. The presented information in Table 2 was collected based on the available references from force-velocity plots or pressure-flow rate characteristics if available. The information which we acquired was spread over a quarter of the century, and the earliest reference included in the comparison was by Spencer et al. [92]. The data set in Table 2 includes the performance data of several automotive MR flow-mode single-tube dampers for which the geometry and performance characteristics were acquired.

The information in the table can be considered only as ballpark figures. The lack of the unified testing procedure, various approaches to presenting the performance information did not make the comparison easy. Therefore, only the data concerning MR valves sharing similar features in the flow path can be reliably compared against each other. Moreover, the configurations used in commercial MR actuators have been the subject of engineering efforts and optimization over the years to enhance their performance. This is not the case for research or academic prototypes. We deemed the inclusion of commercial MR actuator data necessary as they offer excellent performance which is usually achieved with relatively simple means. However, despite the differences and discrepancies, the following statements can be drawn:

(i) The top performers of the analyzed set are the concept valve with radial coils analyzed in [55] (numerical study only, no experimental verification), and the benchmark valve of the train damper [93].
(ii) All commercial MR valves dominate in the upper range of the examined set of MR valves.
(iii) The enhancement in the performance of the valves with the radial coils is due to their active area ratio of nearly 100%.
(iv) The performance of nearly all analyzed valves having one coil assembly, one single flow channel is within the range from 500–1000 kPa for the velocity range $0.1, \ldots, 1.0$ m/s ($\Delta\tau$).
(v) The proposed metric can be a convenient way of characterizing the actuator's force capacity (see Equation (10)).
(vi) Several claims on superior dynamic range of the proposed configuration cannot be substantiated based on their evaluated τ ($\Delta\tau$) metrics.

Table 2. Summary: estimated shear stress metrics for selected MR actuator configurations; the ♡ symbols refer to studies using commercial flow-mode MR dampers.

Ref.	Type	v_{ref}, m/s	$F_{d,ref}$ (min.), N (est.)	$F_{d,ref}$ (max.), N (est.)	τ (min.), kPa	τ (max.), kPa	$\Delta\tau$, kPa
Kubík et al. [94]	Figure 7a	0.20	400	1950	120.1	586.13	466.03
Kubík et al. [95]	Figure 7a	0.30	450	1900	110	570	460
Kubík et al. [93]	Figure 7a	0.2	1894	15,600	291	2397	2106
Spencer et al. [92]	Figure 7d	0.1	495	1350	381.9	1041.7	659. 8
Sohn et al. [60]	Figure 7d	0.60	900	3500	214	833.3	619.3
♡ Gołdasz et al. [78]	Figure 7d	0.52	750	4000	225.2	1200.3	975.1
♡ Gołdasz et al. [79]	Figure 7d	0.52	430	2600	132.3	800.3	668
Elsaady et al. [77]	Figure 7c	0.008	1500	≈8000	84.2	≈450.1	365.9
Gołdasz [67]	Figure 7g	1.0	430	2950	145.6	982.1	836.5
Cheng et al. [85]	Figure 8d	0.06	69	3305	14.2	683.1	668.9
♡ Commercial MR damper (meas.)	Figure 8a	0.52	617	4126	161.1	1077.2	916
Bai et al. [96,97]	Figure 8a	1.0	620	3100	55.7	278.75	223.1
Gołdasz [55]	Figure 8e	0.52	790	4300	374.5	2038.6	1664.1
Trębacz et al. [87]	Figure 8f	0.52	330	3000	88.8	808.9	720.1
Liu et al. [50]	Figure 8e	0.132	215	4320	44.7	898.6	851.9
Hu et al. [83]	Figure 8b	0.05	920	6838	46.8	347.9	301.1
Kim et al. [68]	Figure 8b	0.132	240	3250	53.1	719.4	666.3

5. Summary

The purpose of our analysis was to provide a thorough and detailed review of prior art on MR flow-mode control valves. It is a deliberate attempt to tie particular configurations with the performance envelopes they are capable of providing. We believe that no review on MR valves was provided in this specific context. Over the years, a variety of configurations have been developed to fulfill specific application needs. Manipulating the geometry, selecting adequate material properties, etc., are simple ways of tuning the performance of the valve and adjusting its dynamic range. In addition to that, various arrangements of flow channels, coil assemblies, etc., can be proposed for optimum active surface area and realizations of particular pressure-flow rate envelopes.

The variety of MR valves of different sizes and performance limits makes the direct comparison difficult. Therefore, we propose an easy-to-determine shear stress metric that is based on calculating the force per unit shear area. This allows for a sound comparison of MR valve configurations based on their capacity to generate adequate pressure drop (forces).

Author Contributions: Conceptualization, J.G. and B.S.; methodology, J.G.; software, J.G.; writing-review, editing and validation, B.S., M.K., O.M., W.B., T.S. and A.S.T.; supervision, J.G. All authors have read and agreed to the published version of the manuscript.

Funding: The authors wish to acknowledge the kind support of the Czech Science Foundation (Grantová agentura České republiky—GACR) and the National Science Centre (Narodowe Centrum Nauki—NCN, Poland)—grant IDs: GACR 21-45236L (CZ) and 2020/39/I/ST8/02916 (PL).

Institutional Review Board Statement: Not applicable.

Informed Consent Statement: Not applicable.

Data Availability Statement: Data can be obtained from the corresponding author upon request.

Conflicts of Interest: The authors declare no conflict of interest.

References

1. Rabinow, J. The magnetic fluid clutch. *Electr. Eng.* **1948**, *67*, 1167. [CrossRef]
2. Lam, H.F.; Liao, W.H. Semi-active control of automotive suspension systems with magnetorheological dampers. In Proceedings of the Smart Structures and Materials 2001: Smart Structures and Integrated Systems, Newport Beach, CA, USA, 16 August 2001; SPIE: Bellingham, WA, USA, 2001; Volume 4327, pp. 125–136.
3. Gołdasz, J.; Sapiński, B. *Insight into Magnetorheological Shock Absorbers*; Springer: Berlin/Heidelberg, Germany, 2015.
4. Harris, D.C. History of magnetorheological finishing. In Proceedings of the Window and Dome Technologies and Materials XII, Orlando FL, USA, 27–28 April 2011; SPIE: Bellingham, WA, USA, 2011; Volume 8016, pp. 206–227.
5. Kumar, M.; Kumar, A.; Alok, A.; Das, M. Magnetorheological method applied to optics polishing: A review. *Proc. Iop Conf. Ser. Mater. Sci. Eng.* **2020**, *804*, 012012. [CrossRef]
6. Choi, S.B.; Gołdasz, J. Controllable actuators utilizing smart MR materials and ER suspensions. In Proceedings of the ACTUATOR 2018—16th International Conference on New Actuators, Bremen, Germany, 25–27 June 2018; pp. 1–9.
7. Ulrich, S.; Böhme, G.; Bruns, R. Measuring the response time and static rheological properties of electrorheological fluids with regard to the design of valves and their controllers. *J. Phys. Conf. Ser.* **2009**, *149*, 012031. [CrossRef]
8. Jolly, M.; Bender, J.; Carlson, J. Properties and applications of commercial magnetorheological fluids. *J. Intell. Mater. Syst. Struct.* **1999**, *10*, 5–13. [CrossRef]
9. Goncalves, F.; Carlson, J. An alternate operation mode for MR fluids—Magnetic gradient pinch. *J. Phys. Conf. Ser.* **2009**, *149*, 012050. [CrossRef]
10. Goldasz, J.; Sapinski, B. Magnetostatic analysis of a pinch mode magnetorheological valve. *Acta Mech. Autom.* **2017**, *11*, 229–232. [CrossRef]
11. Sung, K.G.; Choi, S.B.; Lee, H.G.; Min, K.W.; Lee, S.H. Performance comparison of MR dampers with three different working modes: shear, flow and mixed mode. In *Electrorheological Fluids And Magnetorheological Suspensions (Ermr 2004)*; World Scientific: Singapore, 2005; pp. 645–651.
12. Yazid, I.I.M.; Mazlan, S.A.; Kikuchi, T.; Zamzuri, H.; Imaduddin, F. Design of magnetorheological damper with a combination of shear and squeeze modes. *Mater. Des.* **2014**, *54*, 87–95. [CrossRef]
13. Abdul Aziz, M.; Muhtasim, S.; Ahammed, R. State-of-the-art recent developments of large magnetorheological (MR) dampers: A review. *Korea-Aust. Rheol. J.* **2022**, *34*, 105–136. [CrossRef]
14. Sapinski, B. Energy-harvesting linear MR damper: Prototyping and testing. *Smart Mater. Struct.* **2014**, *23*, 035021. [CrossRef]

15. Zhu, X.; Jing, X.; Cheng, L. Magnetorheological fluid dampers: A review on structure design and analysis. *J. Intell. Mater. Syst. Struct.* **2012**, *23*, 839–873. [CrossRef]
16. Abd Fatah, A.Y.; Mazlan, S.A.; Koga, T.; Zamzuri, H.; Zeinali, M.; Imaduddin, F. A review of design and modeling of magnetorheological valve. *Int. J. Mod. Phys. B* **2015**, *29*, 1530004. [CrossRef]
17. Rahman, M.; Ong, Z.C.; Julai, S.; Ferdaus, M.M.; Ahamed, R. A review of advances in magnetorheological dampers: Their design optimization and applications. *J. Zhejiang Univ.-Sci. A* **2017**, *18*, 991–1010. [CrossRef]
18. Ismail, A.F.; Ibrahim, M.F.; Osman, K.H. A technological review of magnetorheological dampers. *Int. J. Adv. Res. Eng. Innov.* **2021**, *3*, 8–15.
19. Aziz, M.A.; Aminossadati, S.M. State-of-the-art developments of bypass magnetorheological (MR) dampers: A review. *Korea-Aust. Rheol. J.* **2021**, *33*, 225–249. [CrossRef]
20. Mohtasim, S.M.; Ahammed, R.; Rahman, M.; Rashid, M.M.; Roy, R.; Aziz, M.A. Recent developments of regenerative magnetorheological (RMR) damper: A review. *Korea-Aust. Rheol. J.* **2021**, *33*, 201–224. [CrossRef]
21. Oliver, M.; Kruckemeyer, W.; Bishop, T.; Jensen, E. Magnetorheological Strut Piston with Compression Bypass. U.S. Patent US6637557B2, 28 October 2003.
22. Deng, H.; Gao, Y.; Hu, R.; Zhao, S.; Han, G.; Lian, X.; Ma, M.; Zhong, X. Self-sensing automotive magnetorheological dampers for low frequency vibration. *Smart Mater. Struct.* **2021**, *30*, 115015. [CrossRef]
23. Gavin, H.P. Annular Poiseuille flow of electrorheological and magnetorheological materials. *J. Rheol.* **2001**, *45*, 983–994. [CrossRef]
24. Carlson, D.; Chrzan, M. Magnetorheological Fluid Dampers. U.S. Patent US5277281A, 11 January 1994.
25. Namuduri, C.; Alexandridis, A. Magnetorheological Fluid Damper with Optimum Damping. U.S. Patent US6390252B1, 21 May 2002.
26. Gołdasz, J. Theoretical study of a twin-tube magnetorheological damper concept. *J. Theor. Appl. Mech.* **2015**, *53*, 885–894. [CrossRef]
27. Jensen, E.; Kruckemeyer, W. Twin-Tube Magnetorheological Damper. U.S. Patent US20020139624A1, 3 October 2002.
28. Zhang, X.; Li, Z.; Guo, K.; Zheng, F.; Wang, Z. A novel pumping magnetorheological damper: Design, optimization, and evaluation. *J. Intell. Mater. Syst. Struct.* **2017**, *28*, 2339–2348. [CrossRef]
29. Oakley, R. Twin-Tube Magnetorheological Damper. European Patent EP1908985A1, 9 April 2008.
30. Anderfaas, E.; Banks, D. Magnetorheological Damper System. U.S. Patent US6953108B2, 11 October 2005.
31. Shin, D.K.; Phu, D.X.; Choi, S.; bok Choi, S. An adaptive fuzzy sliding mode control of magneto-rheological seat suspension with human body model. *J. Intell. Mater. Syst. Struct.* **2016**, *27*, 925–934. [CrossRef]
32. Idris, M.H.; Imaduddin, F.; Ubaidillah; Mazlan, S.A.; Choi, S.B. A concentric design of a bypass magnetorheological fluid Damper with a serpentine flux valve. *Actuators* **2020**, *9*, 16. [CrossRef]
33. Nam, Y.J.; Park, M.K. Performance evaluation of two different bypass-type MR shock dampers. *J. Intell. Mater. Syst. Struct.* **2007**, *18*, 707–717. [CrossRef]
34. Bai, X.X.; Zhong, W.M.; Zou, Q.; Zhu, A.D.; Sun, J. Principle, design and validation of a power-generated magnetorheological energy absorber with velocity self-sensing capability. *Smart Mater. Struct.* **2018**, *27*, 075041. [CrossRef]
35. Hou, S.; Liu, G. Research on Theoretical Modeling and Parameter Sensitivity of a Single-Rod Double-Cylinder and Double-Coil Magnetorheological Damper. *Math. Probl. Eng.* **2020**, *2020*, 5489896. [CrossRef]
36. Mao, M.; Hu, W.; Choi, Y.T.; Wereley, N.M. A magnetorheological damper with bifold valves for shock and vibration mitigation. *J. Intell. Mater. Syst. Struct.* **2007**, *18*, 1227–1232. [CrossRef]
37. Zeinali, M.; Mazlan, S.A.; Choi, S.B.; Imaduddin, F.; Hamdan, L.H. Influence of piston and magnetic coils on the field-dependent damping performance of a mixed-mode magnetorheological damper. *Smart Mater. Struct.* **2016**, *25*, 055010. [CrossRef]
38. Hong, S.; Wereley, N.; Choi, Y.; Choi, S. Analytical and experimental validation of a nondimensional Bingham model for mixed-mode magnetorheological dampers. *J. Sound Vib.* **2008**, *312*, 399–417. [CrossRef]
39. Maharani, E.T.; Ubaidillah, U.; Imaduddin, F.; Wibowo, K.; Utami, D.; Mazlan, S.A. A mathematical modelling and experimental study of annular-radial type magnetorheological damper. *Int. J. Appl. Electromagn. Mech.* **2021**, *66*, 543–560. [CrossRef]
40. Imaduddin, F.; Mazlan, S.A.; Zamzuri, H.; Yazid, I.I.M. Design and performance analysis of a compact magnetorheological valve with multiple annular and radial gaps. *J. Intell. Mater. Syst. Struct.* **2015**, *26*, 1038–1049. [CrossRef]
41. Chen, C.; Liao, W.H. A self-powered, self-sensing magnetorheological damper. In Proceedings of the 2010 IEEE International Conference on Mechatronics and Automation, Xi'an, China, 4–7 August 2010; pp. 1364–1369.
42. Grunwald, A.; Olabi, A.G. Design of magneto-rheological (MR) valve. *Sens. Actuators A Phys.* **2008**, *148*, 211–223. [CrossRef]
43. Choi, Y.T.; Wereley, N.M. Flow mode magnetorheological dampers with an eccentric gap. *Adv. Mech. Eng.* **2014**, *6*, 931683. [CrossRef]
44. Foister, R.; Nehl, T.; Kruckemeyer, W.; Raynauld, O. Magnetorheological (mr) Piston Assembly with Primary and Secondary Channels to Improve MR Damper Force. U.S. Patent US8327984B2, 11 December 2012.
45. Lee, T.H.; Kang, B.H.; Choi, S.B. A quasi-static model for the pinch mode analysis of a magnetorheological fluid flow with an experimental validation. *Mech. Syst. Signal Process.* **2019**, *134*, 106308. [CrossRef]
46. Oliver, M.; Kruckemeyer, W. Electrical Coupling Assembly for a Magnetorheological Damper. U.S. Patent US6345706B1, 12 February 2002.
47. Nehl, T.; Gopalakrishnan, S.; Deng, F. Direct Flux Control System for Magnetic Structures. European Patent EP1868214B1, 26 January 2011.

48. Bai, X.X.; Wereley, N.M.; Choi, Y.T.; Wang, D.H. A bi-annular-gap magnetorheological energy absorber for shock and vibration mitigation. In Proceedings of the Active and Passive Smart Structures and Integrated Systems 2012, Online, 22–26 March 2012; Sodano, H.A., Ed.; SPIE: Bellingham, WA, USA, 2012; Volume 8341, pp. 646–666. [CrossRef]
49. Bai, X.X.; Wereley, N.M.; Hu, W. Maximizing semi-active vibration isolation utilizing a magnetorheological damper with an inner bypass configuration. *J. Appl. Phys.* **2015**, *117*, 17C711. [CrossRef]
50. Liu, S.; Feng, L.; Zhao, D.; Huang, H.; Shi, X.; Chen, L.; Jiang, J. The development of an outer multi-pole magneto-rheological damper with high dynamic range of damping force. *Smart Mater. Struct.* **2018**, *27*, 115025. [CrossRef]
51. Bai, X.X.; Shen, S.; Wereley, N.M.; Wang, D.H. Controllability of magnetorheological shock absorber: I. Insights, modeling and simulation. *Smart Mater. Struct.* **2018**, *28*, 015022. [CrossRef]
52. Zheng, J.; Li, Y.; Wang, J. Design and multi-physics optimization of a novel magnetorheological damper with a variable resistance gap. *Proc. Inst. Mech. Eng. Part C J. Mech. Eng. Sci.* **2017**, *231*, 3152–3168. [CrossRef]
53. Hu, G.; Long, M.; Huang, M.; Li, W. Design, analysis, prototyping, and experimental evaluation of an efficient double coil magnetorheological valve. *Adv. Mech. Eng.* **2014**, *6*, 403410. [CrossRef]
54. Elsaady, W.; Oyadiji, S.O.; Nasser, A. A one-way coupled numerical magnetic field and CFD simulation of viscoplastic compressible fluids in MR dampers. *Int. J. Mech. Sci.* **2020**, *167*, 105265. [CrossRef]
55. Gołdasz, J. Electro-mechanical analysis of a magnetorheological damper with electrical steel laminations. *Przegląd Elektrotech.* **2013**, *89*, 8–12.
56. Sassi, S.; Cherif, K.; Mezghani, L.; Thomas, M.; Kotrane, A. An innovative magnetorheological damper for automotive suspension: from design to experimental characterization. *Smart Mater. Struct.* **2005**, *14*, 811. [CrossRef]
57. Badri, Y.; Syam, T.; Sassi, S.; Hussein, M.; Renno, J.; Ghani, S. Investigating the characteristics of a magnetorheological fluid damper through CFD modeling. *Mater. Res. Express* **2021**, *8*, 055701. [CrossRef]
58. Sassi, S.; Sassi, A.; Cherif, K.; Tarlochan, F. Magnetorheological damper with external excitation for more efficient control of vehicles' dynamics. *J. Intell. Mater. Syst. Struct.* **2018**, *29*, 2919–2932. [CrossRef]
59. Kubík, M.; Mazurek, I.; Roupec, J.; Strecker, Z.; Macháček, O. Design of semi-active magnetorheological valve with non-magnetic bypass. *Trans. Electr. Eng.* **2015**, *4*, 20–23.
60. Sohn, J.W.; Oh, J.S.; Choi, S.B. Design and novel type of a magnetorheological damper featuring piston bypass hole. *Smart Mater. Struct.* **2015**, *24*, 035013.
61. Oh, J.S.; Kim, K.S.; Lee, Y.S.; Choi, S.B. Dynamic simulation of a full vehicle system featuring magnetorheological dampers with bypass holes. *J. Intell. Mater. Syst. Struct.* **2020**, *31*, 253–262. [CrossRef]
62. Kim, B.G.; Yoon, D.S.; Kim, G.W.; Choi, S.B.; Tan, A.S.; Sattel, T. Design of a novel magnetorheological damper adaptable to low and high stroke velocity of vehicle suspension system. *Appl. Sci.* **2020**, *10*, 5586. [CrossRef]
63. Kuzhir, P.; Bossis, G.; Bashtovoi, V. Optimization of magnetorheological fluid valves. *Int. J. Mod. Phys. B* **2005**, *19*, 1229–1235. [CrossRef]
64. Bai, X.X.; Wang, D.H.; Fu, H. Principle, modeling, and testing of an annular-radial-duct magnetorheological damper. *Sens. Actuators A Phys.* **2013**, *201*, 302–309. [CrossRef]
65. McLaughlin, G.; Hu, W.; Wereley, N. Advanced magnetorheological damper with a spiral channel bypass valve. *J. Appl. Phys.* **2014**, *115*, 17B532. [CrossRef]
66. Namuduri, C.; Alexandridis, A.; Madak, J.; Rule, D. Magnetorheological Fluid Damper with Multiple Annular Flow Gaps. U.S. Patent US6279701B1, 28 August 2001.
67. Gołdasz, J. Study of a magnetorheological fluid damper with multiple annular flow gaps. *Int. J. Veh. Des.* **2013**, *62*, 21–41. [CrossRef]
68. Kim, K.; Chen, Z.; Yu, D.; Rim, C. Design and experiments of a novel magneto-rheological damper featuring bifold flow mode. *Smart Mater. Struct.* **2016**, *25*, 075004.
69. Potnuru, M.R.; Wang, X.; Mantripragada, S.; Gordaninejad, F. A compressible magneto-rheological fluid damper–liquid spring system. *Int. J. Veh. Des.* **2013**, *63*, 256–274.
70. Hu, G.; Long, M.; Yu, L.; Li, W. Design and performance evaluation of a novel magnetorheological valve with a tunable resistance gap. *Smart Mater. Struct.* **2014**, *23*, 127001. [CrossRef]
71. Coulter, J.P.; Weiss, K.D.; Carlson, J.D. Engineering applications of electrorheological materials. *J. Intell. Mater. Syst. Struct.* **1993**, *4*, 248–259.
72. Duclos, T.G. Design of devices using electrorheological fluids. *SAE Trans.* **1988**, *97*, 532–536.
73. Nguyen, Q.H.; Choi, S.B. Optimal design of a vehicle magnetorheological damper considering the damping force and dynamic range. *Smart Mater. Struct.* **2008**, *18*, 015013. [CrossRef]
74. Nguyen, Q.H.; Choi, S.B.; Wereley, N.M. Optimal design of magnetorheological valves via a finite element method considering control energy and a time constant. *Smart Mater. Struct.* **2008**, *17*, 025024.
75. Soong, W. Sizing of electrical machines. *Power Eng. Brief. Note Ser.* **2008**, *9*, 17–18.
76. Miller, T.J. *Brushless Permanent-Magnet and Reluctance Motor Drives*; Office of Scientific and Technical Information: Washington, DC, USA, 1989.
77. Elsaady, W.; Oyadiji, S.O.; Nasser, A. Magnetic circuit analysis and fluid flow modeling of an MR damper with enhanced magnetic characteristics. *IEEE Trans. Magn.* **2020**, *56*, 1–20.

78. Gołdasz, J.; Sapiński, B. Nondimensional characterization of flow-mode magnetorheological/electrorheological fluid dampers. *J. Intell. Mater. Syst. Struct.* **2012**, *23*, 1545–1562. [CrossRef]
79. Gołdasz, J.; Sapiński, B. Verification of magnetorheological shock absorber models with various piston configurations. *J. Intell. Mater. Syst. Struct.* **2013**, *24*, 1846–1864. [CrossRef]
80. Nehl, T.; Alexandridis, A.; Foister, R.; Kruckemeyer, W.; Deng, F. Magnetorheological Fluid-Based Device Having a Magnetorheological Piston Assembly. U.S. Patent US8286763B2, 16 October 2012.
81. Jeniš, F.; Kubík, M.; Macháček, O.; Šebesta, K.; Strecker, Z. Insight into the response time of fail-safe magnetorheological damper. *Smart Mater. Struct.* **2020**, *30*, 017004. [CrossRef]
82. Yang, X.; Chen, Y.; Liu, Y.; Zhang, R. Modeling and Experiments of an Annular Multi-Channel Magnetorheological Valve. *Proc. Actuators. Multidiscip. Digit. Publ. Inst.* **2022**, *11*, 19. [CrossRef]
83. Hu, G.; Liu, H.; Duan, J.; Yu, L. Damping performance analysis of magnetorheological damper with serial-type flow channels. *Adv. Mech. Eng.* **2019**, *11*, 1687814018816842. [CrossRef]
84. Abd Fatah, A.Y.; Mazlan, S.A.; Koga, T.; Zamzuri, H. Increasing effective region in magnetorheological valve using serpentine flux path method. In Proceedings of the 2013 World Congress on Advances in Structural Engineering and Mechanics, Jeju, Korea, 8–12 September 2013; pp. 8–12.
85. Cheng, M.; Chen, Z.; Xing, J. Design, analysis, and experimental evaluation of a magnetorheological damper with meandering magnetic circuit. *IEEE Trans. Magn.* **2018**, *54*, 1–10. [CrossRef]
86. Jensen, E.; Kruckemeyer, W. Magneto-Rheological Damping Valve Using Laminated Construction. U.S. Patent US6481546B2, 19 November 2002.
87. Trebacz, P.; Bańkosz, W. Novel magnetorheological flow-mode valve. 2022, *unpublished*.
88. Dominguez, G.A.; Kamezaki, M.; Sugano, S. Proposal and preliminary feasibility study of a novel toroidal magnetorheological piston. *IEEE/ASME Trans. Mechatron.* **2016**, *22*, 657–668. [CrossRef]
89. Dominguez, G.A.; Kamezaki, M.; He, S.; Sophon, S.; Schmitz, A.; Sugano, S. Design optimisation and performance evaluation of a toroidal magnetorheological hydraulic piston head. In Proceedings of the 2016 IEEE/RSJ International Conference on Intelligent Robots and Systems (IROS), Daejeon, Korea, 9–14 October 2016; pp. 350–355.
90. Nehl, T.; Alexandridis, A. Magnetorheological Devices with Permanent Magnet Field Bias. U.S. Patent US8651250B2, 18 February 2014.
91. Boese, H.; Ehrlich, J. Performance of magnetorheological fluids in a novel damper with excellent fail-safe behavior. *J. Intell. Mater. Syst. Struct.* **2010**, *21*, 1537–1542. [CrossRef]
92. Spencer, B., Jr.; Dyke, S.; Sain, M.; Carlson, J. Phenomenological model for magnetorheological dampers. *J. Eng. Mech.* **1997**, *123*, 230–238. [CrossRef]
93. Kubík, M.; Strecker, Z.; Jeniš, F.; Macháček, O.; Prikryl, M.; Spalek, P. Magnetorheological Yaw Damper with Short Response Time for Railway Vehicle Boogie. In Proceedings of the ACTUATOR—International Conference and Exhibition on New Actuator Systems and Applications 2021, Online, 17–19 February 2021; pp. 1–4.
94. Kubík, M.; Macháček, O.; Strecker, Z.; Roupec, J.; Mazurek, I. Design and testing of magnetorheological valve with fast force response time and great dynamic force range. *Smart Mater. Struct.* **2017**, *26*, 047002. [CrossRef]
95. Kubík, M.; Gołdasz, J. Multiphysics Model of an MR Damper including Magnetic Hysteresis. *Shock Vib.* **2019**, *2019*, 3246915. [CrossRef]
96. Bai, X.X.; Hu, W.; Wereley, N.M. Magnetorheological damper utilizing an inner bypass for ground vehicle suspensions. *IEEE Trans. Magn.* **2013**, *49*, 3422–3425. [CrossRef]
97. Bai, X.X.; Hu, W.; Wereley, N.M. Analysis and testing of an inner bypass magnetorheological damper for shock and vibration mitigation. In Proceedings of the Active and Passive Smart Structures and Integrated Systems 2013, San Diego, CA, USA, 10 April 2013; SPIE: Bellingham, WA, USA, 2013; Volume 8688, pp. 292–309.

MDPI
St. Alban-Anlage 66
4052 Basel
Switzerland
Tel. +41 61 683 77 34
Fax +41 61 302 89 18
www.mdpi.com

Applied Sciences Editorial Office
E-mail: applsci@mdpi.com
www.mdpi.com/journal/applsci

www.ingramcontent.com/pod-product-compliance
Lightning Source LLC
Chambersburg PA
CBHW061454210926
51850CB00008B/1713
* 9 7 8 3 0 3 6 5 8 4 9 1 1 *